Selling Places

The City as Cultural Capital, Past and Present

Policy, Planning and Critical Theory

Series Editor: **Paul Cloke**
St David's University College, Lampeter, UK

This major new series will focus on the relevance of critical social theory to important contemporary processes and practices in planning and policy-making. It aims to demonstrate the need to incorporate state and governmental activities within these new theoretical approaches, and will focus on current trends in governmental policy in Western states, with particular reference to the relationship between the centre and the locality, the provision of services, and the formulation of government policy.

Also available in this series

Policy and Change in Thatcher's Britain
Paul Cloke

Forthcoming titles in this series include

The Global Region: Production, State Policies and Uneven Development
David Sadler

People, Place, Protest: Making Sense of Popular Protest in Modern Britain
Michael J Griffiths

Policy and Planning for International Agriculture
Richard Le Heron

Gender, Planning and the Policy Process
Jo Little

Uneven Reproduction: Economy, Space and Society
Andrew Pratt

Selling Places
The City as Cultural Capital, Past and Present

Edited by

Gerry Kearns
University of Madison-Wisconsin, USA

and

Chris Philo
St David's University College, Lampeter, UK

PERGAMON PRESS
OXFORD · NEW YORK · SEOUL · TOKYO

UK	Pergamon Press Ltd, Headington Hill Hall, Oxford OX3 0BW, England
USA	Pergamon Press Inc, 660 White Plains Road, Tarrytown, New York 10591-5153, U.S.A.
KOREA	Pergamon Press Korea, KPO Box 315, Seoul 110-603, Korea
JAPAN	Pergamon Press Japan, Tsunashima Building Annex, 3-20-12 Yushima, Bunkyo-ku, Tokyo 113, Japan

First edition 1993

Library of Congress Cataloging-in-Publication Data

Selling places: the city as cultural capital, past and present / edited by Gerry Kearns and Chris Philo.—1st ed.
p. cm.—(Policy, planning, and critical theory)
Includes bibliographical references.
1. Cities and towns. 2. Civic improvement. 3. Public relations—Municipal government. 4. Urban renewal.
I. Kearns, Gerard. II. Philo, Chris. III. Series.
HT155.S45 1993
307.76—dc20   93–7681

ISBN 0-08-041385-4 (hardcover)
ISBN 0-08-041384-6 (flexicover)

Printed in Great Britain by BPCC Wheatons Ltd, Exeter

Contents

Contents

Contributors

MARK BILLINGE
Department of Geography
University of Cambridge, UK

DARREL CRILLEY
Department of Geography
Queen Mary and Westfield College, London, UK

ANDREW DAVID FRETTER
Head of Economic Development
Gwent County Council

PAUL GLENNIE
Department of Geography
University of Bristol, UK

MARK GOODWIN
School of Geography & Environmental Science
University of North London, UK

BRIAVEL HOLCOMB
Department of Urban Studies
Rutgers University, New Brunswick, NJ 08903, USA

GERRY KEARNS
Department of Geography
University of Madison-Wisconsin, USA

ERIC LAURIER
Department of Geography
St. David's University College, Lampeter, UK

Contributors

MICHELLE LOWE
Department of Geography
University of Birmingham, UK

CHRIS PHILO
Department of Geography
St. David's University College, Lampeter, UK

LAURA REID
Department of Geography
Rutgers University, New Brunswick, NJ 08903, USA

DAVID SADLER
Department of Geography
University of Durham, UK

NEIL SMITH
Department of Geography
Rutgers University, New Brunswick, NJ 08903, USA

NIGEL THRIFT
Department of Geography
University of Bristol, UK

RACHEL WOODWARD
Department of Geography
Centre for the Study of Public Order, Leicester, UK

Preface

THIS VOLUME is about *cultural capital*, where 'capital' refers both to money and to 'capital' or sizeable cities, and attention is paid to the complex and often contested processes whereby the managers of large urban areas – local authorities and entrepreneurs of various kinds – manipulate cultural resources for capital gain: whether by converting them into 'commodities' that can be bought and sold in their own right, or by using them as a lure to inward investment from industrialists, tourists and shoppers. These processes are fairly obvious dimensions to the phenomenon of *selling places* (and are now considered in various literatures), but attention is also paid here to the more intangible phenomenon whereby cultural resources are mobilised by urban managers in an attempt to engineer consensus amongst the residents of their localities, a sense that beyond the daily difficulties of urban life which many of them might experience the city is basically 'doing alright' by its citizens.

The objective of bringing a diverse set of essays together in this one volume is hence to open a number of different windows on the selling places phenomenon: not so much to list the techniques of how places are sold, as to examine the 'discourses' that sustain the practice of manipulating culture in the selling of places, and to tease out the material contexts (in terms of the national and local economies, polities and societies) that are generating this practice as a key feature of urban governance in the late-twentieth-century Western world. At the same time, the intention has been to lay out in detail the contents of culture – concrete manifestations such as theme parks and exhibitions; symbolic codings present in advertisements, monuments and other human products; ideologies of what constitutes a 'good' or 'proper' way of living – that are being manipulated more or less consciously by urban managers, and to investigate what it is that these managers are endeavouring to do with these contents of culture. It has further been intended that contributors should reflect upon moments of resistance to the selling of places, asking about how and why such resistances appear, and about the classed, gendered and racialised identities of the peoples commonly involved in these acts of resistance. In addition, it is hoped to display a sensitivity to history: to consider the phenomenon of selling places as it has occurred in past times, but more importantly to inspect the various uses that are made of history – through historical references; through evoking memories of past events, peoples and places – in the striving of the place managers to sell their places.

The volume grew out of a one-and-a-half day conference session that ran as part of the Annual Conference of the IBG held in Glasgow in January 1990. This session was jointly sponsored by the Social and Cultural Geography Study Group and by the Historical Geography Study Group of the Institute of British Geographers (IBG), and was convened by ourselves along with Jacquie Burgess. For the year of 1990 Glasgow assumed the mantle of 'European City of Culture', and as such it seemed appropriate to organise a session that dealt at length with the interconnections of culture and capital whose dimensions and implications were inevitably heightened in Glasgow at this time (Boyle and Hughes, 1991), but which are currently becoming central to the existences of many urban centres both great and small. The session was adjudged sufficiently successful to warrant bringing the papers delivered together in a single published collection, although for various reasons we have not been able to include all of the original papers that we would like to have done, and have therefore commissioned four additional essays (by Mark Billinge, Mark Goodwin, Rachel Woodward, and Eric Laurier) as well as preparing an introductory essay of our own. All of the contributions have been specially written up for the volume, then, and the papers originally delivered at the conference have all been rewritten to a greater or lesser extent. Some delay has occurred in the preparation of the volume, we must admit, but we think that taken as a whole the volume can still mark a significant intervention in – and hopefully some new perspectives on – the debate over the phenomenon of selling places.

We would like to thank Geraldine Billingham at Pergamon for both her patience and her encouragement, and also to acknowledge the advice of Paul Cloke (the Series Editor) during the planning and execution of the volume. Various colleagues and friends have played an important supportive role as well, notably Hester Parr in helping to prepare the index, but our most important thankyous must undoubtedly go to our contributors for the care which they have shown in writing their chapters.

GERRY KEARNS
CHRIS PHILO

May 1993.

1

Culture, History, Capital: A Critical Introduction to the Selling of Places[1]

CHRIS PHILO AND GERRY KEARNS

An Obscure Script?

'The bored tourists who pay their nine francs at the desk or are admitted free on Sundays may believe that elderly nineteenth-century gentlemen – beards yellowed by nicotine, collars rumpled and greasy, black cravats and frock coats smelling of snuff, fingers stained with acid, their minds acid with professional jealousy, farcical ghosts who called one another *cher maître* – placed these exhibits here out of a virtuous desire to educate and amuse the bourgeoisie and the radical taxpayers, and to celebrate the magnificent march of progress. But no: Saint-Martin-des-Champs had been conceived first as a priory and only later as a revolutionary museum and compendium of arcane knowledge. The planes, those self-propelled machines, those electromagnetic skeletons, were carrying on a dialogue whose script still escaped me' (Eco, 1989, p. 8).

Like Casaubon, the narrator of Eco's novel *Foucault's Pendulum*, we must often wonder what precisely is going on when we find ourselves surrounded by bits and pieces of the past that all manner of people – and certainly not just 'elderly nineteenth-century gentlemen' – seem to think it appropriate that we encounter. Casaubon finds himself in a museum of technology, the Conservatoire des Arts et Métiers at Saint-Martin-des-Champs in Paris, attempting to make sense of the bewildering variety of antique machines displayed there: but in principle his situation is similar to that of any tourist confronted by a portion of the past packaged up before them – the sights of a town heritage trail; the walk along a castellated city wall; the reconstruction of a Roman town house; the renovated cotton mill – or indeed of anybody confronted by a range of historical references (commonly architectural ones) deliberately inserted into the built landscape all around them. Casaubon reflects upon the *raison d'être* of the museum, immediately noting the coins that exchange hands at the admissions desk. In this instance the coins may genuinely go to the upkeep of the museum, but is it not so often the case that the past on show in a museum, a theme park, an anniversary celebration or whatever has become a commodity designed to make money for somebody else? Casaubon then speculates that the museum's exhibits could have been put there to educate the Parisian bourgeoisie in the history of science, and in so doing to hint at a broader history of progressive civilization, and he acknowledges too that this process of education might further serve the purpose of amusement as Parisians laugh at the curious contraptions put together by peoples remote in time and space. This explanation may seem obvious, but does it not disguise a further series of questions about the assumptions that both informed the founders of the museum and become internalised by its patrons: assumptions about the superior vantage-point of Parisians (and, more generally, of urban Europeans) supposedly at the apex of the histories of science and civilization? And does not the museum therefore serve the purpose of, to anticipate a phrase from later in this chapter, educating people in the basic ideological commitments of a specific society? There are several additional aspects of the Paris museum that are relevant here − notably the fact that the museum itself was opened in an historic place, the old priory, and also the role of French 'revolutionary' ideals in shaping our engagement with history (see the chapter by Kearns) − but it will suffice to identify the logics of *economic gain* and *social control* that Casaubon gestures to as he contemplates the Conservatoire. We would want to insist at the outset that there may be other logics at work here beyond the economic and the social,[2] but it is these two logics embedded in the obscure script being 'spoken' by the strange museum exhibits − the two sets of connections perceived by Casaubon between culture, history, capital and (by implication) the city − that remain the principal ones under scrutiny in this volume.

The focus of the collection is what we are calling the practice of *selling places*: a phenomenon which involves and has implications for a range of economic and social activities − often explicitly formulated *policies* − pursued by those individuals and organisations who 'manage' places, most obviously large urban areas but also on occasion rural areas or whole (often quite heterogenous) regions. We are principally thinking about this phenomenon as it affects the 'developed' Western world of the late-twentieth century, or as it affects those numerous cities in 'developing' or Eastern countries which have become Western-ised in function and appearance, but this is not to deny that it is possible to find rather different sorts of places both historically and geographically adopting promotional strategies to boost their own well-being (Cosgrove, 1990; Robinson, 1991). At bottom, the practice of selling places entails the various ways in which public and private agencies – local authorities and local entrepreneurs, often working collaboratively – strive to 'sell' the image of a particular geographically-defined 'place', usually a town or city, so as to make it attractive to economic enterprises, to tourists and even to inhabitants of that place (see similar claims in Harvey, 1989a). The chief ambitions are to encourage economic enterprises (and notably footloose high-technology industries) to locate themselves in this place and to entice tourists to visit the place in large numbers, and both of these ambitions obviously tie in with the attempts that all sorts of localities in Britain, North America and elsewhere are making to secure inward capital investment, a degree of local job creation and hence *local economic (re)generation*. Here we arrive at the essentially economic logic of selling places, of course, but there is also a more social logic at work in that the self-promotion of places may be operating as a subtle form of *socialisation* designed to convince local people, many of whom will be disadvantaged and potentially disaffected, that they are important cogs in a successful community and that all sorts of 'good things' are really being done on their behalf.

Central to the activities subsumed under the heading of selling places is often a conscious and deliberate manipulation of *culture* in an effort to enhance the appeal and interest of places, especially to the relatively well-off and well-educated workforces of high-technology industry, but also to 'up-market' tourists and to the organisers of conferences and other money-spinning exercises. In part this manipulation of culture depends upon promoting traditions, life-styles and arts that are supposed to be *locally* rooted, and in this respect the selling of places has what the humanistic geographers might call an 'authentic' quality spawned by the cultural life of the places themselves (Relph, 1976), but in part too this manipulation can involve using a range of loosely 'cultural' motifs, events and exhibitions (such as Casaubon's museum) that have no necessary associations with the places concerned and that might thereby be adjudged 'inauthentic'. At once it is possible to see how the manipulation of culture can give rise to tensions and potential conflict, then, given that many people within a place may feel that the cultural materials drawn upon by the 'place marketeers' are inappropriate: and this is particularly true if the marketeers are claiming certain cultural practices to be local in origin and spirit, whereas the majority of people in a place reckon them to be unfaithful and unwanted representations of what local cultural life is actually all about. The situation quickly becomes more complicated when there are several different visions of local culture, perhaps associated with different class, gender or ethnic groupings in the vicinity, all of which arguably

3

should be accorded respect by the marketeers. But this reference to conflicts over cultural representation is to run ahead of the argument, since for the moment we are simply noting the pivotal role that culture has come to perform for capital in the selling of places, principally as a resource for economic gain (through attracting inward investment) but also as a device for engineering social consensus (although, and as just indicated, the emotive quality of culture can often shatter consensus politics).

An additional key ingredient to be stirred into this mix is that of *history*, and in so doing we introduce into the discussion of selling places the whole terrain of debate amongst historians and cultural critics (Hewison, 1987; Wright, 1985), as well as amongst historical geographers (Datel and Dingemans, 1984; Hardy, 1988) and archaeologists (Baker and Thomas, 1990), concerning the ways in which the past is appropriated in the present. It is evident that the culture of a place, however this might be understood, is intimately bound up with the history of that place and with the histories (which may *not* always be locally-rooted) of the peoples who have ended up living in that place. To paraphrase Wright (1985, p. 3), the past is indeed a 'cultural presence' in the present, and the way in which we approach the present is inevitably – and often in a manner scantly recognised or reflected upon – shaped by the understandings that we have of what went on (what was right and what was wrong; who was winning and who was losing) in the history (or, better, histories) of the peoples and places around us. Wright concentrates on a sizeable place, all Britain, in explaining and critiquing the components of the 'national past', but there can be little doubt that similar processes of history being bound into contemporary self-understandings occur in the context of smaller places such as specific towns and cities. Indeed, the continual fascination with local history must derive at least in part from a desire of local people to gain a sort of self-knowledge through knowing what their predecessors (maybe but not necessarily their biological ancestors) have been doing over the centuries in the particular corner of the world that is their home place. White neatly captures what is involved here, whilst also elaborating upon the peculiar significance of a place and its constituent physical structures in the 'making' of local history:

'Consciousness of the past is a living experience at the local level in a way that it rarely is in any other sphere. It is forced on us every day of our lives by the physical change in neighbourhoods – streets and buildings that were there one, five, ten years ago, but are now gone; by the changing functions of buildings and places – the street market now a car park, the cinema now a warehouse, the factory now a wasteland; by the change in the people around us, the coming and going of our neighbours and our families and our friends. Change has no meaning without comparison with what used to be, and those comparisons in things near at hand are a part of the daily thought processes of all of us in recognising just where we are' (White, 1981, p. 34).

4

Given this centrality of local history to processes of self-identification in the present, it is surely inevitable that all manner of people will pick up on such history, and will seek to work with it in a variety of ways to serve a variety of ends. And this means that any manipulation of culture involved in the selling of places will tend also to be a manipulation of history: an attempt to tap historical resources, whether these be 'relic' features in a landscape such as a castle or associations with an 'historic' event such as the beheading of a queen, in the course of marketing the image of a place both externally and internally.

However, it should immediately be pointed out that the implication of history in this selling phenomenon is in no way restricted to the obvious level of the castle in the town wall or the industrial heritage museum. Rather, we would want to identify several further dimensions to the role of history in selling places, one of which occurs in the shape of an anniversary such as the Bicentenary in Paris when a place-specific 'historic' event is turned into a money-making extravaganza in its own right *and* into a vehicle for the more general promotion of what the place can offer to enterprises and to tourists. In this respect the anniversary celebrations serve a similar function to equally grand events such as Vancouver holding a 'world fair' (see Ley and Olds, 1988) or Barcelona staging the Olympics, and it might be added that the situation is not entirely dissimilar for smaller and more frequent events such as carnivals and festivals, many of which combine cultural and historical referencing with momentary surges in local capital accumulation. The financial objectives here are easy to detect, but what some commentators argue is that these events are also significant in pacifying local peoples whose everyday lives beyond the event in question are monotonous and unrewarding: and the suggestion is that we see in operation a formula of social control sometimes referred to as that of 'bread and circuses' (a term adopted by Harvey, 1989a, 1989b) which says that you give the disaffected a taste of bread and a day of fun, and then (and for somewhat longer) they will forget about their problems and believe in the benificence of the system. As in the previous examination of culture, it quickly becomes apparent that this manipulation of history – along with the deceptively straightforward showcasing of history in the castle or museum – is going to create tensions and conflicts, since the same event or object will mean different things to different people (class, gender and ethnicity will again be vital sources of difference, as too will sexuality, age, religion and politics) whilst these different people may consider entirely different events and objects more worthy of celebration or display.

This account of contesting history is again to run ahead of the

5

argument, though, and for the moment it will suffice to mention three further ways in which history is implicated in selling places, each of which is scrutinised in several of the chapters that follow. The first of these occurs when local authorities and local entrepreneurs self-consciously draw upon the economic and social history of a particular place as a source of pride and inspiration for the present, and the ironic thing is that place marketeers are often able to turn the most negative of historical associations – consider the black, grimy, smokestacked landscapes of the mind conjured up by the mention of old industrial cities and regions – into a powerful statement about a local history full of initiative and enterprise which projects into an exciting and prosperous local future (see in particular the chapter by Lowe, who discusses this use of a region's industrial past in selling the 'Black Country'). The second and related possibility here entails the use of 'heroic' imagery surrounding specific historical processes as a lever for money-making and persuasion in the present, and we are thinking at this point about how the heroism – a highly contentious heroism – of the people involved in processes such as revolution or colonisation is more or less consciously mobilised to cast a favourable light upon processes occurring today. The French Revolution is again a good example (see the chapter by Kearns), but so too is the use of the North American 'frontier myth' as a means of legitimating the current economic colonisation of the inner-city 'wilderness' being undertaken by the forces of gentrification (see the chapter by Reid and Smith). The third possibility here involves the planned adoption of all manner of historical references, particularly architectural references, in the fabric of the built environment, so as to foster the 'cosy' ambience of a place that is basically familiar – unlike the constructions of modernism that allegedly ignore history, culture and place, and thereby alienate people from their surroundings – and which is likely as a result to encourage inward investment from enterprises and tourists whilst also securing the loyalty of local residents (see the chapters by Crilley and by Laurier). Once addressing these latter sorts of questions we inevitably engage with the debate about postmodernism in the city, a debate that is very much concerned with what happens to culture and history in the selling of places, and to this debate we will return shortly.

The concern of the collection hence becomes *culture, history and capital*, and the complicated fashion in which these elements (which are in any case difficult to define in isolation) tangle together in the phenomenon of selling places. Leading from Casaubon's encounter with the museum of technology – a museum whose contents are at once scientific, cultural and historical; a museum which earns money and socialises the populace; a museum whose own historical character and presence in a specific cultural-historical *milieu* helps in selling it to the public – we have now introduced in outline the key issues of concern for the volume. In the remainder of this chapter we will develop some of these issues further. We begin by signposting arguments about the historical emergence of cities as elite-bourgeois centres in which culture and capital are spun together in ways that reflect in part the working of specifiable economic and social logics; and then we consider those 'other peoples' who have come to stand outside of these elite-bourgeois equations and to pose something of a threat to them. Having sketched in these *critical historical geographies*, we use them as an optic through which to view the contemporary *promotion of cities* (and of other places); and more specifically we proceed by examining the 'New Right' discourses of selling places, the connections between postmodernism and selling places, and the relationships between 'history' and 'memory' implicated in selling places.

Critical Historical Geographies

Centralising the Surplus, the Power and the Glory

An essential prerequisite of the existence of cities is an agricultural *surplus* which can release some individuals from direct food production, and most writers on the origins of urbanism agree that the emergence of urban centres as dense concentrations of people – few of whom directly produce in the sense of growing crops or tending livestock – must have depended upon the systematic appropriation of a surplus from the surrounding countryside (Harvey, 1973, p. 216). Upon this food surplus all of the commercial and cultural activity of cities is dependent and most of their secondary and tertiary economic functions founded, and radical theorists might go further to claim that this 'social surplus'[3] extracted by the city from the country is actually the prime basis for exploitation within human society. Such a claim now requires qualification, particularly given feminist arguments about the gender dimensions to exploitation (see below), but Marx was surely correct to identify a key 'division of labour' within society as that between agricultural labour (people working the land) and industrial and commercial labour (people turning nature into products and then organising the exchange of these products): a division that leads to the separation of town and country and to an inevitable clash of interests between them.[4] Moreover, and whilst we need to be careful here given what he also claimed about property relations in the constitution of class, Marx clearly supposed that the appearance of an *urban elite* – an elite which was non-productive in that it could not feed itself – was a crucial first step in the splintering of society into social classes defined through their relationship to the processes of production.[5] Here is how he and Engels encapsulated their thinking about surplus, urbanism and class in *The German Ideology*:

'The antagonism between town and country begins with the transition from barbarism to civilization, from tribe to state, from locality to nation, and runs through the whole history of civilization to the present day. . . . The existence of the town implies, at the same time, the necessity of administration, police, taxes, etc.; in short, of the municipality, and thus of politics in general. Here first becomes manifest the division of the population into two great classes, which is directly based on the division of labour and on the instruments of production' (Marx and Engels, 1970, p. 69).

The appropriation of a surplus from the countryside hence allows the creation of an elite class, and the conjoint *centralisation* of both the surplus and the elite was what brought the city into being and – as the above quote emphasises – ushered into history a whole new set of geographically-focused practices to do with administration, policing, taxation and the suchlike through which the wheels of surplus extraction and class formation could be further oiled.

This argument about the city as a mechanism for ensuring the economic gain of a bourgeois class is an orthodoxly Marxist one, of course, and it is vital immediately to recast its economically determinist outlines by discussing other perspectives on the part that cities play in the socio-spatial landscape (and on why they should have appeared at all). Marx's own references to practices such as administration signpost these other perspectives, since the tendency in much recent thinking is to stress the many and various forms of *power* that can be exercised through the resources of the city by an elite class who become not just 'managers' of the city but 'managers' of whole societies. This ensnaring of the city in historical geographies of power is something that Mumford came to appreciate sooner than most:

'The city, as Mumford has shown brilliantly, may be regarded as a special form of "container", a crucible for the generation of power on a scale unthinkable in non-urban communities. In his words, "[t]he first beginning of urban life, the first time the city proper becomes visible, was marked by a sudden increase in power in every department and by a magnification of the role of power itself in the affairs of men [and women]"' (Giddens, 1981, p. 96).

An economic component to the power relations hinged around the city must have been present in ancient times, just as it obviously is today, but for many writers on urban origins it was less economic power *per se* – less the simple possession of wealth and commercial arrangements – and more other sources of power that enabled urban elites to exert their will on peoples in surrounding countrysides and farther afield. Indeed, it is argued by some that economic gain was more a *consequence* of power being deployed in 'non-economic' ways than a cause or enabler of this deployment:

'For the economic power generated by the early cities seems of lesser significance than political and military power, centred first in theocratic and later in monarchical control, which in the vast majority of cases appears to have been most consequential in their formation and subsequent development. Mumford's view is that ancient cities were above all "containers" of religious and later royal power, the temple and the palace. It was these, he argues convincingly, which (by fair means or foul) attracted people from a distance, including traders: the drawing power of the city brought the merchants, rather than *vice versa*' (Giddens, 1981, p. 100).

A writer such as Mann pays special attention to the political and administrative power channelled through the earliest cities, and to give just one example he describes the Greek city-state as a 'political power unit, centralising and co-ordinating the activities of [a] small territorial space' (Mann, 1984, p. 224); and in related fashion Giddens explains that cities became powerful because of their centrality in 'the retention and control of information or knowledge' (Giddens, 1981, p. 94). This latter point is a crucial one, hinting as it does at an important association between cities and *writing*: notably the writing down in lists of regularised official knowledge about people, possessions, resources and finances, but also the writing of a literate class now able to ensure through the written word some continuity of cultural forms.

This observation reintroduces the matter of culture into our chapter, and a key assertion for us is that the birth of the city – which freed a group of people from the drudgery of working the land, giving it the opportunity to develop human activity in very new directions – both opened up a space for culture and made available new horizons through which an elite class could exert power over others. There is a danger of overstating this argument, and of slipping into the overly simple equation of urbanism with culture or (more broadly) civilisation that (dis)figures much thinking about the city, but there does seem to be warrant for regarding the city as a site of culture and civilisation 'insulated from nature' (Renfrew, 1972, p. 74) where what we could term cultural or ideological power has become crystallised and disseminated. Renfrew actually specifies three 'insulators' in the sealing off of civilisation from nature: one of these is the city itself, and it is revealing that the other two are writing (see above) and ceremony or, to be more precise, ceremonial centres. This is an intriguing specification, since there is much historical-archaeological evidence to suggest that many of the earliest cities were at least in part ceremonial centres with cosmological significance – their very location and lay-out apparently being deeply imbued with cosmic symbolism (see Wheatley's remarkable research on ancient Chinese cities: Wheatley, 1971) – and whose elite residents constituted a priesthood regarded as mediators with the gods. Dodgshon reasons that the boundaries between spiritual and secular were probably very blurred in all of this, and he examines 'the assumption of divine status by chiefs and tribal kings' (Dodgshon, 1987, p. 120) which meant that the theocratic and 'monarchical' elements described by Mumford occupied structurally similar positions in the life of the early cities. Furthermore, and as Dodgshon goes on to state with reference to centres as different as the small mortuary sites of European prehistory and the massive religious complexes (containing palaces, temples, shrines and ceremonial platforms) of the Middle and Far East:

'The extent to which the power of elites became imbued with sacro-religious authority and meaning is more visibly demonstrated by the way in which centres of ritual practice took on a more elaborate and lavish appearance. Through their command of tribute, *corvée* and slave labour, they were able to initiate the construction of monuments that were patently meant to impress' (Dodgshon, 1987, p. 120).

A more nuanced account would be appropriate here, but what should be clear is that this distinctively cultural geography – this coming together of ceremony, priesthood, kingship and glorious architectural constructions – was in no sense some 'nice' add-on to the operations of early urban systems. Rather, it was through this cultural geography that power was exerted over whole societies: it was because the priests and the kings had more or less consciously taken upon themselves this cosmological significance, and had become in the eyes of the 'masses' pivotal to the very functioning of creation and thereby legitimated in their dispensing of justice, demanding of monies and enforcing of duties, that they were so powerful and so able to command (if not always easily) massive territories from the tiny pin-pricks in the map that indexed the locations of their home 'cities'. And, to reiterate an earlier remark, the mobilisation of this power certainly led to economic gains for the urban elites – it allowed them to collect the surplus, to channel it in through the city gates – but it was cultural or ideological power, and not economic power, that justified and made possible these gains. Having followed this reasoning, then, we have now managed a first 'cut' through the tangled web of what cities are all about: demonstrating, chiefly by reference to historical materials, that the city as cultural capital is dependent upon the extraction of a surplus, but that cultural resources are mobilised to sanction this extraction and also to organise its collection.

Urban Culture and Social Control

The city is hence to be seen as a focus of power exerted by an urban elite over the peoples of surrounding territories, and in this sense we are talking chiefly about power radiating out *beyond* the city to shape the activities of those not actually in the city. But in what follows it is important to recognise that the urban elite has also looked to exert power *within* the city itself, with the aim of shaping the activities of all of those *other peoples* – those other non-elite peoples – who have, as it were, 'turned up' in the city for a variety of reasons, perhaps to serve the elite, perhaps to avoid persecution or perhaps merely to escape rural poverty. We must again avoid too crude a perspective on what is involved here, but we still want to suggest that with the passage of history – as ancient times passed into feudal times and then into the age of capitalism – the urban elite, which we can refer to now as an urban-based *bourgeoisie* (see **Endnote 5**), has continued to manipulate cultural resources in a dual attempt to command 'other peoples' both beyond and within the city.

Weber saw the history of early agrarian societies as the domination of the countryside by the town, as had Marx before him, but Weber went further and examined the way in which the emergent bourgeoisie forged an urban culture that created 'space' for political activity against the constraints of the feudal order. The precise articulation of town and country in feudal times remains a subject of contention, as too is the extent to which urban areas could be both linchpin and solvent of the feudal order, and numerous writers now regard these questions as basic to the whole matter of the transition from feudalism to capitalism (see Dunford and Perrons, 1983; Langton and Höppe, 1983; Merrington, 1975). The debate hinges in part on the position of the emerging bourgeois class between feudalism and capitalism, but most writers are in agreement that there *was* a distinct social stratum in the towns and cities that organised its activities along capitalistic lines (involving the co-ordination of production and consumption for monetary gain). What Weber realised, and what needs to be made more central to the transition debate, is that the self-consciousness of this particular social stratum took the form of a distinctive urban culture (Weber, 1958, p. 96). The bourgeoisie or, rather, bourgeoisies (since it is important not to overlook differences between the bourgeois communities of different urban centres) identified themselves with their own urban settlements and organised their lives in terms of specific urban corporations, and in their own name they defined an urban culture against both the broader regional feudal powers[6] and the poorer, propertyless 'other peoples' of their towns and cities. Against the feudal magnates, the air of the town was said to set people free; against the disorderly poor, the increasingly powerful 'city fathers' began to lay down a social and spatial grid which marked the city out as essentially a bourgeois arena.

This self-aware definition of an embryonic bourgeois urban culture probably first occurred in the city-states of Renaissance Italy,[7] but Varey (1990) has produced an account of something similar occurring in the eighteenth-century English city of Bath,[8] and Borsay (1989) has described the more general links between social control and spatial order that arose in the Early Modern British provincial town. As the urban bourgeoisie gained greater purchase in national politics, so they increasingly expressed their power in the form of town plans and city architecture: the city hence became in itself a mechanism of social discipline. The distinction between public and private space was renegotiated, as Stallybrass and White (1986) explain so compellingly, both to assert bourgeois values against a 'corrupt' aristocracy (the old feudal powers) and to put the lower sorts of the city 'in their place'. The bourgeoisie projected its private understandings of how social life should be conducted upon the public spaces of the city – the streets, which were now rigorously policed to ensure that middle-class conceptions of appropriate public behaviour were adhered to; the public parks, with their own distinctive codes of behaviour (Malchow, 1985) – and Harvey neatly evokes this development in the context of a nineteenth-century Parisian elite striving to turn the boulevards of the French capital into 'bourgeois interiors':

'Paris experienced a dramatic shift from the introverted, private and personalised urbanism of the July Monarchy to an extroverted, public and collectivised style of urbanism under the Second Empire. . . . The boulevards, lit by gas lights, dazzling shop window displays, and cafés open to the street (an innovation of the Second Empire), became corridors of homage to the power of money and commodities, play spaces for the bourgeoisie. When Baudelaire's lover suggests the proprietor might send the ragged man and his children packing, it is the sense of proprietorship over public space that is really significant, rather than the all-too-familiar encounter with poverty. . . . It was for this reason that the reoccupation of central Paris by the popular classes [under the Commune] took on such symbolic importance. For it occurred in a context where the poor and the working class were being chased, in imagination as well as in fact, from the strategic spaces and even off the boulevards now viewed as bourgeois interiors' (Harvey, 1985, p. 204).

The Parisian bourgeoisie hence sought to colonise all of Paris with its blend of culture and capital, a set of understandings in which cultural items became commodified and capital accumulation was aestheticised, but at the same time it also made efforts either to discipline those social elements which were non-conforming or to oust them from the public places of the capital altogether (a strategy of exclusion elsewhere described as one of spatial 'purification': see Sibley, 1989). A new form of socio-spatial order was being nurtured, then, and the character of this order was expressed most overtly in the political and indeed cultural revolts against its limitations: herein lay the divisions between Haussmann and the Commune,[9] and also the connections between Arthur Rimbaud and Auguste Blanqui (Ross, 1988).

It is worth considering in a little more detail how urban form itself was mobilised to act as a tool of social control disciplining and ousting the 'other peoples' of the city. An obvious example is the way in which the exclusion of working-class people from certain parts of the city – notably the parts which are politically vulnerable, the financial and military centres – was expected to vaccinate cities against revolution, as Harvey has argued for Paris and as Harrison (1988) has suggested for British cities. But these bourgeois cities were meant to discipline in a creative as well as a restrictive sense: the architecture of public buildings was intended to impress with its magnificence, and to make the political order seem immortal through its classical references (and a thread of continuity can be traced here from the 'glory' of the ancient ceremonial city: see the essays by Woolf and Grant in Cosgrove and Daniels, 1988). Height (as with Sacre Coeur in Harvey's account: see Harvey, 1979), arrangement (as with those ancient Chinese cities studied by Wheatley: see Wheatley, 1971) and naming practices (think of the renaming of cities, squares and streets during the collapse of Communist Eastern Europe): all of these have been manipulated so as to present the current social hierarchy as natural and permanent, and in this respect we may speak of urban form as an ideological project.

Urban culture is thus the active project of the urban bourgeoisie, and the city is an important instrument in their social dominance.[10] Marx and Weber saw cities as central to the emergence of class, as explained above, and both also regarded class as a key social divide around which exploitation has come to be organised. Yet class is most definitely not the *only* divide around which exploitation arises in and through the city, and it is important to recognise that rather different networks of exploitative social relations are also knitted into the workings of culture, capital and the city. Marx and Engels may have recognised that this was the case when speaking of the family as the basis of private property and then of property as the basis for exploitative patriarchal relations, but Marxist urban theorists have only recently woken up to the way in which the bourgeois organisation of the city into public and private realms not only puts the proletariat 'in its place', it also puts women into theirs. It may be true that the urban bourgeoisie has often sought to turn public city spaces into its own private play spaces, but it must be remembered that in a very real sense the bourgeoisie was (and is) male and so its cultural projection on to the city stemmed more from the male sociability of the drawing room than from the female domesticity of the hearth. Privacy and inner-directed domesticity have increasingly become equated, though, and when these two states came together to define new suburban spaces or new arenas of leisure in the late-nineteenth-century city these bourgeois spaces were clearly gendered. Suburbanisation was admittedly in part the spatial expression of class-distance, of the bourgeoisie escaping from physical association with an inner-city proletariat, but the sharp division that ensued between home and work also delimited a domestic space which parcelled off the supposedly 'natural' site of women's being and doing from the supposedly 'natural' site of men's wheeling and dealing. For women to operate outside of this space was far from easy, and for bourgeois women it was to break social codes and for working-class women it was to expose them to the risk of being preyed upon in a variety of ways (see Davidoff and Hall, 1983; McDowell, 1983; Women and Geography Study Group, 1984). As Massey underlines in the course of deconstructing Harvey's essay (1985) on Second Empire Paris, men and women here were spatially 'separated' and entertained rather different possibilities for spatial 'penetration':

'[Harvey] discusses suburbanisation at a number of points, but does not mention the separation of the spheres. Or again, he discusses how Frédéric Moreau, [bourgeois] hero of Flaubert's *L'Éducation Sentimentale*, "glides in and out of the differentiated spaces of the city, with the same sort of ease that money and commodities change hands . . .". Comments Harvey: ". . . it was the possession of money that allowed the present to slip through Frédéric's grasp, while opening social space to casual penetration. Evidently, time, space and money could be invested with rather different significances, depending upon the conditions and possibilities of trade – between them" (Harvey, 1985, pp. 263–264). Well, yes, nearly but not quite. Frédéric, as he casually penetrated these social spaces, did have another little advantage in life too' (Massey, 1991, p. 49).

And this is why Wolff (1985) concludes that the so-called 'soft city' (Raban, 1974) of the urban *flâneur*, the urban stroller, was (and is) inaccessible to women, certainly to 'respectable' women but also in many ways to 'non-respectable' women (the barmaids, dancers and prostitutes) whose 'movement would still be effectively restricted by the threat of male violence' (Massey, 1991, p. 47: see also Valentine, 1989): in short, there were (and are) no *flâneuses* and much of the city was (and is) for women a 'no-go area' or at least an area fraught with danger.

The City and its 'Others'

It is not only women who have had their lives moulded and constrained through the cultural assumptions of the bourgeois city, though, since many human groupings definable on a variety of grounds (both by themselves and by society-in-general) have been regarded as standing outside of the orthodoxies of bourgeois existence and therefore as irritations and possibly threats to be countered. This has been most obviously the case when divides along ethnic lines have been identified and acted upon, a process that has often been enforced by legal means and through deliberate policing tactics. Some scholars refer to this process as the fundamental characteristic of the 'colonial city', and argue that the intense socio-spatial separations of the South African 'Apartheid city' merely mark the most recent and formalised expression of a development, for example, the small boroughs of Medieval Wales by royal statute, which were designated English places of living, working and trading where native Welsh people should not be found nor possess any legal rights (Carter, 1989; Christopher, 1983: see the essays in Griffiths, 1978). A few, though by no means all, of these centres hence became (to borrow a metaphor) English islands in Welsh seas, and the resulting separations were sanctioned as much by the cultural mores of the English urbanites 'planted' in these early boroughs as by any more basic repulsions triggered by skin colour or physiognomy. Once the European conquests of the world had exported these arrangements to distant lands and therefore given rise to the 'colonial city' proper, the separations involved became more transparently rooted in 'racial' differences and structured around the socio-spatial exclusions laid down by the hegemonic white colonists, but there can be no doubt that the supposed superiority of white culture – the notion that this was indeed the apex of civilisation compared to the barbarism of non-white peoples – was deeply implicated in driving and legitimating this particular chapter in world history (see King, 1976).

There is also a sense in which the 'colonial city' has now 'come home', of course, in that the global redistributions of human populations from the nineteenth century onwards have contributed to most Western urban areas becoming ethnic mosaics scarred in all too many instances by segregationist and prejudicial attitudes on the part of white 'host' groupings. The mechanisms leading to 'immigrants' being discriminated against include the operations of labour and housing markets – operations through which capital uses and abuses ethnic minorities in the city – but energising this practical discrimination, both flowing into and feeding upon it, has been the complex discourse of 'white supremacy' that is at once cultural, social and political (Smith, S., 1988, 1989). From the nineteenth-century North American city where a mental map of ghetto and slum was given an ethnic dimension and explained in cultural terms (Ward, 1989), through to the late-twentieth-century British city where inner-city problems from unemployment to lawlessness are ascribed to the cultural inadequacies of black and Asian communities (Smith, S., 1988, 1989), bourgeois urban culture specifies who is 'correct', who should be where and who needs policing within the context of this urban ethnic mix.[11]

In setting out and policing the spaces for their 'own' culture, and in also apportioning other spaces for other cultures – spaces that ideally they would probably be happy to obliterate, but on whose presence they may also be economically dependent – the urban bourgeoisie duly makes a statement about gender and ethnic relations as well as about those of class. And this account could easily be extended to consider many other social groupings in the city whose failure to square with bourgeois cultural values leads them to stand outside of the 'normal' spaces of urban living, often because they are actively excluded but sometimes as a conscious strategy allowing them to avoid the 'normalising' pressures of bourgeois expectation. Think in this respect of such groupings as diverse as children and elderly people, Gays and lesbians, people with physical or mental disabilities, people with unusual political or religious beliefs, tramps, travellers, criminals: all of these overlapping groupings – together with workers, women and ethnic minorities – constitute the 'other peoples' of the city who for one reason or another depart from, irritate and threaten the spaces of bourgeois culture. We acknowledge certain drawbacks that arise from thinking in terms of 'otherness' – and in using dualisms such as 'same' and 'other', 'inside' and 'outside', 'inclusion' and 'exclusion' – given that these terms and concepts cannot avoid being locked into a Western tradition of rational thought, complete with its 'colonial, bourgeois and phallocentric economies of meaning' (Rose, 1991, p. 346: see also Corrigan, 1991; Young, 1990) that ascribe value in some directions but not in others, and yet we *do* want to suggest that it is these 'other peoples' of the city that furnish the critical edge for our perspective here (and indeed throughout the volume) on the specific phenomenon of selling places.

What we want to emphasise in this respect is that these 'other peoples' have relationships with the city – or, to be more precise, with the particular city places in which they live, work, rest, play and dream; often the places 'left over' after those with power have chosen theirs – that differ (often quite dramatically) from both the relationship lived by the bourgeoisie and the 'respectable' relationship intended for them by the bourgeoisie. And what we further want to claim is that these 'other peoples' hence possess *other attachments* to the city that differ from the arguably superficial attachments of the bourgeoisie – those to do with property-ownership and fancy possessions, the surface badges of cultural capital – and that alternative attachments of this sort are always shot through with meanings, though not necessarily in all that reflected upon a fashion, to the extent that these 'other peoples' will feel their lives to be seriously compromised if their particular city places are mistreated in any way. We are even tempted to argue that these other attachments are *more* vital to the peoples concerned than they are to their bourgeois counterparts, in large measure because such peoples tend to be more firmly 'fixed' in place than are the bourgeoisie thanks to greater financial limitations on their ability to get up and go, but this is not to slip into a whimsical or romantic vision of people naïvely 'loving' their places and only seeing good things in their everyday workings. In fact, we would argue that the 'other peoples' of the city can often be highly critical of their own places – recognising (say) the dangerousness of the inner-city park or the unhealthiness of the Victorian tenement block – but will nonetheless continue to feel deep place attachments which colour their whole lives, and will routinely draw upon the pros and cons of local life in these local environments to inform their cultural, social and political conduct. The arguments that we are hinting at here arise from an intriguing convergence between humanistic geography (which has sometimes been guilty of sentimentalising place) and a more critical social-cultural geography that recognises the importance of the shared meanings that people can ascribe to and derive from their places (see Cloke *et al.*, 1991, chapter 3), and what is more the importance of such meanings in the contest over contemporary capital restructuring is also beginning to be taken seriously in the debates of political-economic geography. Here is Hudson exploring the tensions that arise when a locality is treated simply as a *space* by capital, as a location whose profitability to enterprise is what enters into the capitalist's equations and beggar the people who may actually live there:

'[T]he point is that for these people the locality is *not just* a space in which to work for a wage but a *place* where they were born, went to school, have friends and relations etc.; places where they are socialised human beings rather than just the commodity labour-power and, as a result, places to which they have become deeply attached. These localities are places that have come to have socially endowed and shared meanings for people that touch on all aspects of their lives and that help shape who they are by virtue of *where* they are. At the same time, however, these localities may also lead a precarious existence, vulnerable to the disinvestment decisions of capitals (and, it must be acknowledged, national states)' (Hudson, 1988, pp. 493–494: see also Hudson and Sadler, 1986).

We do not want to understate the complexity of what is involved here – the 'other peoples' of the city are so many and various; the places identified with are so diverse and yet often overlapping; the exact attachments are so differently constituted and differently impelling

– but we are certain that here, in this discussion of the 'other side' to the city's human geography that squidges out from beneath the economic and social logics isolated above, we have found the key to why the city's cultural capital can never be manipulated as consensually as the place marketeers would like. The marketeers have assumed that the places being sold are the spaces of bourgeois culture, and in a way they are right: the problems arise because in the process the marketeers also try to sell places that mean other things to the 'other peoples' of the city, who thereby resist the form that the selling takes (along with its primarily economic motivation) and who also resist the 'bread and circuses' element of this selling (its attempt to exert social control by convincing people of truths that are not their own).

The Promotion of Cities

The Discourses of Markets, Competition and Individualism

Selling places is now a well-known feature of contemporary urban societies: it is discussed by all sorts of commentators, particularly in the more applied sense of how the 'marketing' of places does and should occur (for a recent statement see Ashworth and Voogd, 1990), as well as in the context of debates over how North American cities seek to achieve 'growth' through self-promotion (see Logan and Molotch, 1987, along with more radical critiques such as Harvey, 1989a; Leitner, 1990; Mair and Cox, 1988). It is wholly unsurprising that the phenomenon has become couched in the prevailing languages of 'New Right' capitalism – and here 'Thatcherism' and 'Reaganism' still cast long shadows – such that places are not so much presented as foci of attachment and concern, but as bundles of social and economic opportunity *competing* against one another in the open (and unregulated) *market* for a share of the capital investment cake (whether this be the investment of enterprises, tourists, local consumers or whatever). In this discourse places do indeed become 'commodified', regarded as commodities to be consumed and as commodities that can be rendered attractive, advertised and marketed much as capitalists would any product (see in particular the chapter by Goodwin), and we hence need to appreciate the constituents of this discourse – considering as we do the various turns that the discourse takes for different 'actors' in the process – and their varying practical consequences. At one level this means recognising what Martin describes as the attempt of Thatcherite 'New Conservatism' to build a fresh 'one nation politics' rooted in a brave new programme designed to arrest long-term economic decline:

'This programme was to proceed on two main fronts. The first and most urgent was to purge the economy of the inflation that had become endemic during the 1970s and to revive the regenerative powers of efficient capitalism, by controlling the money supply, limiting state spending, and fostering free market competition and a climate of entrepreneurial innovation. The second objective has been to forge a new social and ideological consensus, a new *Weltanschauung*, consistent with this free enterprise economy: a "cultural revolution" based on the virtues of individualism, self-help, private property and respect for law and order' (Martin, 1988, p. 410).

It is not difficult to see how the specific discourse about selling places fits into this 'cultural revolution' (and it is revealing that Martin talks of 'culture' in this respect): and neither is it difficult to see the compatibility here with a second level of arguments about how the problems of individual localities can best be solved not through the conventional mechanisms of local authority planning, but by mobilising local resources of private capital, entrepreneurship and self-help in the scramble for rewards available in the free (place) market. As Duncan and Goodwin (1985, 1988) explain, the whole terrain of thinking about local economic policies (and about local political forms more generally) is being shifted, and the result is that a range of local institutions – including local authorities, but also including other bodies such as chambers of commerce and Urban Development Corporations – now internalise the idea that the interests of a place are best served by lifting the 'dead hand' of regulation and by opening it to the sway of market forces.

The specific discourse in which places are converted into products to be sold in competitive markets clearly embraces these ideas, as can be seen from virtually all of the contributions to this volume, and Fretter's contribution is interesting because it is written very much from *within* the discourse. This is unsurprising, of course, given that Fretter's everyday task is to find ways of rendering places such as Birmingham and Gwent attractive to inward investment (ways that are indeed very much bound up with the self-conscious manipulation of local and sometimes not-so-local cultures and histories). It is important to be careful in assessing such a contribution, however, since it would be easy from an unthinking radical position to cast the arguments involved in a negative light, and to imply that place marketeers possess no real concern for the places that they promote using the vocabularies of market and competition. Would such a criticism be fair, and is it not important to leaven the radical critique with a more humanistic sensitivity to the care for a place that may be quite genuinely felt by members of local authorities and by local entrepreneurs? In her chapter Lowe encounters this question as she considers the activities of the Richardson brothers, the Black Country 'regional entrepreneurs', who clearly subscribe to a free market vision of local economic development but who also appear to treat the region

19

as something more than simply a resource to be milked dry (as a space to be exploited by capital until it ceases to yield profits). And once we begin to tackle substantive matters such as these, some of the more abstract claims offered above and below about the superficiality of the bourgeois relationship to places do inevitably begin to look overstated and in need of qualification. This being said, we would insist that a critical edge still be retained when interpreting the (ab)use of culture by capital in the phenomenon of selling places, an (ab)use that is in a sense demanded by the imperatives of capital accumulation operating beyond the control of individual place marketeers, and Sadler provides just such a critical edge when dissecting in detail the hegemonic assumptions and exclusions – the things that should not be said, which presumably would include most of this book – associated with place marketing in Britain. Holcomb does something similar with respect to the 're-visioning' of old industrial places accompanying North American place marketing, and it is further the case that a number of the chapters (notably those by Goodwin, Reid and Smith, and Lowe) contrast the idealised and 'heroic' images conjured up by the place marketeers with the less ideal experience on the ground of what 'sold' places offer to local people (the 'other peoples' of the city) in terms of employment and quality of life.

An additional concern to flag here is a certain irony that attaches to the way in which the 'New Right' discourse of individualism – the notion of individual people, enterprises, localities or whatever striving to earn their just rewards using their own resources – does not contribute in the way that might be expected to an appreciation of the *individuality* which renders one 'thing' different from any other. With respect to people, for instance, it is clear that individualism does not mean acceptance of the many different sorts of individuals present in Western societies: indeed, what it really does is to posit as 'correct' a certain sort of individual (bourgeois, male, white, straight, able-bodied) who internalises certain sorts of values (about respect-ability, advancement, efficiency), and as such it is really desiring not variety but *sameness*. With respect to other 'things' the logic is repeated, and so for places the idea is not so much that they be genuinely different from one another but that they harness their surface differences in order to make themselves in a very real sense nothing but 'the same': to give themselves basically the same sort of attractive image – the same pleasant ensemble of motifs (cultural, historical, environmental, aesthetic) drained of anything controversial – with basically the same ambitions of sucking in capital so as to make the place in question 'richer' than the rest. Both Fretter and Sadler recognise this contradiction, pointing out the absurdity that writers of promotional literatures find themselves extolling the supposedly 'unique' qualities of supposedly 'unique' places using an actually quite universal vocabulary of 'better, bigger, more beautiful, more bountiful' and so on. The implication here must be that to this way of thinking the individuality of different places matters far less than cultivating an image of a certain sort of place with certain sorts of attributes, and it might even be argued that a great many 'poorer' and 'uglier' places thereby end up suffering the same fate as those individual people who have the 'misfortune' to depart from bourgeois norms. 'New Right' individualism in no way sanctions a tolerance let alone a celebration of diversity, then, whether we are thinking about people, places or whatever: and the practice of selling places may even generate sameness and blandness despite its appearance of bringing geographical difference into the fold of contemporary economic and political discourse.

Play, Politics and Postmodernism

This latter claim brings us neatly into another set of themes that run through several contributions to this volume, and these revolve around the evaluation of what the 'postmodern city' entails. There are various arguments currently being pursued about this phenomenon called 'postmodernism', with some commentators asserting that we are now living in a distinctively different 'postmodern' epoch or period marked by very new ways in which human affairs are organised – very new ways in which economies, polities, societies and cultures are constituted internally and connected externally; very new ways in which time and space are implicated in these constitutions and connections – and with other commentators supposing that our intellectual reflections upon this world are now tending towards a 'postmodernist' rejection of all of the certainties and truths of 'modernist' thought as initially forged in the crucible of the European Enlightenment (see Cloke *et al.*, 1991, chapter 6). But more specifically, it is suggested by many that the contemporary Western city now warrants being described in a new vocabulary (perhaps that of 'postmodernism') because of the great changes that have taken place in its workings and appearances – changes that have allegedly involved a dramatic fragmentation into smaller, more diverse yet more complexly (even chaotically) interwoven socio-spatial units – and which have revolutionised everything from the economic functions of the city through to the experienced forms of the city. And more specifically still, it is in the forms of the city – in the designs of houses and offices; in the lay-outs of streets and parks – as manipulated by contemporary architects and planners that a self-conscious notion of creating a 'postmodern landscape' has begun to emerge, a landscape shaved in particular of that excessive modernist growth which was Le Corbusier's concrete high-rise slab-block for living and working. This postmodern architectural vision is one in which difference and diversity come to dominate; one in which the austere homogeneity of the white monolith is replaced by the colourful playfulness of architects mixing and matching all manner of styles, references and materials:

'It has become commonplace in postmodernist circles to favour a reintroduction of multivalent symbolic dimensions into architecture, a mixing of codes, an appropriation of local vernaculars and regional traditions. Thus Jencks [the best-known advocate of postmodern architecture] suggests that architects look two ways simultaneously, towards the traditional slow-changing codes and particular ethnic meanings of a neighbourhood, and "towards the fast-changing codes of architectural fashion and professionalism"' (Huyssen, 1984, pp. 14–15: see also Jencks, 1987).

This description neatly encapsulates much of the 'spirit' of post-modernism, not just as it becomes inscribed in the city but as it permeates whole fields of practical and intellectual existence in the late-twentieth century, and what the quotation also underlines is the way in which the resources of culture and history readily get stirred into the postmodern melting-pot.

It should be easy to see how the postmodern turn as described here is embedded in the practice of selling places under investigation in this volume, and the linkages are spelled out explicitly in the chapters by Kearns, Holcomb, and Crilley, whilst being very much present in the other contributions. Indeed, there is the direct sense of selling the 'postmodern city' which entails the deliberate creation of cultural-historical packages – the more-or-less obvious lumping together of cultural and historical elements to produce marketable pastiches such as theme parks (Sorkin, 1992), simulations of rituals or past events, hyped-up monuments, and so on – and then there is the more indirect sense which entails the subtle 'playing' with cultural and historical materials in the production of what are supposed to be attractive, pleasing and uplifting environments. Both of these manoeuvres obey an economic logic by luring capital into the city, either through people paying to encounter the packages or through enterprises and their workforces deciding to settle in the locality, but they also obey a social control logic in being more-or-less consciously orchestrated to persuade people of the 'good things' on offer in the city (the exciting spectacles and the enticing surroundings). As Crilley explains with respect to postmodern architecture, the objective is to provide a 'warm', 'comfortable' and 'friendly' built environment – one which manipulates the known design features of popular urban locations, and which in consequence manipulates the cultural and historical associations (and spatial configurations) of old town squares, markets, courtyards and the like – and in so doing the aim is to manufacture an environment that will secure the acceptance and even the affection of peoples who might otherwise rebel against it. The suggestion is hence that the ideas and practices of postmodern architecture give an accommodating face to grand urban redevelopment schemes such as London Docklands, and thereby not only lure in big capital (the enterprises and the 'yuppies') but also defuse much of the antagonism that might otherwise have surfaced from indigenous local populations. Seen in this light, postmodern architecture appears less as an 'architecture of the people' (cf. Ley, 1989) and more as just another mutation in the 'architecture of power', with power no longer being so nakedly expressed in city form as in the days of massive skyscapers 'pricking' the skies but being cunningly disguised amidst the homeliness of cultural, historical and (above all) local remembrances.

The critical response to this state of affairs, and more specifically to this aspect of selling places, objects strongly to the playfulness and to what is arguably an 'irreverence' to the realities of cultures, histories and localities which declares 'mastery' over these realities by dismissively rearranging them into new patterns that might earn money and acceptance (Harvey, 1989b; Sorkin, 1992). 'Modern' architecture and urban planning is of course criticised for such sins, as is evident from the revulsion of Prince Charles to the monstrosities that he considers to be so 'disrespectful' of the historical urban fabic of central London, but postmodern architecture and planning is equally assaulted for being just as 'disrespectful' if more subtle in its processes and effects. Many writers hence take the 'postmodern city' to task for what is actually an excessive *decontextualisation* rather than the avowed sensitivity to cultural, historical and local contexts: and the argument here is that, by playfully scissoring elements out of their contexts ready to glue them back together any-old-how, these elements and their contexts are deprived of the meanings that they initially acquired by dint of being integral to (of being the very moorings for) the lives of 'real' peoples seeking to survive and to thrive in these 'real' places. There is an argument to be made about the superficiality of the postmodern engagement with the city, then, as well as the argument that a postmodernism which talks of contexts whilst decontextualising them parallels in its hypocrisy (possibly in a manner less than accidental) a 'New Rightism' which talks of individuals whilst de-individualising them (see above).

The further step that is taken by critics of the 'postmodern city', paralleling that taken by critics of postmodernism in all of its manifestations, is one that latches on to the apparent draining of politics out of the urban arena which occurs once culture and history are severed from their contexts and hence depleted of meaning. A nineteenth-century quayside where casual dockworkers laboured in appalling conditions for ridiculously low wages was a context rich in meaning and political tinder, for instance, but such a quayside done up as the backdrop for postmodern warehouses-turned-into-apartments occupied by a mobile new middle-class has been stripped of its original meanings and political resonances (except in the sense that some groupings may endeavour to retrieve these meanings and resonances as almost an 'imaginary' site of resistance). Similarly, nineteenth-century mills demanded an alienation of labour both from itself and from nature that was a lived reality and a likely spur to reaction, but one of these mills converted into a postmodern industrial heritage museum (often containing relics of industrial activity and lifestyle that have little to do with the mill itself) can convey little of this alienation nor of this impulsion to political unrest (and the pictures often portray a smiling labour force working happily in the great project of industrialising a great society).

The summation of the above arguments is duly a suspicion of the 'postmodern city' and of the various ways in which the managers of this city play with its cultural and historical legacies – decontextualising them and sucking out of them all political controversy – so as to sell its places both to outsiders who might be sources of inward investment and to insiders who might otherwise feel alienated or encounter encouragements to political defiance. As such, our line of reasoning closely parallels that given by Harvey in his treatment of *The Condition of Postmodernity* (1989b), although we would depart from Harvey in not seeing everything about postmodern society as a knee-jerk response to the most recent transformations of the capitalist mode of production (principally, the shift to 'flexible accumulation') and in granting more autonomy of thought-and-action to an urban elite-bourgeoisie that has for centuries (from the ancient ceremonial complex to the postmodern shopping mall) been installing cultural practices which yield both economic reward and social control. And where we would also depart from Harvey, and at the same time hook up our thoughts to a radical version of postmodernism as a way of thinking about the world (see Cloke *et al.*, 1991, chapter 6), is in paying more attention to those many 'other peoples' of the city discussed earlier: in not regarding the voices of these 'other peoples' simply as variants on or as only properly articulated through the voice of the working class, and in recognising that the involvements of these 'other peoples' with matters of culture, history and indeed locality are not so much ephemera as important strands in the development of an oppositional politics which proceeds from a multiplicity of 'centres' to challenge the hegemony of the bourgeois treatment of places (and, most notably, its selling of them).

Memory and History: Conflation and Contestation

As explained above, then, there are various tactics that the place marketeers adopt when trying to anticipate and to negotiate any conflicts that might arise over their policies: indeed, they may mobilise a 'New Right' discourse that depicts as reactionary and unhelpful any oppositional positions (and they probably 'shout loudest' in the process: see the chapter by Sadler), or they may adopt a deliberate coding of 'friendly', consensual and locally-rooted cultural and historical references into the built environment so as to secure acceptance of present transformations (one feature of the 'post-modern city': see the chapters by Holcomb, Crilley, and Laurier). But such tactics do not always work as well as their authors would like, with the result that conflicts *do* frequently arise because the manipulation of culture and history by the place marketeers runs against the understandings of local culture and history built into the daily encounters with city spaces of the city's 'other peoples' (as well as sometimes impacting upon these peoples in very direct and material ways, perhaps by causing them to be evicted or to be neighbours of alien developments: see the chapters by Goodwin, Crilley, and Woodward).

We have already addressed this theme of conflict from various angles, and in closing we want to spiral back to it by recasting our discussion in the vocabulary of 'memory' and 'history' that is suggested by Nora (1984a, 1984b) when claiming that the latter seeks a *critical* and the former an *uncritical* relationship with the past. In this view 'history' (with a small 'h') is concerned with the evaluation of claims about the past and their re-examination in the light of both current historiography and extant evidence, and as such it is essentially provisional in its interpretations. 'Memory', on the other hand, is concerned with the celebration of a certain account of the past which seeks to legitimate certain developments in the present, and as such it can easily become arrogant, chauvinist and exclusionary (particularly in the sense of using the past to legislate on who should be where and doing what in the present). It is crucial to appreciate that there are many sorts of memories, and that it is out of the spectrum of different memories associated with different peoples remembering the past in different ways that *contestation of the past in the present* emerges. Our specific source of disquiet here, however, concerns the transformation that all too commonly occurs whereby one particular form of memory – namely, that of the bourgeoisie – becomes the officially-sanctioned History (with a capital 'h') of a given territory, nation or city.[12] Integral to this transformation is the use of those many props from heritage centres to historical anniversaries which can be employed (more or less consciously) to create the impression of a truthful History with unavoidable lessons for the present, and the whole process is hence deeply implicated in the bourgeois project of selling places: it is both a support for this project, providing materials for the place marketeers to utilise, and a process that is itself secured through the infrastructure and happenings put 'in place' by the marketeers. To put matters another way, the act of selling places clearly conflates history and memory in the course of meeting economic needs (the immediate need to attract inward investment), but this conflation is also engineered by the bourgeois 'managers' of places to reaffirm the ideological commitments of a society (or, at least, the ideological commitments that society's most powerful groups would like to be hegemonic). It is relatively easy to detect the conflating of history with bourgeois memory in cases such as the celebrating of the Bicentenary in Paris or the staging of the Great Exhibition in London (see the chapters by Kearns and by Billinge), since in such cases it is only *specific* anniversaries or objects – and then only *specific facets* of these anniversaries or objects – that are chosen for *re*-presentation: the choosing is evidently far from random and answers more-or-less clearly to the needs of organised lobbies for economic gain and social control. The further point here is hence to bring what Nora means by 'history' as a critical and on-going intellectual excercise to bear upon the potential distortions of memory, and upon the status of History as bourgeois memory frozen into truth, and thereby to use history as a judge of 'whose' memory is most appropriately being re-presented (with the best remembering and the least forgetting as far as can be ascertained from the findings of careful scholarship) in any particular instance of the place marketing project.

There is a significant dimension to this discussion that should be noted, however, in that we are claiming that the exercises of historical re-presentation undertaken in the name of an urban-based bourgeoisie tend to be ones of memory which possess *limited* correspondence to the past, whereas the exercises of historical re-presentation under-taken in the name of those 'other peoples' of the city mentioned earlier usually operate with situated and rooted memories – ones impregnated, as we have already suggested, with a vital relationship to the specific places of these peoples and their forebears[13] – which possess considerably *greater* correspondence to what really happened in past times. We are fully aware of the pitfalls inherent in such an argument: we recognise that to talk of more or less accurate representations of a 'real' past is to risk charges of academic naïvety, and that to champion one account as more historically correct or faithful than another is to risk charges of playing judge and jury with the evidence. But up to a point the consensus of this volume *is* that something akin to Nora's 'history' must, can and does remain as a court of appeal when considering alternative re-presentations of culture and history in the present: the consensus appears to be that some accounts simply *do* have a greater veracity than do others through getting closer – in a manner that is more alert to evidence, but also in a manner that is very much 'nearer to the ground' (more involved, more lived) – to the processes, struggles and meanings that have shaped the existences of peoples (usually those 'other peoples' again) in past periods and places.

And this means that the cultural and historical appropriations of the bourgeoisie frequently *do* appear untenable when put alongside the alternative cultural and historical appropriations made by the 'other peoples' of the city, whether by themselves – and note the significance in this respect of 'cultural representors' indigenous to such groups[14] – or by enlisted intellectuals. This also means that the sympathies in this volume tend to lie, although not in an overly romantic or uncritical way, with the oppositional or merely forgotten 'other pasts' of 'other peoples' that tend to be deformed or discarded by an urban bourgeoisie (or at least a bourgeoisie whose roots of power and money are still very much in the urban arena) looking for attractive cultural and historical packages to use in the selling of places to enterprises, tourists and residents. We think it important that these 'other pasts' of 'other peoples' should not be so easily sidelined, and that strenuous efforts should be made to aid in their rediscovery – complete with the stories about past struggles, past wrongs, past victories that they often contain – as a tool to use in the contestation of selling places in the present, particularly when this practice of selling places leads (for instance) to massive redevelopment schemes such as London Docklands that are so damaging to pre-existing communities and their livelihoods.

Let us now conclude this section with a brief example touching upon the practice of selling places, one that encapsulates the concerns of the preceding discussion. This example hinges around a small urban space, the Piazza del Popolo or Town Square of the Italian town of Terni (fifty miles north of Rome), which is central to the lives of the townspeople:

'Some of the most dramatic events in the history of Terni took place there: from the killing of five workers by the police in 1920, to the barricade after the lay-offs [from the steelworks] in 1953. "That point was like a nerve centre; it was a division between the bourgeoisie and the proletariat", says Arnaldo Lippi' (Portelli, 1991, p. 88).

Moreover, a carnival ritual known as the *merancolata* grew up around this square during the early years of this century, the core of which involved groups of working-class and middle-class youths lining up on either side of the square and pelting each other with *merancola* (a stone-hard hybrid of orange and lemon). Portelli (1991, p. 88) suggests that folk recollections of events in the Square, when coupled with this ritual which 'represent[ed] a discrete urban space crossed by an ideal barrier', served to cement popular conceptions of Terni as divided both socially and spatially along class lines. But during the 1940s the *merancolata* was gradually substituted by another ritual, that of the *Cantamaggio*, which contained within itself a rather different set of meanings:

'This was originally a rural wassail ritual, which the local petit bourgeoisie revived at the turn of the century as a symbol of their identity which was threatened by industrialisation. It soon began to gravitate towards the town, and, in the Fascist period, was appropriated by the authorities. It became a representation of the town's new identity, from rural roots to industrial present, in the form of a parade of decked floats and a vernacular song festival. The ritual now culminates, every May Day, in the symbolic crossing of the old border [in the Square] in front of the authorities. The class division of the *merancolata* is thus replaced by the continuity of urban space, representing the continuum of a fluid citizenry instead of the discrete [*sic*] of conflicting social classes' (Portelli, 1991, pp. 88–89).

The bourgeois manipulation of a place's cultural and historical associations is here transparent, and it is evident that the current treatment of the past in this context – partly for the economic gain to be gleaned from a commodified ritual; chiefly for the social control to be exerted over the minds of the town's non-bourgeois residents (Terni's 'other peoples') – stems from a selective, even dishonest, appropriation of the past (one proceeding from bourgeois memory) rather than from a more measured 'historical' evaluation of it (which would arguably be closer to the spirit of the folk recollections that can be recovered and their truth-content inspected using the techniques of oral history).[15] Indeed, what seems to have occurred here is a clear schism between what might be termed a bourgeois 'colonisation' of a place's past and what might be termed a 'lived' feeling for that place's past possessed by peoples debarred from the ranks of the bourgeoisie.

Summary

As a final note, then, what we might argue is that this chapter has told a 'story' – and in parts it *is* something of a 'story', moving rapidly over vast swathes of complex theoretical and historical material, and operating with somewhat caricatured depictions of both the urban elite-bourgeoisie and the city's 'other peoples' – but we would insist that this 'story' still reveals something important about how cultural and historical resources have long been manipulated by society's powerful (and often urban-based) groupings in their own economic and social self-interest. And, moreover, it reveals something about the most recent mutation in this process of manipulation whereby urban bourgeoisies are seeking to mobilise segments of culture, history and locality in the competitive selling of their places both to outsiders (to attract capital) and to insiders (to legitimate redevelopment). We have outlined something of the longer-term historical geographies here, stressing the ramifications for those 'other peoples' whose urban experience has so often been moulded and constrained by the activities of their bourgeois 'superiors', and have then traced several quite specific themes (to do with 'New Right' discourses, postmodernism in the city, and the meeting of 'history' and 'memory') which shed light on the continuing efforts of the urban bourgeoisie to serve its own self-interests through the practice of selling places. And in a number of different ways we have paid attention to how this practice (complete with its effective denial of individual, contextual and historical difference) obliterates both deliberately and on occasion more accidentally the lives of the city's 'other peoples'; and we have emphasised instead the need to appreciate the many 'other attachments' – the different and arguably more vital attachments that these 'other peoples' have and feel for their city places – which can be threatened by, but which can also serve as a powerful reservoir for opposition to, any insensitive commodification of places. What our volume may hence be leading towards is a form of 'geography' much like Nora's 'history': a form of geographical inquiry which seeks constantly to examine and to re-examine claims about places – claims about their cultures and their histories; claims about who lives and works in them, and why and with what implications; claims about inequalities; claims about meanings – so as to question the appropriateness of a mentality that trades only in stereotypes of places with a view to enhancing their marketability.

Notes

1 This essay seeks to introduce some (though by no means all) of the themes that run through the volume: in so doing the other contributions to the volume are referenced, but no attempt is made to explore systematically or in detail the connections and differences between them. Our intention is that this essay should 'stand in its own right', offering certain thoughts and arguments of its own. Various people have read and commented upon a draft of the essay, but we would particularly like to thank Felix Driver for his careful and insightful criticisms.

2 Indeed, there is a further sense in which Eco supposes there to be an obscure script being 'spoken' by the exhibits: a complex semiotic logic whereby the exhibits

as 'signs' (as ensembles of signifiers and signifieds) are locked into a network of meaning whose interpretation by human subjects is always unstable and variable, and which maybe only the objects themselves can ever properly 'know'! Related claims about the construction of meaning through museums and exhibitions that display past or exotic objects are now being made in various texts (Jackson, 1991; Mitchell, 1989; Pred, 1991; Vergo, 1989), and it is interesting that such claims – even if not working expressly with semiotic tools – often consider the extent to which messages are coded into objects on display through the spatial disposition of these objects relative to one another (and to entrances, walls, windows, lighting) in the museum or exhibition (see the chapter by Billinge).

3 Considerable controversy rages over the appropriate definition of 'the surplus' – can it be identified in an 'absolute' sense applicable to any society, or must it always be identified 'relative' to the particular requirements and organisational arrangements of different societies? – but here we follow Harvey's definition of the 'social surplus' as 'the quantity of labour power used in the creation of product for certain specified social purposes over and above that which is biologically, socially and culturally necessary to guarantee the maintenance of labour power in the context of a given mode of production' (Harvey, 1973, p. 238).

4 It is important to note that, whilst it is appropriate to identify such a 'division of labour' in abstract and analytical terms, in practice it may often have been the case that people working on the land were *also* people working as miners, petty commodity producers and even as merchants co-ordinating trading activities (and think about the findings of detailed historical studies dealing with dual economic systems and 'proto-industrial households': see Kriedte, Medick and Schlumbohm, 1981).

5 In this essay some distinction is made between 'urban elites' as the powerful people (priests, chiefs, kings) of Ancient and Medieval cities, and 'urban bourgeoisies' as the powerful people (merchants, industrialists) of Early Modern and Modern cities. We want to suggest that in both cases these powerful people have pursued goals of economic gain and social control through the purposeful manipulation of culture, but we would tend to agree with a Marxist history of urbanism that stresses the increasingly determinant role of *economic relationships* – those organised around the production, circulation and consumption of 'value' – in the constitution of urban bourgeoisies post-*circa* 1700. Our historical geography in this respect is taking a grand historical sweep, and one consequence is to depict the elites and the bourgeoisies as occupying structurally similar positions in their respective periods and places, and also to regard the latter as in effect 'heirs' of the former (and of their elite position at the economic, political, cultural and ideological apex of urban society).

6 Note that in feudal societies considerable powers resided *outside* of urban areas in the hands of landlord-nobles whose wealth apparently derived from agricultural production in rural regions, and debate therefore occurs over just how autonomous this power base was from that of the urban 'bourgeoisie' and from the trading arrangements which this bourgeoisie arguably controlled.

7 In a paper delivered at the original conference on which this collection is based, Cosgrove (1990) claimed that city self-promotion can be traced back at least as far as sixteenth-century Venice, when a number of devices – the use of ritual as a focus of 'leisure' and even 'tourism'; the reworking of myth to suggest the 'favouring' of Venice historically; the elevation of the visual image over the written text; the propagation of Venetian architecture as a pastiche of other styles and references – were deployed by the city patriarchs in a fashion anticipatory of what Harvey (1989b) and others regard as the postmodern hallmarks of place marketing. Cosgrove was therefore led to interpret sixteenth-century Venice as 'postmodern', and in so doing he scrambled the time-bound conception of postmodernity as the distinctively 'here and now' which colours much writing about the city. He also noticed, furthermore, that these features of Venetian self-promotion in the sixteenth century apparently embraced the goals both of securing economic gain and of exerting social control (of convincing ordinary Venetians of the city's greatness and holiness, notwithstanding its day-to-day miseries).

8　More speculatively perhaps, it is suggested by some writers (see the chapter by Thrift and Glennie) that the supposedly distinctive quality of the contemporary selling process whereby our personal and group identities are intimately bound into the process of consumption – with the things that we buy and with the things that advertising convinces us we ought to buy; with the persistent barrage of information conveyed to us through the signs and symbols of places such as the urban shopping arcade – is not really so novel at all, and can actually be traced to older 'discourses of consumption' such as those that surfaced in late-seventeenth- and early-eighteenth-century British urban areas.

9　Writing of the events of the Commune, Harvey suggests that 'much of it had its roots in the processes and effects of the transformation of Paris in the Second Empire' (Harvey, 1985, p. 217). Haussmann's dramatic spatial engineering in central Paris, which saw the opening up of the inner city through the insertion of the boulevards, was of course a key achievement of the Second Empire (see also Scott, 1980, pp. 204–209).

10　Consider Seed and Wolff's account of culture in general: '[c]ulture was and is a crucial arena in the development of classes – their identification, segregation, and relations of dominance, subservience and incorporation. More than this, the cultural activities and concerns of social groups is part of their formation and self-recognition' (Seed and Wolff, 1984, p. 38).

11　It might be argued that the black inner city becomes the focus of what, following Said's example, can be termed an 'imaginative geography'. In the course of dissecting the components of 'Orientalism', the various discourses that have grown up in the West about the peoples and civilisations of the East, Said (1978, pp. 54–55) talks of an 'imaginative geography' through which the Western mind uses characterisations of distant places 'to intensify its own sense of itself by dramatising the distance and difference between what is close to it and what is far away'. In other words, what is projected on to *both* the people and their place is indeed a sense of 'distance and difference': a sense perhaps of being alien, wild, promiscuous, irreligious, unrestrained and diseased that says as much about the 'darker' side of Western society (a side that is here being conceptually exiled) as it does about any 'real' Oriental peoples and places. And it may be appropriate to think of the Western mind creating in its collective (bourgeois?) heads a similar 'imaginative georgaphy' through which the peoples and places of the black inner city are imagined as alien, wild and so on, and where the physical decay of the built environment is taken as a reflection of – and maybe as being actually caused by – the cultural shortcomings of the black inhabitants (and this is a line of thought suggested by Burgess, 1985; Jackson, 1989, chapter 6: and see too the arguments about socially constructing black inner cities as 'no-go areas' where law and order has broken down in Keith, 1988).

12　It is in this respect that place attachments can easily weave their way into a form of memory (and often a form of History) with strong nationalistic or 'localist' commitments, and the result can often be a chauvinistic sense of 'us and them' that precludes co-operation with different people in other places and can lead to different peoples being excluded from (and maybe actively chased out of) 'our' places. In this respect we agree with a commentator such as Harvey (1989b) when he worries about the *negative* and 'reactionary' character of such place attachments, and we would want to signpost the dangers inherent in the unthinking translation of such attachments into both historical re-presentations and related political programmes.

13　It is in this respect that we would speculate about certain place attachments weaving their way into rather different forms of memory to the chauvinist ones referenced in **Endnote 12** (and perhaps doing so in a manner that would be looked upon more favourably by the critical scrutiny of history). In this respect we would therefore want to stress the *positive* and 'progressive' character of certain place attachments, a claim that Harvey (1989b, esp. p. 351) does himself offer, but to stress that such attachments should be reflected upon carefully by the peoples involved and with a deliberate attempt being made not to close a place's boundaries around a particular sort of 'people' (a nation, a race, a class) in such a way that alliances cannot be made with different peoples elsewhere or tolerance cannot be extended to different peoples 'at home'. If such conditions can be met, then historical re-presentations of the local past are unlikely to be as insensitive as their bourgeois counterparts, and at the same time the possibilities are opened up for a more emancipatory politics rooted in a considered engagement with culture, history and locality.

14 Colls (1990), in a paper delivered to the original conference, discussed the role of such indigenous 'cultural representors' – writers, singers, artists – in the case of working-class communities from the North East of England. And this was a theme echoed by Rose (1990: also in a paper delivered to the original conference) when describing the 'maps of meaning' imagined by people in Old Docklands as they strive to counter the imagery pedalled by the London Docklands Development Corporation, and when outlining the constructs of 'history', 'community' and 'land' fostered quite deliberately by local community arts groups (the indigenous 'cultural representors' in this case) as they criticise the ripping of Docklands places away from their cultural and historical contexts.

15 Note that Portelli himself uses terms such as 'memory' and 'history' in a rather different fashion to Nora, and hence considers his own project to be one of using oral history sources to trace what happens to an 'experienced event' (the *évenement*) – in this case the death of Luigi Trastulli, a young steel-worker from Terni – as it accretes symbolic, legendary and imaginative qualities in 'the *longue durée* of memory and culture' (Portelli, 1991, p. 1).

2

Historical Geographies of Urban Life and Modern Consumption[1]

NIGEL THRIFT AND PAUL GLENNIE

Introduction

The theorisation of contemporary consumption has been a striking growth area right across the social sciences in recent years. In this chapter we seek to advance two contentions. The first is that many theorisations of current consumption are deficient because they inaccurately characterise the history (and geography) of modern consumption. As a result, they fail to clarify which facets of contemporary consumption represent continuities with earlier practice(s), and which are genuinely novel. Our second contention, on which the first rests, is empirical, and concerns the emergence of modern consumption. We situate this emergence not in industrialised mass production, as is usual, but during the phase of relatively rapid English urbanisation of the century or more that *preceded* factory industrialisation. It is ironic that analysts of modern consumption have employed the industrialised-mass production oriented historical account of consumption just as it has become less widely accepted among historians. In addition, we emphasise dissimilarities between English and North American consumption patterns which invalidate the notion that US consumption is paradigmatic of modern consumption in general.

The paper divides into two parts. The first discusses aspects of English and North American consumption before the twentieth century. We focus attention on the timing of new consumption practices and on the contexts in which they emerged. These historical observations inform the second part of the paper, which identifies the ways in which contemporary consumption does seem to be undergoing important changes. Throughout the chapter, we place urban contexts at the heart of the development of new consumption practices.

Historical Debates

The Emergence of Modern Consumption

We can distinguish two major constellations of views about the chronology, the setting, and the determinants of the emergence of modern consumption. The first places the emergence of a consumer society in Britain firmly in the late-eighteenth century, following McKendrick, Brewer and Plumb's *The Birth of a Consumer Society* (1982), and has proved particularly influential beyond history's disciplinary boundaries.

McKendrick and his co-authors viewed consumer society as spawned by mass production, and shaped by advertising which directed and manipulated emulative instincts intrinsic to human nature. It is not merely the purchase of particular sorts of goods which is portrayed as having diffused through society as a result of emulation. It is also 'consumer culture': that is to say, ideas about how to consume, and about what to consume, and about the links between individual character and consumption behaviour (in other words, how people's appearance, conduct and domestic setting are 'read'). 'Consumer culture' is portrayed as originating among middle-class urbanites, 'educated' largely by images of commodities promulgated through capitalist advertising. These consumption attitudes supposedly diffused to the more prosperous sections of the labouring classes as they acquired disposable income in the late-nineteenth century, becoming virtually universal in the modern West since the 1920s and 30s, when household appliances and cars became the paradigmatic modern consumption goods (Miller, 1991). In this respect McKendrick parallels other writers who have stressed the diffusion of consumption patterns from aristocratic to middle-class consumers (Thornton, 1987; McCracken, 1987; Earle, 1989).

Before highlighting problematic areas in the McKendrick thesis, it is important to identify two major areas in which his approach is considerably more sophisticated than much earlier, and some later, work. The first is in situating consumption behaviour within social interaction patterns, which centre on establishing identity partly through consumption. The second is in seeing consumption as an ongoing process, not as a single moment of purchase. Thus consumers of clothing were consuming whenever they wore or displayed particular items (or even thought about them), not just in the act of buying.

However, insofar as McKendrick offers an explanation for a new consumer culture, it is simply that new sources of trade and manufacturing wealth enabled greater social mobility, which made consumption an important weapon in battles for status. And, whilst Brewer and Plumb link new consumption habits to wider changes in society and politics, including the commercialisation of leisure and of childhood, they likewise offer little comment on what consumption could have meant to participants and observers *other* than in the rather mechanical terms of emulation (Fine and Leopold, 1990). Moreover, this stress on the late-eighteenth century as a formative period now looks more eccentric and overstated than it did even five years ago. Work on both England and New England – what Breen (1986) has styled 'the Anglo-American Empire of Goods' – points to considerable continuities in consumption practices through the eighteenth century, and also to some radical-looking breaks in the decades *prior* to 1700.

On both sides of the Atlantic, substantial changes occurred in both the actual consumption of, and more importantly the propensity to consume, a wide variety of consumer goods: household fabrics, pewter, brass, ironware, ceramics, furniture, ornaments, clothing (Weatherill, 1988; Shammas, 1990; Glennie, 1992). These relatively mundane household items were fashion goods, in the sense that they sold at least in part upon their appearance, and that specifications of appearance changed frequently, often from year to year (Styles, 1992). Yet their significance has long been obscured by the focus of art and furniture historians at the top end of the market, and recognition of their significance has arisen only with the broadening of these academic specialisms from connoisseurship to social history (Carr and Walsh, 1988).

The importance of this second chronology of modern consumption is that it also involves different sets of views on the causes and contexts of new consumption habits and discourses, and on the relationship between changes in consumption on the one hand and in production on the other. The most important differences in interpretation are three. First, that modern consumption emerged on the basis of artisan and proto-industrial production, prior to the mass production associated with factory-based industrialisation. Secondly, that new consumption practices were closely linked to particular kinds of urban settings, rather than to mass production, and indeed were associated with the revival of notions of urbanism as a distinct *way of life* embracing consumption, architecture, and leisure skills: in other words, the practical knowledge for urban living (Borsay, 1977, 1989; Corfield, 1989). Thirdly, that a new account is required given dissatisfaction with emulation as a sufficient explanation of what appear to be clear changes in the meaning of both goods consumed and consumption as an activity (Fine and Leopold, 1990). This second chronology hence relates the emergent modern consumption to a specific conjunction of factors, to which we now turn.

The Contexts of New Consumption Practices

In several respects, the Early Modern English economy was unusual by contemporary European standards. A first point of comparison is simply the high level of, and the rapid growth in, urbanisation in England. English urban population had soared in the century after about 1650, very much against the European trend, and even while national population was stagnating (De Vries, 1985; Wrigley, 1984). A second point is that even beyond the major urban centres, rural English society was permeated by market transactions to a remarkable extent. The level to which English men and women were routinely involved in 'the market' (focused on a dense network of small towns), and the extent to which their clothing and domestic goods came from distant producers, were by-words amongst both domestic and foreign commentators. A third point is that the English 'middling sort', that is commercial farmers, wealthier traders and artisans, were both numerous and wealthy. International comparisons of the size and social shape of the market for traded consumer goods are at an early stage, but it is clear that more people lived comfortably in England than elsewhere in Europe (De Vries, 1974, 1976; Schuurman, 1980; Weatherill, 1988). And, fourthly, it should be noted that the spending patterns of the English 'middling sort' diverged from that of their less numerous continental equivalents. English consumers amassed very considerable quantities of relatively inexpensive goods, rather than luxuries such as gold, silver, plate, paintings, jewellery, fur, and so on. English consumption affected high-volume industrial production rather than low-volume-but-high-value-added luxury crafts.

Both economic and social changes promoted more open and fluid social relations, particularly in urban contexts. New sources of wealth – both agrarian and non-agrarian – increased scope for individual economic advancement (Clay, 1984). Religious diversification, coupled with the decline of notions of community and collective identity, weakened cultural homogeneities which had previously provided social cohesion. Particular cultural practices were no longer universal, but became characteristic of particular social positions (Wrightson, 1982; Sharpe, 1987). In urban centres, then, a number of developments occurred: the scale of urbanisation created populations much less familiar with one another than in communities of more static composition (Corfield, 1989), and as the old cultural homogeneities broke up, particular consumption practices became cultural markers, creating greater scope than hitherto for self-manipulation of status by choice of consumption practice. At the same time the separation of commodity producers and urban consumers was of considerable significance, since it increased the importance of the consumer rather than the producer in processes of interpreting artefacts (Giddens, 1987). The meaning carried by a particular artefact was less under the control of its producer, and consumers' interpretative powers were correspondingly increased, enabling consumers to foster a partly autonomous social-cultural sphere of commodity interpretation.

That a range of economic and social developments created new contexts for consumption, and changed what could be achieved through consumption, did not inevitably amount to a change in the concurrent meaning of consumption. The crucial element in bringing about new forms of consumption was thus the development (in the late-seventeenth century) and subsequent elaboration of new consumption *discourses*. These discourses developed in relation to specific sorts of goods, and in particular social and geographical contexts, but rapidly became generalised across a wide range of objects and people. By 1750 or so, new vocabularies of description and sets of ideas had diffused from their points of origins in various old and new urban centres to become characteristic, not only of much of the English population, but of much of the population in the North American colonies as well (Styles, 1989; Breen, 1986, 1988; Shammas, 1990).

A pivotal element in these discourses was novelty in tandem with ideas about how to consume novel goods. An immediate stimulus had been the appearance of new types of foreign goods: tea, coffee, sugar, chocolate, tobacco, new fabrics, and various ceramic, wood and metal items. Their very novelty made their consumption more open to interpretation than that of more familiar items. It is important to note that novelty could reside not just in a material object itself, but also in the techniques or the motifs and concepts of design involved in its making (Mukerji, 1983). Particularly relevant too is that several of these key goods were associated with new social practices: think, for example, of tea and coffee drinking, or of smoking. Around the consumption of novel items developed consumption discourses which prized novelty as part of the experience of consumption, and also as a source of pleasurable experience in itself. As Styles notes:

'the kinds of receptiveness to visual novelty and differentiation normally associated with the Victorian middle classes was already present at relatively humble levels of the domestic market from the late-seventeenth century' (Styles, 1991, p. 37).

Even as they developed, these discourses were being extended to apply to familiar objects as well as new ones. Novelty as an aspect of goods thereby became applied to familiar materials: textiles, leather goods, wooden furniture, metalwares, as well as to hitherto unfamiliar and exotic items. Crucially, these included a wide range of mundane non-luxury goods. The power of novelty-centred discourses was strikingly evident from the response to them of both merchants and shopkeepers, and also in the practices of both artisan and proto-industrial producers. For traders and manufacturers, the ability of a good to offer novelty became a prime design and trade consideration (Styles, 1988, 1992). Novelty-centred interpretations of goods gave a new impetus to fashion. Novelty entered into how consumer goods were 'read', both by purchasers and by consumers, as well as by observers. In short, fashion was not important because of emulation, but rather the reverse. Emulation became a more influential process because of the rise of mass fashion, but the rise of fashion was itself dependent on consumption discourses which prioritised novelty. These were already well-formed, and were spreading rapidly by the early-eighteenth century.

Urban contexts were central both to the learning of new consumption practices, and to their pursuit (Borsay, 1988; Corfield, 1989). Knowledge of consumption was essentially practical knowledge, not acquired through instruction or advertising, but from the experience of participating in the activities comprising the dense interaction and information networks of urban life. Quasi-personal contact and observation in the urban throng were the main ways of learning for many people in Early Modern England (Clark, 1983; Fawcett, 1990). Observation at fairs, markets, inns, shops, recreations and entertainments as well as in the streets, and at home, provided the basis for constructing interpretations of individuals based on fragmentary images and indications.

It is true that perceptions of identity must always have been particularly fragmented within large towns, but the implications of this fragmentation would have depended heavily on specific intra-urban contexts. In late-seventeenth- and early-eighteenth-century England these contexts furnished unprecedented but varied scope for social mobility, a vast array of new objects and ideas through which identity could be stated and interpreted, and new ideas about the very activity of consumption itself. Furthermore, urban consumption was not a uniform process from one town to the next, and there were striking variations around the country, and between different types of urban centre, in the extent to which the above-mentioned characteristics were manifest (Weatherill, 1988). The contrasting supply and diffusion patterns of different commodities strikingly illustrates such inter-town variations: the distribution of dealers in tea and coffee followed the 'old' urban hierarchy based on centres of provincial and county administration, for instance, whereas watch and clockmakers were distributed as widely through new and rapidly-growing commercial-manufacturing towns, and in port-based networks, as they were through traditional urban centres (Glennie, 1990). Comparisons between pairs of neighbouring towns of different character reinforce the point: York and Leeds (Looney, 1989); Chester and Liverpool; Shrewsbury and Stoke (McInnes, 1988, 1990; Borsay, 1990).

Consequences of Industrialised Mass Production

The main topics to be considered now are the impact of industrialised manufacture; the social extension and diversification of the consumer goods market; emulation and the relationship between middle-class and working-class consumption discourses; and advertising and the sources of consumer knowledge.

If mass consumption and the development of modern consumption discourses *pre-date* factory-based mass production, what was the contribution of industrialisation to the development of modern consumption? Rather than marking the appearance of mass demand for consumer goods, the late-eighteenth and nineteenth centuries saw the incorporation of factory-produced goods into the evolving (and increasingly socially-extensive) consumption discourses that had developed during the previous century-plus. In the course of this incorporation, mass production profoundly affected the content of consumption discourses in three ways. Firstly, through the 'menu' of physical objects they made available, and at what prices and market conditions. Secondly, by changing the relationships (both economic and cultural) amongst producers, distributors, retailers and purchasers. And thirdly, through the further growth of advertising and the way in which the changing character of advertising reacted to and stimulated socio-cultural changes.

The growing repertoire of consumer goods available, and the growing social range of consumers appearing as working-class expenditure rose, have been widely recognised, but it is common to find the diffusion of purchasing patterns through society treated as a process involving working-class emulation of middle-class 'consumer culture' mediated by advertising images. For example, Richards presents late-nineteenth-century advertising as reaching out to 'discipline' the working class, which previously,

'. . . having never had much experience with consumer goods, was even less disposed to revere the sight of them . . . the system of advertised spectacle would extend to mass periodicals that spoke to the working class only after the consumer economy had begun to reach them' (Richards, 1990, pp. 7–8).

39

We believe such treatments to be badly misguided. Moreover, those who dismiss working-class consumption as unimportant because of its insignificance in terms of overall consumer expenditure are in danger of neglecting the existence and development of strong *working-class* discourses about consumption. These were intimately related to processes of social interaction (Williams, 1987; Johnson, 1988). When real income for working classes eventually rose, their consumption decisions stemmed from consumption cultures which *they* had already created in low-income contexts. Working-class consumers hence did not 'pick up' ready-formed middle-class consumer attitudes as their incomes rose.

Symbols of respectability and of surplus even pervaded expenditure on subsistence items. Beyond keeping bodies fed, housed and clothed, they dominated other key areas of consumption spending: ranging beyond manufactured goods into services such as funerals, holiday outings, mass entertainments, and insurance clubs. In all these areas Yeo's phrase 'visible participation' nicely captures the dimension of social interaction in even modest spending, and Johnson is surely right to link middle-class incomprehension of impoverished families' spending to their insensitivity regarding social dimensions of working-class consumption behaviour. Such incomprehension simply does not fit with the notion that working-class consumption discourses 'trickled down' from middle-class behaviour, then, and the consumption discourses of different classes need to be seen as differing not just in degree (the essence of the emulation approach) but also in kind. Nor do we mean to imply here that the consumption discourses of 'the middling sort' were impervious to modification by, and interaction with, the diverse practices among groups whose consumption was coming to embrace a widening variety of goods (any more than aristocratic consumption choices were entirely independent of those displayed by the *nouveaux riches*). Threads of different consumption discourses continued to evolve, to fragment, to interweave, and to unify, under a host of influences.

Through most discussions of consumption behaviour runs an unposed question: what were the sources of consumer knowledge? In the late-twentieth-century West a central role is obviously played by advertising: 'a set of purposive procedures for producing consumption'. Advertising redefines the scope of what are considered necessities; expands definition of 'necessity' when introducing new products; and attempts to change the ways in which people relate to ordinary articles by making the ordinary seem extraordinary, primarily through the use of spectacle. In this view, advertisers control consumers by the knowledge that they impart. Commentators have used the changing character of advertising since the mid-nineteenth century to trace changes in consumer knowledge. Typically, this is done through dichotomising the primitiveness of early-nineteenth-century selling (passive, dirty, private, hidden) and advertising (sparsely used, personal) on the one hand, and late-nineteenth-century 'modern' selling (active, public, creative) and advertising (use of spectacle, homogenising consumer tastes) on the other. But, whilst such Whiggish views about consumption have become hegemonic, they are not well-grounded.

To treat consumer learning as advertising-dominated is a narrow view, overly modernist about both the nature of selling and of consumer-learning. It illustrates the danger of treating earlier periods simply as weaker forms of those familiar to us today. We instead regard advertising as just one source of knowledge among many, as a point on a continuum of discourses influencing consumers; not so much a locus of authority as a source of potential values and ascriptions: like the state, the Church and other institutions, a provider of mutable and hence subvertible images. The situation of advertising within a wider array of symbols and images is demonstrated by the fact that certain images were widely used across unrelated commodities. There were no advertising agencies to oversee such thematicism, and as yet no advertising 'theories' to inform it.

To summarise our argument, then, we want to suggest that mass production interacted with modern consumption discourses, rather than creating them. The consumer goods market became socially more extensive and diverse, but not through an emulatively-inspired social diffusion of middle-class 'consumer culture', and advertising was not the pre-eminent arena for the acquisition of practical knowledge about consuming. Thus 'modern' consumption discourses neither appeared during this period, nor were they static. Consumption practices developed between the early-eighteenth and late-nineteenth centuries, and stood in a different set of relationships with new forms of production. In addition, we want to stress that to highlight interactions between various consumption discourses held by different classes or social groups, and to emphasise that advertising interacted with other sources of imagery, is to underscore modern consumption's *adaptability*. This adaptability is indeed one of its key characteristics, and this 'evolutionary' capacity is both a defining characteristic of modern consumption and one of its most enduring features.

England and America in the 'Anglo-American World of Goods'

Because modern consumption is essentially a dynamic concept, its historical components need to be specified chronologically, geographically, and socially, and yet it is not uncommon for contemporary analysts to treat the experience of the United States as the general Western experience. This we believe to be incorrect, and we wish to emphasise the divergence, or rather the *further* divergence, of consumption in England and North America over the past three centuries. In the seventeenth century, Britain's North American colonies were not only largely populated from Britain (at this stage particularly from England), they were also bound into English production and distribution systems for manufactured goods. The range of goods supplied from England in the late-seventeenth and early-eighteenth centuries is truly astonishing (see, for example, Ligon, 1673). For obvious reasons, English goods sent to North America and the Caribbean were usually sent in large batches, and were often of standard design. Moreover, they were usually finished rather than semi-finished goods, an important distinction because of its consequences for relationships between retailers and consumers, a point missed in the important book by Shammas (1990). This means that the consumer goods available in North America and the Caribbean were more standardised than those which reached English consumers. To this claim we should add the commonplace observation that North American settler society was less stratified by status and wealth than that of European motherlands (Lemon, 1973), and was therefore less riddled by complex negotiations over status that entailed deploying small differences in the goods consumed as 'marks' of differential status. These are deep-seated contrasts between the character of Old English and New English consumer societies, and they are important to the historical geographies being sketched out here.

These important contrasts between developments in consumption on either side of the Atlantic were magnified and elaborated during the nineteenth and early-twentieth centuries following industrial growth in the United States. Firstly, US production involved far more standardisation of consumer goods than was the case in the UK. For historians such as Samuel, the prominence accorded to design differentiation in Britain constrained mass production of uniform items (Samuel, 1977). Secondly, there was the contrasting geographical scale of the respective space-economies within which consumer goods circulated and were obtained. The two systems were characterised by differing means of supply, with obvious contrasts in shopping behaviour both within the urban system (relating to the specialisation of shops and the changing social meaning of 'shopping') and beyond it (best illustrated by the role of mail order in carrying urban imagery into some distinctly non-urban realms). And thirdly, British class distinctions remained more intricate, generating greater concern with refinements in the design of objects as cultural markers. It is predictable to find this pattern reflected upon in middle-class biographies and commentaries of the time, but it was at least as strongly revealed by working-class biographers (Johnson, 1988; Blumin, 1989).

Given these differences, there are clear dangers in universalising the experience of North America (or, for that matter, that of Europe) as paradigmatic of modern consumption as an analytical category. Yet often, and especially in discussions which pivot around dichotomies between 'modern' Fordist production and 'postmodern' flexible production, this is exactly what happens. As a result, it is difficult to identify and to analyse continuities and novelties in current consumption with any reliability.

A Perspective on Modern Consumption

What, then, does seem to be original about late-twentieth-century consumption? We want to fix on five relatively new and distinctive processes which we would count as the most important in the modern meeting of culture, capital and the city.

The first of these processes is the growth of reflexivity: that is, a development in the ability of human subjects to reflect upon the social conditions of their existence. In the past, this reflexivity was predominantly cognitive and normative, but increasingly it is also aesthetic. For example, judgements are now made about the *value* of different social and physical environments, as found most noticeably in the present fascination with history as a 'source' of value in contemporary landscapes:

'[S]uch a cosmopolitanism presupposes extensive patterns of mobility, a stance of openness to others and a willingness to take risks, and an ability to reflect upon and judge aesthetically between different natures, places and societies, both now and in the past' (Lash and Urry, 1991, p. 10).

The growth of aesthetic reflexivity has had important effects on the practices of consumption. For example, because there are so many aesthetic authorities to choose from and to interpret, so it becomes much more difficult for any one person, agency or whatever to control the meaning of commodities. Also, the growth of aesthetic reflexivity has produced a different attitude to authenticity, one which is more bound up with aesthetic allusions (for example, in modern advertising) than with a quest for the real or the deeply spiritual.

A second process has been the massive increase in social interaction mediated by electronic means. This has had three direct results, each of them inter-related to the others. The first has been the operation of a much extended public sphere. The advent of the television has resulted in a faster 'disembedding' of the event (Giddens, 1990), as well as the making of many events more public (indeed many events are now purposively styled for television). Events become more public both directly through those who witness them and indirectly, either through their assimilation in discussion with others (what Thompson, 1991, calls 'discursive elaboration') or their assimilation by other media (what Thompson, 1991, calls 'extended mediazation'). Thus:

'the experience of publicness no longer requires individuals to share a common locale' (Thompson, 1991, p. 244).

Moreover, the time structure of this extended public sphere has changed. This is most apparent since the advent of video, which adds a longer time dimension to the assimilation of events and the possibility of their mechanical reproduction at times now controlled by the consumer.

The third novel process is the greater degree of attention being paid to the fostering of individuality, especially as projects of self-actualisation. This change has happened for a number of inter-related reasons. Firstly, there has been the entry of more and more groups into consumption culture through a mixture of greater spending power and greater power over spending, as indicated by the growth of the middle class, the greater participation of women in the labour force, the greater power of certain ethnic groups, and so on. Secondly, there has been a growth (particularly within the middle class) of various social groupings which are often committed to particular patterns of consumption (the ecological movement, the conservation movement, the women's movement). Thirdly, there has been a massive increase in the scale and power of consumer movements, especially in the 1980s (Mennell, 1985). Fourthly, the business of marketing individuality (through the design of products, shops and so on) has consequently become more reflective, and this niche marketing has been both created by and constitutive of new modes of individuality. Fifthly, the shaping of personality has become a more reflexively systematic concern. As Featherstone (1990) has noted, there has been a gradual shift from a nineteenth-century (and earlier) model founded on ideas of *character* to performative models of *self* in which appearance and bodily presentation comes to 'express' the self. For example, amongst many social groupings commodities like clothing have increasingly become, to use Carlyle's phrase, 'emblems of the soul' (Giddens, 1991): part of a general striving for 'attractive otherness'.

Thus, individual identities have in one sense become more diffuse, becoming attached to discourses emanating from particular social groupings. Yet in another sense they have also become more tightly drawn to particular lifestyles, which consist of normative, cognitive, and aesthetic choices, often made in a quite deliberate fashion. The conclusion to be reached is that individuality is both broadening *and* deepening. Seen in this way, many new 'postmodern' consumer developments like shopping malls, redesigned city centres and the like are not shining spires, drawing consumers to them in a trance-like state of desire like moths to a candle, but rather ways of encompassing the large numbers of different vocabularies of description employed by diverse social groupings. The power of these places, such as it is, lies in their ability – through a combination of the buildings and their design, the commodities on offer, the overall layout, and consumers' reactions – to speak to a number of these different vocabularies at once. In other words, the postmodern production of difference that they are meant to epitomise is simply a way of swimming with increasing social diversity, of fitting into as many modes of group interaction and self-description as possible, of creating appropriate discursive frames (Morris, 1989) which provide the basic props for discourse. On a smaller scale, the diversity of fly-posting on contemporary urban walls, with posters for sales, gigs, political meetings, ecological causes, and so on, also addresses many groups.

The fourth process follows on from this forging of new spaces for both consumption itself and the sustenance of diverse consumption discourses. Developments like new city centres or shopping malls can be seen as part of a more general project to *recompose* the city as a dense information network. This is being done to meet the need to provide different social groups with different sets of information in the same place. And it can also be seen as an attempt to recreate the dense information networks of seventeenth- and eighteenth-century cities in which knowledge was not derived from advertising but from 'the throng'. It is an attempt, in Shotter's (1989) words, to reproduce a 'communicative commotion' of people who are able to gain their identity from interpreting each other (since 'you' needs to be defined by 'others'), and in so doing this new city supports each social group's discourse *and* provides each social group with a means to tap into other social groups' discourses (thus, indirectly at least, supporting a discourse of novelty). Thus, although there may have been through time what Giddens (1991) calls a progressive 'pluralisation' of life-worlds and their associated contexts, with particular social groupings increasingly moving in carefully circumscribed strings of contexts, it remains the case that settings which are common to *many* social groupings are still needed. Indeed, we could argue that such contexts have become more necessary now than ever before to provide the ground for new interpretations. In any case, the importance of interaction is not confined to display of oneself nor to simple observation of others: it is crucial for monitoring the reactions of others to one's own behaviour, for acquiring feedback about others' interpretations of oneself. This function is provided by 'the throng' as it lives, jostles and consumes in the city of everyday human intercourse, and it is a function that cannot be met by an 'automated throng' constructed through mass media.

The fifth change follows on again. One way, perhaps the *only* way left, to provide a joint appeal to this whole unstable, ambivalent mess of social groupings is by providing them with a 'sense of place': a 'sense of place' that invests the practices of buying and trafficking in commodities with localised meanings. After all, it is the case that most commodities are mass-produced and that the thrill of owning or wearing them is more likely to be present in the advertiser's mind than the consumer's. We must be careful to avoid letting what Morris calls 'commodity boudoir talk' about the seductive powers of the commodity escape from the few upmarket shopping malls and designer city centres where bourgeois luxury goods are sold. As Morris puts it (1988c, p. 122), 'it isn't necessarily or always the objects consumed that count in the act of consumption, but rather that unique sense of place', and this sense of a distinctive locality to which a range of social groups belong (complete with their diverse consumption discourses and practices) is something that the city of interactions can still provide against the placeless 'hyperspace' of advertising through the mass media.

Concluding Comments

All of the above is not to suggest that there are no important continuities from the past to be found in modern consumption. By way of conclusion, we wish to identify five such continuities, but these differ significantly from those conventionally emphasised in the literature.

Firstly, it is clear that there was a discourse of wants and needs *before* industrial capitalism, based upon novelty and pleasure, and a very powerful discourse it was too. Industrial capitalism intervened in this discourse. In some ways this discourse was ideally fitted to industrial capitalism, but the two did not share a common point of origin. This discourse still exists, and not just as an effect of industrial capitalism.

Secondly, the power of advertising as a producer of consumer demand has been consistently overestimated, and it is at least debatable as to whether advertising has the effects which are often ascribed to it. Thus, the modern literature on advertising consistently downplays its power to stimulate and shape demand. Furthermore, it is clear that advertising must be seen as only one part of a discourse of consumption, and not as the discourse itself.

Thirdly, it is obvious that the tension between mass-produced and individually-tailored goods has been a continuing one throughout the last three hundred years. Whilst it is clear that the nineteenth century saw the spread of both a mass market and mass consumption, and that the significance of this spread should not be underestimated, it is still the case that individually-tailored goods (often but not always of a luxury nature) have continued to have powerful societal impacts, both by virtue of their scarcity (which often makes them attractive as a means of social division) and by virtue of their ability to act as exemplars for mass-produced goods. In recent years many individually-produced goods have hence become models for mass-produced consumer goods through the emergence of 'retro' and rural styles (Samuel, 1989). This is not even to note the way in which people can, through practices like collecting, transform mass-produced goods into part of their of self-identity (Featherstone, 1991; Willis, 1978, 1990).

Fourthly, it is clear that the identities of people have been built on consumption practices for at least the last three hundred years. The goods that people have owned have consistently been a vital way in which they have effectively constituted their sense of self and of other. The idea that there has been a major shift in the form of identity, from being focused chiefly in production to being focused chiefly in consumption, is too simple, especially among the middle classes.

Fifthly, the idea that individual identities are disappearing into a seductive flux of free-floating signs is questionable. It is, of course, arguable whether individual identities have ever been well-formed wholes, as any reading of a social-constructivist text would attest. It is also questionable whether modes of individualism and individuality (Abercrombie, Hill and Turner, 1986) are currently leading to the disintegration of individual identity. Our thesis is rather that they are leading to more differentiated notions of individualism and individuality based upon a greater number of systems of cultural reference, but this does not imply that individual identity is necessarily disintegrating, only that it is changing its form.

None of this is to suggest that the period from the beginning of the nineteenth century did not see a massive increase in the importance of consumption which has continued down to the present. We do not claim that nothing in the sphere of consumption has changed in recent years. It is to say that the effects of this increase have been exaggerated (by taking a particular date as the baseline for change, by accentuating certain changes and by disregarding evidence to the contrary) and homogenised (by taking the experience of the United States as pivotal). It is for these reasons that paying greater attention to the historical geographies of urban life and consumption – as a variety of discourses and practices tied to different social groups, regions, cities, towns and intra-urban sites – is an essential component of rethinking the standard theorisations of modern consumption.

Note

1 The arguments of this paper are extended in Glennie and Thrift (1992).

3

The City as Spectacle: Paris and the Bicentenary of the French Revolution[1]

GERRY KEARNS

Introduction

I want to describe some of the links between culture, history and politics which we may excavate from the celebration in Paris during 1989 of the Bicentenary of the French Revolution. The celebration was continually contested. This contestability illustrates one of the ways that the significance of history is both nurtured and rejuvenated. The images of the Revolution have not yet broken free from their historical content.

This was certainly a special case. Anniversaries, those arbitrary collisions between calendar and culture, inevitably raise questions of interpretation. Out in the 'global village', context may be recycled as stage-set and events as costume-drama, but on the occasion of anniversaries there at least exists the possibility that they will be reined back to their original content, or to an explicit reflection upon their continuing significance. Cultures interpellate their constituents, and in modern democracies these members are defined territorially rather than in the narrower terms of blood or property. National birthdays, then, attempt to deny difference on the basis of race, colour or creed in the name of citizenship. Yet, to bear this democratic weight it is frequently necessary to do violence to history; to construct a pure national founding myth, a story of the origins of a nation which finds unimpeachable motives at its conception, if not at its birth. National anniversaries, then, raise in a particularly acute way questions about the 'lessons' of history. The second reason why the Bicentenary brought the symbols of the Revolution together with their meaning is that the memory of the Revolution had been scratched into the Parisian urban fabric with significant sites and historical place-names, etched there by subsequent re-enactments of the Revolution (1830, 1848, 1871, 1968) and buffed up by an ideology of urban preservation. If the act of commemoration raised the possibility of re-examining the Revolution, the urban fabric provided a map of signifiers waiting for precisely this rejuvenation.

None of this was inevitable. The continuing significance of the Revolution could have been denied. The cultural capital tied up in the historic sites and images could have been devalued through meaningless circulation in an inflationary spiral of pointless photo-opportunities and human-interest historical soap-opera. In some respects this danger was not averted. Yet François Mitterand, as President, and his advisers in the Socialist Party (PS) were aware of this problem and took steps to avoid it. This is thus the third of our special local circumstances: the Bicentenary celebration was partly directed by a Socialist President with a particular political mission. Mitterand wanted to wed the Socialist Party to a consensual French Republican tradition. Being seen as the group making revolution poses problems for a Left wanting to be perceived as a natural party of government. The notion of permanent revolution appears to put the Left outside the Republican tradition, and leaves the gelder of patriotism in the Right's account. Politically Mitterand wanted the Socialist Party to capture the centre-ground of French politics.

I want to describe some of the events of 1989, set them in their historical context and offer some reflections upon the significance of all this for debates about the 'postmodern condition'. My account of 1989 is organised to draw attention to Mitterand's strategy and to the ways in which it was challenged. I will pay particular attention to the use that was made of the *City of Paris*. First I describe how the Socialists tried to ensure that the occasion of the Bicentenary received attention, and then I will trace how they tried to get their message across and summarise what it was they were saying. I consider the importance of Paris in two ways: in terms of the politics of architecture and in terms of the politics of space. Paris was the context in which Mitterand and his colleagues addressed the heritage of the Revolution with, I shall argue, significant consequences for their approach to architecture and to town planning.

The Politics of Spectacle

The Events of 1989

In 1989 Paris was the European 'City of Culture'[2] and the place where the world's seven richest nations had their annual meeting about the state of the world economy. No doubt if the Olympics or World Cup had been available, the French President might have bid for them too. By making the celebration of the Bicentenary as visible as possible, Mitterand at once aimed at helping the French tourist industry and at raising the stakes for his domestic critics. The central event was the Bicentenary, but these additions raised both profile and cash.

The main events were staged over three days, beginning at midday on 13 July with a simple, even sombre, ceremony dedicated to the 'Rights of Man' on the Esplanade of the Rights of Man at the Trocadero. Thirty-two heads of state or government were present, arrayed around the French President in accordance with the protocol of eighteenth-century France: heads of state before heads of government, and ranked by length of tenure within each group – first, Houphouet-Boigny, the President of the Ivory Coast, then the President of the United States, and way back in the corner mere heads of government such as Margaret Thatcher (Berger, 1989). In addition to the 'Big Seven' (Canada, France, Great Britain, Italy, Japan, United States, West Germany) present for their summit, the other representatives came from Senegal, Djibouti, Cyprus, India, Yugoslavia, Congo, Cameroun, Nigeria, Pakistan, Japan, Uganda, Madagascar, Zimbabwe, Venezuela, Portugal, Togo, Zaire, Mali, Egypt, Uruguay, Ivory Coast, Bangladesh, Philippines, Brazil, Greece, Mexico, Gabon: a spread which was important in showing the universality of the ideals of the French Revolution and in balancing the rich seven against a numerical preponderance of the poor (Gauthier and Valladao, 1989). Like everything at this celebration, protocol was still subject to political adjustment, as when Margaret Thatcher was given new neighbours at the lunch following this ceremony – placing her next to Robert Mugabe and Jacques Delors was considered too provocative (Maigne *et al.*, 1989). Longstanding critics of Mitterand's Bicentenary plans, *Le Quotidien de Paris* thought the ceremony 'flat' rather than 'sombre' (Vernet, 1989), whilst *Le Figaro* said it was 'not a success' since many of the two hundred civil rights organisations invited did not come, their places being filled at the last minute by representatives of foreign military (Delsol, 1989a). *Le Monde* was more generous, finding the ceremony 'simple' but 'moving' and commenting on the 'egalitarian' mixing of rich and poor imposed by the protocol (D. R., 1989).

On the evening of the 13th, the international cast reassembled with French politicians and personalities for the inauguration of the Opéra-Bastille. With June Anderson, Teresa Berganza, Placido Domingo, Barbara Hendricks and Ruggero Raimondi under the direction of Georges Pretre in sets designed by Bob Wilson and some in dresses by Lacroix, Ungaro, Kenzo, Tarlazzi, Givenchy, Lanvis, Dior and Yves Saint Laurent, it was certainly a glittering occasion. *Le Monde* was amused that, in Mitterand's presence, the orchestra opened with 'Et Satan Conduit le Bal' ['And Satan Leads the Ball'] from Gounod's 'Faust' (Lonchampt, 1989). The press liked the music and acoustics, with even *Le Figaro* conceding that it had passed its inaugural test with flying colours (Doncelin, 1989). The exclusiveness of the event and the screening of the building from the crowd with security shields, allied with a long-standing dissatisfaction with the architecture ('a camembert in steel and glass'), prompted some criticism (Braitberg, 1989). But the following evening there was a free performance of Berlioz's 'Te Deum' at the Opera and for a week there were cheap concerts there. The balancing act was evident.

After their night at the Opera, the President's guests were entertained to dinner in the Musée d'Orsay where each table was named after a revolutionary hero, and Margaret Thatcher found herself at table with Rajiv Ghandi, Benazir Bhutto and Gabriel Garçia Marquez under the sign of Abbé Grégoire (regicide, clergyman and soon to be Panthéonised). The morning of the 14th brought the traditional military parade. The guests, behind the bullet-proof glass shield demanded by George Bush, joined 800,000 other spectators in admiring the men and women of the French armed forces trundling their *materiel* along the Champs-Élysées. Mitterand's guests then left for their garden party at the Élysée, the presidential palace, before the Big Seven detached themselves from the rest and began their Summit with a photo-opportunity at the new Pyramid at the Louvre. Then followed dinner, with the rich and poor re-united, at Mitterand's insistance: same menu (Guilbert, 1989b), but separate tables (Laurent, 1989). And then, it was back to the site of the morning's parade for Jean-Paul Goude's 'Marseillaise'. Eighty foreign television companies took the pictures to 700 million viewers, while 3,500 journalists and technicians were accredited to report it, the Summit and the rest of the Bicentenary for their newspapers (Schwartzenberg, 1989; Grundmann, 1989).

Goude's commission was to celebrate the French national anthem, 'La Marseillaise', along the Champs-Élysées whilst representing the French provinces (Vincendon, 1989). Goude knows little history, however, disliked the period costumes used at the centenary of the Statue of Liberty and argued that World Music was the genuine contemporary revolution and that in harmonising different cultures, sensibilities and races, it appropriately expressed the international aspirations of the Rights of Man (Nataf, 1989). So the march brought the costumes of a comic-book 'King Solomon's Mines' to the rhythms of world music; topped and tailed by more weighty matters, led in silence by Chinese students with pictograms of liberty, equality, and fraternity on their headbands, and moving to a silent conclusion in the Place de la Concorde where Jessye Norman, draped in the tricoleur, stepped forward to sing the national hymn. The rain stayed away. The French press, at least, were transported. The about-turn was dramatic. *Le Quotidien de Paris* celebrated the greatest Parisian crowd since the Liberation from the Nazis (*'La foule . . .'*) while *Le Figaro* pronounced the Bicentenary a success thanks to Goude, who, Giesbert (1989a) believed, had expressed all of the contradictions of the Revolution; and even Delsol (1989b) was moved by Norman. Criticism shifted to the coverage on French television, which was considered 'disgraceful' by *Le Monde* (Lacan, 1989) and accused of verbal diarrhoea by *Libération* (Darmaillacq, 1989). Criticism was also diverted to those right-wing politicians who had boycotted (*'L'opposition . . .'*) a parade that they thought reeked of 'Mitterandolatorie' (*'M. Juppé . . .'*). *Le Monde* expressed what was more or less a consensus when it said that the worst error of the Bicentenary was Chirac's absence from 'the greatest 14 July since the Liberation, in the town of which he is the mayor' (Rollat, 1989b).

The morning after, the world's press were out at the new Arch at La Défense where a short opening ceremony sent images of the Cube around the world before the Seven got down to their Summit again. Agreement on the joint declaration came quickly enough for the meeting to finish the following morning, the 16th, and for the final dinner at the Élysée to be cancelled (*'Les travaux . . .'*). In keeping with Mitterand's wish to celebrate 1789 as the most significant international point of (re)reference in the diffusion of their democratic commitment to the Rights of Man, the Summit's communiqué on the state of the world economy began: '[i]n 1789 the Rights of Man and Citizen were solemnly proclaimed. Forty years ago the General Assembly of the United Nations adopted the Universal Declaration of the Rights of Man . . .' (*'Droits . . .'*). The Seven went on to commit themselves to support democracy and the human rights of a pluralist society, and to underline their belief that democracy was basic to economic development. Repression should stop in China and foreign forces should withdraw from the Lebanon, agreed the Big Seven, but they stopped short of sanctions. Much of the text was concerned with the need to recognise the environmental implications of economic development, but apart from monitoring drought in the Sahara and expressing their intention to phase out CFCs by the end of the century, no definite commitments were made. On the Third World, they stressed the use of World Bank and International Monetary Fund (IMF) funds to sustain the mountain of private commercial debt in Latin America, more precisely Mexico. In reply to Mikhail

Gorbachev's request to Mitterand to put the case for the Soviet Union being a full party to the Summit talks, the Seven indicated that the Soviet Union was not yet democratic enough to join 'the club' and that only Hungary and Poland were sufficiently democratic or market-oriented even to receive direct aid from the Seven. The rest of the East was to be encouraged along the path away from Communism. The implementation of these policies of the rich towards Eastern Europe was left in the hands of the European Community, whose president, Jacques Delors, was a full participant at the Summit. This was enough for most of the French press. *Le Figaro* said Mitterand could be well pleased with the appearance of the Summit, for it mentioned all of the right things (Marchetti, 1989). Although noting that he had secured little for the Third World, *Le Figaro* recognised that he had brought the world to admire the new Paris and that he had gained his cherished statement about democracy being the basis of economic development (Guilbert, 1989c). Elsewhere Mitterand was praised for his realistic courting of US opinion after his earlier attacks on Anglo-Saxon egoïsm (Miliesi, 1989a), and his critics even tried to take pride in having brought Mitterand to a conciliatory tone for the successful celebration of the Bicentenary (Spitéri, 1989).

As spectacle, the Bicentenary was clearly a success. It received massive press coverage, and most of Mitterand's critics in the French press reversed their position by the 13th, leaving right-wing politicians grinding their teeth and *Le Canard Enchaîné* chuckling at the U-turn of the right-wing press. There was little satire in *Le Canard*'s summary:

'So Dantonton [Mitterand] has given the people their festival . . . and it was well worth the effort. Its success has swept all criticisms aside. The power of the symbol, the "meeting of History and the calendar", the merry-go-round of ceremonies, the sheen of the new Parisian monuments, the pom-pom-pom of the military parade, the nocturnal magic of the Goude opera, and finally the participation, the good humour, the popular pleasure, all went to make the Bicentenary a success despite contrary expectations. And that's without even mentioning the Arch Summit with its consensual and friendly conclusions' (*'Le mur . . .'*).

The Political Use of Symbols

From the beginning of his presidency, Mitterand showed a clear commitment to the politics of spectacle. At the very first press conference following his election he announced a series of *grands projets* for the capital which left his critics agape at the cost, began the persistent criticism of his monarchical style, and sent commentators back to one of his volumes of autobiography (*The Wheat and the Chaff*) where he wrote that: '[t]hroughout the City, I feel like an emperor or architect, I carve up, I decide, I arbitrate' (quoted in

Bonnet, 1989a). He also began immediately to use spectacle to reinforce the legitimacy of his position as both a socialist and a republican. Northcutt (1991, p. 158) notes how Mitterand 'has adroitly used symbols to weaken what was once the political and cultural legitimacy of the Right in France'. For Northcutt, this has had two dimensions. First, in an attempt to 'legitimise the Socialists as having deep roots in the French past' and, second, to show that 'the President and the PS have a vision for the future'. The dimension of history in the Bicentenary spectacle is obviously important for the first of these, whilst the *grands projets* and their architecture speak to the second.

Mitterand wasted no time in mobilising history as spectacle. As Jossin (1983) remarks:

'De Gaulle's France had been wedded to history; that of Pompidou and Giscard d'Estaing to contemporary times. But the French do not like living in some faceless joint-stock company, even if it is prosperous. Old rival of the General, François Mitterand did not forget this. The very day of his investiture as President, he went to the Panthéon, rose in hand, to get the sovereign blessing of history.'

The rose is the symbol of the PS and Mitterand laid one on each of three tombs: Juarès, the pragmatic socialist; Moulin, the Gaullist Resistance leader; Schoelcher, the liberator of the slaves in the French colonies (Northcutt, 1991, p. 144). The choice was significant, as was the dignity of the staging:

'The bright roses, the cold stones, the crowd outside the empty crypt; a solitary man walks into the resting place of the great men; the staging of 21st May 1981 which François Mitterand had planned as a memorial ceremony, a replanting of French history, a regrouping of the national community around its great men. In choosing the Panthéon, however, he elected an old debate about a contentious place, could the Panthéon keep alive the memory of these great men and could they be a rallying point for national unity?' (Ozouf, 1984, p. 139).

In 1986 he repeated the exercise, this time to move to the Panthéon the ashes of another hero from the Resistance, Jean Monnet.

There can be no gainsaying the role of the Left in the Resistance, nor the fact that the Liberation from the Nazis is a powerful founding myth of modern France. Indeed, it is the photographic memory of the Nazis on the Champs-Élysées that largely animates the patriotism of modern 14 July military parades there. Memory works by reinvesting places with new accretions of significance, or it fades. In staging spectacles at the Panthéon, Mitterand is rescuing the place from becoming little more than a non-controversial but unimportant place, increasingly given over to dead scientists, as well as making a statement about his right as a Socialist president to define what the Republican tradition comprises. In 1989 he went beyond his personal credibility as a Resistance fighter and used the occasion of the Bicentenary to move to the Panthéon the ashes of two revolutionary figures, Condorcet and Abbé Grégoire, as well as those of Monge, the founder of the École Polytechnique.

These ceremonies were staged by Jack Lang, Mitterand's Minister of Culture, and by Lang's colleague, Christian Dupavillon. They have been working together since 1966 when they put on a World Student Theatre Festival in Nancy (Jonquet, 1989). Lang was quite explicit about the purposes of spectacle during his period at Nancy (1963–71). Taking its inspiration from Brecht, the theatre festival was to challenge bourgeois assumptions about the world, to educate its audience, to give youth a sense of its collective power and, most importantly, to stimulate creativity in others (Looseley, 1991). It is understandable, as Looseley remarks, that Lang should find the move from political theatre to theatrical politics so attractive. It is, Lang might suggest, just as radical to redefine the national consensus from within the establishment as it is to point up its false unity from without. Certainly, Mitterand's attempt to speak for the nation in 1989 was heavily contested from both Right and Left.

The Right's Challenge to Mitterand's Bicentenary Spectacle

From the Right, Jacques Chirac and his colleagues tried to clip the wings of Mitterand's *grands projets* and Bicentenary celebrations. Mitterand had first stood for President in 1965 when he lost to Charles de Gaulle, and his accession to office in 1981 interrupted a run of right-wing presidents stretching back to the beginning of the Fifth Republic in 1958. He thought it important to mark this Socialist breakthrough with spectacular audacity, but as Mayor of Paris Chirac could frustrate some of Mitterand's efforts. This was a neat reversal of the situation a century before when a left-wing mayor had imposed the Centenary celebrations on a reluctant right-wing government (Ory, 1984). Chirac and Mitterand were clearly competing for the position as master of ceremonies in Paris on 14 July 1989. The first casualty of their competition was Mitterand's proposed World Fair. Seeking to parallel the Centenary celebrations of 1889 – when a World Fair had helped to express French industrial might, the recovery of its pride after the disastrous Franco-Prussian war of 1870, and the self-confidence of its republican virtue – Mitterand projected an international event which would bring 50 million tourists to France. Under the pretext of the traffic problems that it would occasion, Chirac insisted that the World Fair be held outside of the city (Bonnet, 1989a). It was dropped in 1982; a result which may also have pleased Mitterand's Socialist colleagues, who had passed from the reflationary economics of their first year in office to deflationary austerity measures by their second (Derbyshire, 1988, p. 77). In return Mitterand was able to withhold state support from Chirac's pet projects, and it was difficult for Chirac to raise private sector finance without such state guarantees (Bacqué and Murat, 1989). A compromise was reached: the State would celebrate the 14 July and the City would host the festivities on the hundredth birthday of the Eiffel Tower, 17 June 1989.

The other political body with which Mitterand had to deal was the National Assembly. When he became President in 1981 he faced a right-wing parliament which still had two years of its five-year mandate to run. He dissolved it and was relieved to see the popularity of his election as President drive his left-wing colleagues to power. Yet their term of office ran out in 1986, and the loss of support to the Right in the following elections more or less obliged Mitterand to invite Chirac as leader of the largest party of the Right to form a government. The period of cohabitation had begun. Now Chirac had a legitimate right to involve himself in the planning and funding of the Bicentenary. Mitterand appointed Regis Debray as his personal adviser on the Bicentenary, but the choice for the official Mission for the Bicentenary was a Chirac/Mitterand compromise choice, Michel Baroin: a conservative figure and member of a Masonic lodge who immediately consulted Warren Burger, President of the United States Supreme Court, on how the US proposed to celebrate the Bicentenary of the US constitution (Spitéri, 1986). The French knew what the importance of planning was. It had been calculated that the Bicentenary of American Independence (1976) had increased tourism by 20% (Vernet, 1986). In the face of a balance of payments deficit on manufactures, the French needed the receipts on services provided by tourists, yet Chirac refused to release money for an event over which Mitterand would obviously preside.[3] By April 1988 virtually nothing had been done, and only FF15m had been released to the Mission for the Bicentenary: this compared poorly with the Australians, who had their bicentennial ready two years before the event, and with the Americans, who spent one thousand times more on theirs than the French seemed prepared to do (Stenli, 1988). Debray was livid, calling Chirac's lack of preparation 'disgraceful' (Reynaert, 1988). In March 1987 Baroin drowned, and Edgar Faure was appointed to his post. Faure promised a festival of national reconciliation, but reserved most of his energy for the dedication of the Arch at La Défense to the Rights of Man (*Le Monde*, 12 March 1987; Bonilauri, 1987). This at least advanced one of Mitterand's *grands projets*, to Chirac's chagrin (Kajman, 1987). Faure appointed Jean-Michel Jarre his artistic adviser and promised a Jarre laser spectacle for the 14th (Reynaert, 1987). The crucial matter, though, as Dominque Jumet (1986) understood, was who would win the 1988 presidential election because the winner could virtually determine how the Revolution was to be remembered.

Mitterand defeated Chirac in the presidential election, dissolved parliament and again secured a Socialist government, although this time a minority one. Faure died, and Mitterand appointed Jean-Noël Jeanneney President of the Mission for the Commemoration of the Bicentenary ('*M. Jean-Noël . . .*'). Jeanneney expressed an interest in the relations between political history and the operation of the mass media. Michel Rocard, as prime minister, immediately increased the Mission's budget to FF125m, which at least averted the danger that, having committed FF100m of the City's money, Chirac, as Mayor of Paris, would put the President in the shade during the Bicentenary (*Le Quotidien de Paris*, 8 July 1988; Liffran, 1988; '*Un grand . . .*'). Lang went back to the Ministry of Culture and immediately dropped Jarre. Dupavillon claimed that they thought a laser show too static (Lombard, 1989). Chirac picked up Jarre, promising FF10m from the City's coffers to stage the show in the Place de la Concorde. At this point Mitterand played his trump card: the Summit. In June 1988 the Seven met in Toronto and, it being France's turn to host the next meeting, fixed the date for 14 July 1989 in Paris (Camé, 1989b). Initially, it was intended to have the Summit out at La Défense, but by September Mitterand had decided to house the world leaders in Tuileries, which meant that for reasons of security the streets of the west end of Paris would have to be at the disposal of the President rather than the Mayor; and this included the Place de la Concorde. Mitterand instead offered Chirac the night of the 16th (a Monday) for the Jarre concert. Spitting tacks, Chirac said that it was the 14th or nothing, since by the 16th Parisians would certainly have left on holiday and Chirac further advised them to be away over the crucial weekend itself (Varenne, 1988; '*Bicentenaire: Jarre . . .*'). So, there was no Jarre concert, although with astonishing good grace he did release an LP entitled *Revolutions* as his contribution to the festivities (Bouly, 1989). To relieve Parisians' disappointment, Lang and Dupavillon volunteered a spectacle of their own for the 14th, and they commissioned the Goude parade within the Presidential corral in the west end of the city. The Summit had one other important consequence, in that it meant Mitterand could spend money from the presidential purse in preparing Paris for his quests: '[i]n one sense, it is the Summit of the Seven which sponsored the Bicentenary' (July, 1989b).

The Left's Challenge to Mitterand's Bicentenary Spectacle

The Summit itself became the focus of the Left's attack on Mitterand's spectacle. Taking the message of the Revolution to be the need for the poor to struggle against oppression, the Left found the sight of the leaders of the world's seven richest nations at the head of the Bicentenary festival disgusting. They equated the Third World today with the Third Estate of 1789, and presented the world's poor as modern *sans-culottes* demanding justice from the rich (*'Fedéra-tion . . .'*; Begue, 1989). The singer Renaud came up with the slogan '*Ça suffat comme çi*' (enough is enough), and the writer Gilles Perrault called for a demonstration against the Summit to be held on 8 July 1989 at the Place de la Bastille (*'Gilles . . .'*). Renaud organised a concert at the Place de la Bastille featuring, among others, Johnny Clegg. 15,000 people marched around the east of Paris in the afternoon and 100,000 came to the concert in the evening. The 8th July demonstration drew attention to three issues: '[w]e are here today to say no to debt, to apartheid, to colonialism. This is a meeting of those excluded from the official Bicentenary commemorations' (*'Bienvenue . . .'*). This was followed up by a conference on these issues at Maubert-Mutualite and The Other Economic Summit (held to coincide with the 'Big Seven's' meeting each year since 1984), which brought representatives from the world's poorest seven nations to Paris (*Le Monde*, 15 July 1989, p. 4). The poorest seven asked the United Nations to set up a conference on world debt (*Le Monde*, 15 July 1989, p. 5).

These demonstrations against the Summit clearly angered Mitterand, who was trying to present himself as an advocate of the Third World in the courts of the rich. Furthermore, the demonstrations were taking place in what the President liked to consider his own backyard, the east end of Paris (Guilbert, 1989a). Although some commentators tried to present this as little more than party politics with the Communists (PCF) out to embarrass the Socialists, in fact the PCF was far less prominent than the trotskyists and anarchists (*Le Canard Enchaîné*, 12 July 1989, p. 2). And Renaud himself went out of his way to focus attention on the failure of the 'Big Seven' rather than on the President himself: 'I remain, despite everything, an anarcho-Mitterandiste. I have confidence in him. I don't question his humanism, his third-worldism' (*'Renaud . . .'*). Speaking for The Other Economic Summit, Alain Lipietz also put the focus on the Summit saying that he believed the 'Big Seven' listened to the criticisms of the poor (Colson, 1989).

Jacques Attali, the 'sherpa' who represented Mitterand at the preparatory meetings for the July Summit, responded to these attacks in an interview. The heads of state of Senegal, Egypt, India and Venezuela formally requested Mitterand to create a mechanism for regular North–South meetings to pick up again the dialogue which they felt had gone silent since Cancun (1981) (Fabra, 1989). Attali explained that it would take years to organise a North–South summit, claimed that France had always been prominent as the voice of the poor among the Seven, and suggested that agreement would be easier to reach among a group of seven than at a larger summit of 150 (Camé, 1989b). This was an unusually diffident performance from the normally cool Attali, and reflects the genuine despair among the President's advisers (G. B., 1989). In what is almost an official account of the Summit, Camé (1989a) noted that by June 1989 the persistent criticisms of both the Bicentenary and the Summit had left Attali 'ill at ease' during the last preparatory meetings.

Mitterand replied to his critics on two occasions: initially in an interview to *L'Express* and then in his traditional television interview of the 14th. On the latter occasion he said that he also would have joined the demonstration of the 8th if he had been as poorly informed as Renaud and his colleagues, who acted from the best of motives but in ignorance ('*Cap . . .*'). Whilst he dealt courteously with Renaud, he was certainly bitter about criticisms of his failure on the question of Third World debt, telling *L'Express*: '[f]or eight years, I have struggled with the debt problem. It will be the principal subject at the Summit. For me, the problems of the Third World are more serious for humanity than nuclear weapons. They at least are in responsible hands, but poverty is contagious' (Haski, 1989). At Cancun, Mitterand had annulled the debts that the poorest African countries owed the French government. With his veiled references to secret diplomacy and possible breakthroughs, he led people to expect some significant moves on debt at the Summit. 'Wait-and-see' also deflected criticisms past the 14th to the closure of the Summit on the 16th. In the meantime, he went out of his way to emphasise the presence of Third World leaders at the Bicentenary, giving Cory Acquino special attention as a symbol of the triumph of democracy. Acquino, in what *Le Figaro* termed the 'theatre of solidarity' (Guilbert, 1989b), was the only political figure that Mitterand went to meet at the airport (hers was a state visit rather than that of a mere guest), and the French government immediately announced a FF350m loan for joint Franco-Philippine development projects ('*350 millions . . .*'). Moreover, with their rich and poor guests together for the Bicentenary, the French organised bilateral meetings on the morning of the 13th: Mrs Thatcher, for example, met the presidents of Mexico and Uruguay (Tricot, 1989). They also placed the leaders of the four countries calling for a North–South summit together with Bush, Mulroney, Thatcher and Kohl at breakfast on the 14th (*Le Monde*, 15 July 1989, p. 3). Mitterand presented the plea for a North–South summit to the other six, and Attali arranged to meet the leaders of The Other Economic Summit ('*La ruche . . .*').

The Summit did indeed attend to Third World debt, no thanks, as Germain-Robin (1989) pointed out, to Mitterand's charisma but simply because the debt was threatening the stability of Western commercial banking. The Summit also considered Mitterand's request, but, as James Baker rather contemptuously put it, for 'less than ten minutes' (Laurent, 1989). The United States claimed a North–South conference would politicise and dilute the purely technical questions of debt and ecology (C. T., 1989). The US Brady Plan for dealing with Latin American private debt was adopted. Mitterand got nothing at all for Africa (Kalfleche, 1989). The world's bankers breathed a sigh of relief that bicentenial fervour had produced no grand gestures in favour of the poor (G. M., 1989). The day following the Summit the dollar rose on the world market; business as usual (J.-F. R., 1989). Attali kept his promise to meet the leaders of The Other Economic Summit, but the location was kept secret until the last minute ('to avoid embarrassing the rich?', asked *Le Monde*, 15 July 1989, p. 4) and Attali gave them all of eight minutes (*Le Canard Enchaîné*, 19 July 1989, p. 3). If Mitterand had believed that the presence of the poor would shame the rich into concessions, a sort of global 4th August with privileges voluntarily renounced, his hopes came to nought. Perrault was surely justified in concluding that nothing had been delivered (Samson, 1989).

The ideals of the Revolution left the 'Big Seven' unmoved, and unlike in domestic politics Mitterand could not assemble a workable majority at the Summit. Bush, with his pre-Summit visits to Poland and Hungary, was keen for the US to displace West Germany in leading the West's response to the collapse of Communism in the East. Bush gloatingly spoke of 'a great victory for the West, for democracy, for free markets', whilst Gorbachev in his letter to Mitterand spoke of a world without ideological victory, a pluralist, multilateral system of state relations (Hauter, 1989). West Germany and France resisted this US triumphalism, but, with Bush and Thatcher implacable in their commitment to rubbishing Communism in the communiqué, the best they could achieve was a statement that the EC should manage the West's economic relations with the East. This satisfied Bush since it would perhaps cost the US less (and he had already refused to let Hungary and Poland see the colour of his money), and it pleased Kohl as he had no wish, given right-wing gains in recent European elections in West Germany, for West Germany to be seen making all the running on Eastern Europe (de Kergolay, 1989). Bush, seconded by Thatcher, was equally resistant on the question of Third World debt. Any move on preferential treatment for African commerce or on the annulment of elements of inter-nation public debt would have redeemed Mitterand's third-worldist pretensions, but the US – and to a lesser extent the remaining five – had no intention of giving France this symbolic victory. By shifting the focus to Latin America, to Mexico, in promising to use his influence to get commercial banks to reduce Mexico's debts by 35%, and in securing for Mexico $7bn of the $12bn that the World Bank and the IMF had set aside for global debt restructuring, Bush set right the reversal that the US had experienced at Cancun and put the US back in the driving seat on the question of Third World debt (Miliesi, 1989a, 1989c). On the whole, though, the French press came to take Mitterand's point about there being three times as many representatives of the poor as the rich in Paris for the Bicentenary celebrations. Shortly after the Summit, the French, this time in co-operation with Canada, West Germany and (perhaps surprisingly) Great Britain, secured an extension of the Brady Plan to some African countries.

Political Spectacle and Spectacular Politics

Evaluations of the Bicentenary as spectacle were, as mentioned, generally favourable. The summary in *Libération* was only slightly more effusive than most:

'The planet has seen media fireworks before, at sports events and, more recently, at the Bicentenary of the Statue of Liberty, the example which Mitterand seems to have wanted to surpass. But never before has there been such a mixture: commemorative, aesthetic and fashionable pomp; new architecture; spectacular and secret diplomacy; military and civil parades, popular balls and internal politics, classical opera and the opera of the street, displays of police technology and an excuse for new weapons. It only required that a spaceship be launched from the City of Science at La Villette for the whole gamut of national pride to be complete. . . . The presidential will transformed Paris into a gigantic theatre where, with the help of enormous crowds, an armada of police, and thirty-four heads of state a complete play was staged which began in 1789 and ended with Third World debt, drugs and the destruction of the ozone layer, in other words the great dramas of today. He only needed to announce with the Choir of Seven the abolition of Third World debt at the end, to attain political nirvana' (July, 1989b).

And in political terms the spectacle gave Mitterand the patriotic platform he wanted. Indeed, the residual hostility of some on the Left may have done Mitterand little harm as he tried to manoeuvre the Socialist Party into the centre of French politics, a move which required the repudiation of the tradition of permanent revolution. In this respect, just as the sight of the French military recalls the marching Nazis only to erase it once again, so the sight of the Socialists on the streets recalls May 1968 and the commitment of the Left to unconstitutional revolution only, at least in this case, to deny insurrection and to embrace reform. Fontaine's summary in *Le Monde* is significant:

The City as Spectacle: Paris and the Bicentenary of the French Revolution

'The opposition, the Mayor of Paris in particular, must be biting their knuckles at sulking away from the greatest festival, and the most successful, which the capital has known in a long time. The commemoration of the Revolution has never been less revolutionary: harmony rather than civil war reigned. Twenty-one years after May '68, when for several weeks only the red of revolution and the black of anarchy were to be seen, France has spent three days under the tricoleur and the *Marseillaise*' (Fontaine, 1989).

Claiming the platform is crucial and, as I have shown, contested. The importance of the theatre of solidarity in this is clear. As Dupin remarked:

'From Friday the 13th July, things began to move in favour of the Bicentenary-Summit. Cory Aquino arrived in France. The French realised that it was not only the rich who had been invited to the festivities of the 14th July. . . . The simultaneity of the celebration of the Bicentenary and of the Summit of the Seven shines like the summit of Mitterand's own career. By its attitude the Right has given Mitterand the sole right to express patriotic feelings 200 years after the storming of the Bastille' (Dupin, 1989).

The Politics of Memory

History and the French National Consciousness

A stratified sample of 1,010 French adults conducted in 1983 for *L'Express* showed that the Revolution itself no longer divided the French into Republicans and Royalists: '[i]n 1983 the Revolution appears as the foundation myth of the national consciousness: 70% of the French think of it as an event with positive long-term consequences and two-thirds of them think it important to celebrate the Bicentenary in 1989' (Jossin, 1983). Two-thirds of the sample were interested in history, and the statistics on visits to museums, sales of history books and amateur genealogy bore this out, argued Jossin. When asked to name the two most important events of recent French history the Liberation (51%) and the Nazi occupation (31%) topped the list ahead of both the return of de Gaulle to power in 1958 (29%) and the end of the war in Algeria (26%). May '68 (21%) was a good way down the list, and the creation of the Common Market (14%) and the election of Mitterand in 1981 (14%) were put on a par below this. What is significant about this is the importance of the military and of General Charles de Gaulle in the leading events. Moreover, May '68 was primarily significant as a challenge to de Gaulle's paternalism and Mitterand's election as the return of the Left to power after a very long absence, but the general conservative pre-eminence seems clear. Mitterand had plenty of time in opposition to think about the importance of history, and about the extent to which de Gaulle defined the French collective memory.

Aside from de Gaulle's dominance, the other feature of French historical preference which cast a shadow over Mitterand's Bicentenary plans was the extent to which the Terror was almost unequivocally rejected by the French. Danton (26%) and Robespierre (40%) received the highest proportion of critical votes in a list of figures from the Revolutionary period. The 'anti-Terror' stance of La Fayette secured him the highest positive vote (43%), marginally ahead of Bonaparte (39%), and a negative vote (6%) much less than than that of the still-controversial Emperor (21%). Jossin's explanation of La Fayette's popularity is important: '[t]hrough him, the French salute a hero of American independence, a champion of the Rights of Man, a symbol of national unity and liberty, but also an enemy of violence and intolerance'.

History and the Celebration of the Centenary of the Revolution

A comparison with the Centenary is instructive. In 1875 the Republic was officially declared the French Constitution, and when Jules Grevy was elected President in 1879 the theory became fact. The symbols of the Revolution were embraced by the state. The phrygian bonnet, banned in 1878, was legalised again in 1879 and appeared increasingly on statues and in paintings, recalling the symbolism of the First Republic. In 1879 a statue was erected to commemorate the new victory of Republicanism, and it was placed in the east end of Paris in the re-named Place de la République, a site in a working-class area north of the Place de la Bastille (Imbert, 1989). In 1880 the government took up Hugo's poetic suggestion of 1859 that 14 July become the official national holiday, making it the focus of the Republican year and – although the Republican nature of France was fiercely contested between 1880 and 1914 – giving it the passionate significance for Republicans that Easter had for Catholics (Amalvi, 1984). In 1879 'La Marseillaise' was adopted as the national anthem (Vovelle, 1984). Whereas Haussmann, as Prefect of Paris during the Second Empire, had cleared away with relish the houses of great revolutionaries such as Danton and Marat at Odéon to construct the Boulevard Saint-Germain, the City of Paris decided to memorialise the heroes of the Revolution (Poisson, 1990). The statues were intended to show the Third Republic learning from and building upon the work of their Republican forebears. Amongst the heroes selected with the virtues that they were to represent were: Ledru-Rollin (universalism), Etienne Dolet (free thought), Jean-Jacques Rousseau (democracy), Georges Danton (national defence) and Condorcet (public education) (Groud, 1989).

Danton's statue was unveiled in 1889, and became a focal point of the Centenary celebrations: '[a]nticlerical Republican, Danton did not permit, as the ultra-moderates would have wished, the demonisation of '92, but as a victim of the Terror himself, as the base of the statue proclaimed, he appealed to the conservatives. For the rest, his patriotism commanded the middle ground' (Ory, 1984). This was not a search for the lowest common denominator, explains Ory, but instead a gathering of divergent groups with different interests in Danton. Yet a collection of positives is also an impression of negatives, and it is clear that Danton's association with violence did not carry the sanction that it appears to have done for those polled by *L'Express*. The Centenary was about democracy, 'the major controversial issue raised' (Hobsbawm, 1990, p. 70). In a Europe still largely topped with crowned heads, this had both an international and a domestic aspect: in celebrating democracy the World Fair of 1889 alienated monarchical Europe, which stayed away, threatening to cast a shadow over the festival: '[b]ut it proved exalting when, for example, the student delegations from countries' foreign dominations (Poland, Czechoslovakia, Croatia, . . .) transformed the visits into national celebrations, to the strains of a song, stringently suppressed back home: the "Marseillaise"' (Ory, 1982, p. 29). This sense of facing a hostile Europe also had implications for French attitudes toward the key revolutionary figures, putting the stress on the defence of the Revolution against hostile foreign elements rather than on the civil war itself: 'which is why the prominent figures of the year II, formally placed in the Panthéon in 1889 on the anniversary of the abolition of feudalism, were three men of war, Carnot, Hoche, and Marceau' (Hobsbawm, 1990, p. 71). Domestically the democratic and Republican accents of the regime sought to present 1789 as the crucial break with the old order, the *Ancien Régime*, heralding the new nineteenth-century age with its industry, science and prosperity. Republics ensured intellectual freedom and ensured great science: '[t]he living heroes of 1889 were scientists and technocrats, Edison, Eiffel, Pasteur, all citizens of a republic' (Ory, 1984, p. 552). It proved easier to rehabilitate the Revolution, and to connect 1789 to the great progressive sweep of the nineteenth century, if 1789 was seen as part of a movement of ideas rather than in terms of specific people and events (Ory, 1984, p. 550).

Hobsbawm speaks of 'the refraction of the Revolution through contemporary political prisms', arguing that 'everybody had his or her French Revolution, and what was celebrated, condemned or rejected in it depended not on the politics and ideology of 1789 but on the commentator's own time and place' (Hobsbawm, 1990, p. 69). The relations between historical interpretation and political persuasion should be dialectical. History proclaims itself an empirical science, and Hobsbawm goes on not only to explore the political motivation of reinterpretations of the Revolution, but also to contest the empirical adequacy of various 'revisionist' attempts to deny the central importance of the Revolution for the subsequent social, cultural, political and economic development of France and the rest of the world: and he thus refers to 'the extraordinary historical events that transformed the world two centuries ago. For the better!' (Hobsbawm, 1990, p. 113). That is a conclusion to which 70% of the *Express* poll gave their agreement.

Yet the problems of commemoration are clear. The character of the Revolution is contestable. Its implications for the present are thrown into doubt by the polemic which surrounds its interpretation. Any government aiming at consensus would have to tread warily among the possible potential foci of commemoration. If it uses the past to claim legitimacy, it risks losing that harmony in the face of rival accounts of the founding myths. Governments wish to claim the power of memory as a kind of secular religion, but they may also want to disinfect the object of their veneration. In Nora's terminology the mythical consensus of memory is always subject to critical deflation by history (Nora, 1984b). Furthermore, as Lavau (1989, p. 158) warned, foundation myths usually define an internal or external enemy defeated in the crucial events of 'Year Zero' but periodically revivified by commemoration, thereby hindering the prospects for reconciliation. Yet, as with the case of the Panthéon, the power of symbols rests partly with their contested use. The bloody events and ideological clashes still cling to the Revolution even when, as Ory believes, 'to defuse the power of commemoration, governments choose to remember abstract ideas rather than events' (Ory, 1988, p. 23). Indeed, not only does the defusing often fail, but the abstract principles themselves become material for political debate; and a further area of vulnerable flesh is laid bare.

Historical Appreciations of the Revolution in Modern France

The Centenary celebrations in 1889 gave a great impetus to historical studies of the Revolution, a trend which was encouraged by the government since, at that time, scientific enquiry worked to the benefit of the Left (Ory, 1984, p. 550). No one could have expected as much in 1989. In *L'Express*, Kupferman (1986) looked forward to 1989 from 1986, and saw a triumphant right-wing intelligentsia and little festivity. In 1987, when he announced his team for commemorating the Revolution, the historian who Edgar Faure chose was François Furet ('*Commémoration . . .*'). Interviewed in *Le Figaro* (20 February

1988), Furet stated that the French Revolution was now over since the decline of the PCF nailed down the coffin on the Jacobin tradition in French politics. Furet has argued that it is a peculiarly nineteenth-century frame of mind which sees historical change as a continual process of human fulfilment, and which therefore invests the past with the values that people claim to see emerging in the present. He has claimed to find in structuralism a refreshing emphasis on the historical relativism of systems of thought, and on the historical invariance of human nature. This divests the study of history of its political force, so he argues, since our judgements are seen to be grounded in our own time and because the human spirit is no longer seen as having a historical evolution. We have hence reached the end of ideology (Furet, 1984). In one sense, then, Furet's claim that the French Revolution is over marks no more than the special case of a more general argument. Seeing human virtue lying in democracy, the historian of the nineteenth century, argues Furet, saw the Revolution as a clean break with a pre-democratic past and a pure statement of democratic intentions which must be continually defended against royalist and Roman Catholic reaction. The triumph of democracy with the constitution of the Third Republic denied this history its contemporary resonance, and it was only with the Russian Revolution of 1917 that it acquired another as the pure origin and clean beginning for the idea of a social revolution which must be continually defended against bourgeois and liberal reaction. The decline in the belief that a revolution is necessary or desirable, prompted by the failure of world Communism, has for Furet robbed the French Revolution of this emblematic significance (Furet, 1978). Speaking to *Le Figaro* in July 1989, Furet suggested that it was because the Left cannot separate the idea of democracy from the idea of revolution that the collapse of world Communism posed such problems for it (Giesebert, 1989b).

At one level Furet's argument is not about the French Revolution at all. It takes up the attempt by some on the Left, his Jacobin-Marxists, to use the Revolution as the kernel of a socialist revolution which is still unfinished. It is, as Hobsbawm shows, as much about 1968 as 1789. The failure of 1968 to bear out the expectations of Marxist intellectuals led many, such as Furet, to reach some kind of reckoning with their earlier PCF learnings. The crisis of the French Left in large part took the form of the rejection of any sort of agenda of permanent revolution (Hobsbawm, 1990, pp. 98–101). McGarr, in a brilliant survey of historical studies of the Revolution, remarks that Furet's

'argument has as much to do with current politics as history. In fact, the argument is little more than a nicely dressed up version of Burke's tirade against fundamental social change and revolution of 200 years ago. Furet wants to bury the notion of revolution, and the Jacobins and *sans culottes* along with it' (McGarr, 1989, p. 94).

But, as McGarr further notes, Furet is led also to deny the historical significance of the Revolution for the social, political and economic development of France. Furet does recognise a 'liberal' revolution which reformed the political and economic context in which the rich directed the economy. This elite was primarily a landed one, and it was changing its political and economic circumstances before the Revolution and would probably have continued to do so without it. However, this revolution was 'derailed' by the escalating rhetoric and power struggles of the Jacobins which plunged France into a bloodbath and set back economic development. When Giesbert (1989a) wrote on the front page of *Le Figaro* on 15 July 1989 that 'the trouble with the French Revolution was that it turned out bad', this is clearly an example of what Michel Vovelle termed 'vulgar Furetism' (Spire, 1986). In historical terms, it is clear that Furet's revisionism set the agenda for the Bicentenary.

Furet was to be disappointed. Garcia (1988, p. 12) correctly predicted that 'centrists such as Furet who want a non-controversial memory will be denied it by the refusal of the Right and the power of the Revolutionary ideas'. The closest thing to an official historical celebration was a World History Congress held at the Sorbonne (6–12 July 1989). It was opened by Mitterand, and it focused on the international image of the French Revolution. Furet stayed away: an example, said Vovelle, of his arrogant refusal to attend anything that he had not personally organised (Reynaert, 1989). The squabble was long-standing, and Vovelle reported with satisfaction that Jeanneney, President of the Bicentenary Mission, had publicly criticised Furet for his absence (Genevée, 1989). Vovelle had been given the Sorbonne Chair in the History of the French Revolution on Albert Soboul's death in 1982, and had been in charge of the CNRS (Centre Nationale de la Recherche Scientifique) Commission for Commemorating the Bicentenary since 1986; and in 1988, when Jeanneney succeeded Faure, Vovelle replaced Furet as historian to the Mission for Commemorating the Bicentenary. Where Furet had left the PCF, Vovelle had stayed, and the Central Committee of the PCF advised Communists to reject the consensus view of 1789 which Faure had been promoting (Malaurie, 1988). Vovelle (1988) recognised Furet's revisionism as an attack on the PCF, noting the fear of some that French society retained a potential for revolution, and he was in turn criticised for his own partisan views (Bonilauri, 1988). From the Left, Gallo (1986) claimed Furet was part of a counter-revolutionary tradition now being given greater prominence by a media frightened by the Left's acquisition of the presidency. Vovelle also contrasted the interest of publishers in scholarship with the media's urgent courting of counter-revolutionary ideologies (*Contact*, Vol. 257, March 1988, pp. 11–14).

Vovelle continually pressed his claim that in directing the historical events of the Bicentenary, he was motivated solely by considerations of scholarship whilst also insisting on an interpretation of the Revolution which was at best held by only a substantial minority of his colleagues in the academy. Vovelle embraced Clemenceau's assertion that the Revolution had to be taken *en bloc*; the Terror being required to defend the gains of 1789. Vovelle (1987) argued that the Revolution was violent from the start, and that it is impossible to separate the good bits from the bad. Faure (1989, p. 5) had argued that 'it is a mistake to take the Revolution as a whole, it allows the Left to excuse the Terror with the Rights of Man and the Right to reject the Rights of Man because of the Terror'. Ozouf (1986) noted that all history was revisionist, as changing circumstances change our interest in historical events. In persisting with an old polemic, Vovelle was out of sympathy with Faure's attempt to focus on the Rights of Man and avoid pro- and anti-Robespierre quarrels (Bonilauri, 1987). The attention given to the Rights of Man was an official enthusiasm intended to allow Mitterand to define a usable past that might consolidate the Socialists' hold on the centre-ground of politics, as well as to underline Mitterand's third-worldism.

The Rights of Man as an Official Enthusiasm

When Mitterand was asked by the journalists of *L'Express* whether he was making personal or party-political capital out of the Bicentenary, he replied, somewhat disingenuously, 'I have only my love for France, its history and its message' (Haski, 1989). This message was the 'Declaration of the Rights of Man and Citizen'. One commentator claimed that this concern with formal legal rights reflected Mitterand's training as a barrister (Denis, 1989). Indeed, in wishing to mark the Bicentenary with an initiative of his own, Mitterand proposed a reform which would allow citizens to petition the Constitutional Council about any government act that they considered unconstitutional. The President was keen to be seen as a great democrat, and to take back the initiative from left-wing critics of the Summit (Patoz, 1989). The proposal itself had been PS policy since 1972.

In presenting this proposal in the Bicentenary year, Mitterand was also entering into a debate about the nature of France's revolutionary heritage. It is clear that, as Touraine (1989, p. 22) and Kristeva (1988, p. 21) both remarked, the 'Declaration of the Rights of Man and Citizen' presupposes a set of political institutions through which natural rights are to be secured. Yet, the Declaration did not settle the question of what form these institutions should take. The liberty which the French Revolution stood for was little more than 'an absence of arbitrary power' (Starobinski, 1988). In fact, the Parisian administrators of the Revolution considered their social and political programme so self-evidently correct that they optimistically believed reaction would disappear once the few remaining bad influences were removed (Ozouf, 1988). Thus, they gave far too little attention to the *forms* of popular sovereignty (Roman, 1989), and their optimism encouraged an opportunism which recognised no essential difference between the executive and the judiciary, or the need for judicial checks on the executive. As Howard (1988) has pointed out, the 'Rights of Man' actually have no real constitutional status in France. In the US the 'Rights of Man' are the law, whereas in France they are merely the preamble to the Constitution, little more than a piece of rhetoric. The result is that in the US politics is subject to judicial control, whilst in France the opposite is true. If anything, Mitterand was using the Bicentenary platform to make France a little more American – *Le Monde* saw it as a step in 'the Anglo-Saxon direction' of giving the judiciary precedence over politics (Rollat, 1989). For Mitterand it was not enough to revivify the principles of the Revolution, the French must also pay attention to the forms through which those principles were to be defended. Mitterand avowed: '[l]iberty is fragile. I've said it throughout my political life: it is not a natural but a social construction. To preserve liberty institutions are necessary which arbitrate between interests and passions' (quoted in Haski, 1989). The Revolution, he implied, did not provide appropriate institutions for that arbitration.

To a large extent Mitterand's emphasis on France's 'message', the rights of human beings, did shape the uses made of history during the Bicentenary. Furet's 'liberal' revolution was alive and well: 'the Rights of Man, elections, individual freedom, including that of the market', as he put it. The Bicentenary had moved beyond the Bolshevik appropriation of the Jacobin revolution and had

'rediscovered the principles of 1789, and the democratic universalism so long criticised as nothing more than a cover for bourgeois self-interest. The idea of revolution may be in crisis but the principles of 1789 are more universal than ever. It is this contradiction which gives ambiguity but also richness to this Bicentenary' (Furet, quoted in Giesbert, 1989b).

The dialogue with the principles of 1789, the analysis of the context in which they were put forward and the evaluation of the conditions necessary for their success, hardly testifies to history's abdication, or to the end of ideology. It is certainly clear that this emphasis on the principles of 1789 marginalised debate about the events and person-alities of the Revolution. In making it clear that he would be opposed to the Panthéonisation of Robespierre, and that as an opponent of capital punishment he could not have voted for the execution of Louis XVI, Mitterand distanced himself from the claim that the Revolution must be taken as a whole. He also returned to the patriotic rhetoric of 1889 in singling out Carnot and Danton for praise in defending the Revolution against external enemies (Bertier, 1989). Above all, he more or less succeeded in insulating the discussion of the 'Rights of Man' from polemics about the 'excesses' of the Terror.

Challenging the Official Enthusiasm From the Right

Mitterand's success in divorcing the 'Rights of Man' from the Terror was far from inevitable. Some on the Right wished to keep the events of 1793 on the agenda. The Revolution's attack on religious liberty and the Pope's rejection of a natural basis for human rights ensured that the Catholic Church became the figure of anti-Republicanism; and the place of the Catholic Church as a rallying point for some of the opposition to the Communist regimes in Eastern Europe, particu-larly Poland, gave this matter a renewed importance in 1989. Historians have argued that the Catholic Church became the emblem of a regional revolt in the west of France in the 1790s, precisely because it opposed a Revolution which failed to resolve the economic crisis of peasant agriculture in this region (Petifrère, 1988). But the far-Right has always read the revolt of the Vendée and the guerrilla war of the Chouans as primarily a religiously-inspired battle against a blasphemous Republic. The suppression of the Vendée was meant to be exemplary. Pierre Chaunu calls this genocide, and sees it as the more or less ever-present accompaniment of left-wing revolutionaries everywhere: China, Cambodia, Soviet Union. For Chaunu (1986) the Revolution was a complete waste, and he compares the violence and repression in France with the peaceful development of England over the same period. As Rémond (1989) remarks, Chaunu thinks that all revolutions have exactly the same form. Since all legitimate authority issues from the same source (God), all attacks on it are part of the same crime. Benoist (1989) argued that violence is integral to the logic of revolution, for instance, ensuring a direct line between the Revolution and the Holocaust. The counter-revolutionaries planned to march from Nôtre Dame to the Place de la Concorde on 15 August 1989 to commemorate the 'destruction' of the Vendée in August 1793. At the meeting which called for this march, a right-wing journalist showed some of the things which lay under this particular stone. The message of the Revolution, he said, was 'God to the gutter, France for foreigners, and power to Rothschild' (*'Les contre . . .'*). Philipe de Villiers wrote to Mitterand asking that the name of the general who led the Revolutionary army in its scorched-earth criss-crossing of the Vendée, Louis-Marie Turreau, be removed from the Arc de Triomphe (Gene, 1989). De Villiers invited the Polish Solidarity movement to a mass to be held at Puy-de-Fou, in the Vendée, to protest the Revolution's attacks on the Church. Instead, Solidarity sent its representatives to the Bicentenary in Paris (Henry, 1989), but the Catholic primate of Poland, Glemp, went to the mass and called on John Paul III to visit the Vendée in 1993 on the anniversary of its struggle. He said that liberty, equality and fraternity were inconceivable without a firm religious basis, and compared the Vendée of 1793 to the Poland of today. Yet this sort of attack on the Revolution was almost seen as ritual, and – at least from de Gaulle onwards – the main right-wing group in French politics had accepted the secular Republic. Furthermore, the serious discussions about the future of Eastern Europe were going on at the Summit.

The Revolution as Model for the Left

Although some of the Left insisted on taking up particular events and personalities of the Revolution, this was uncommon. Jean-François Vilar, the author of a novel about Hébert, argued that the true popular heroes of the Revolution were being forgotten. He argued that the Bicentenary should celebrate division and chaos (Simsolo, 1989): what Tonka (1988) termed the 'revolutionary project' of 'madness, play and pleasure'. Henri Guillemin wanted Robespierre, of whom he had written a biography rehabilitated. Robespierre, he insisted, was one of the few revolutionary leaders to have looked outside his class to the needs of the poor (Peuchamiel, 1989). The modern relevance of this lay in pointing out that the Revolution immediately betrayed the universal principles of the 'Rights of Man' in narrowing them to the benefit of propertied, white males, and in urging that a truly social revolution was necessary to ensure human rights. From the more radical wing of the Socialist Party, Jean-Pierre Chevènement (Minister of Defence) wished to see Robespierre in the Panthéon before the third centenary of the Revolution (J. H., 1989).

On the whole, however, the Left took up Mitterand's invitation to celebrate the ideals of the Revolution. Liberty, Equality and Fraternity were used to assess the state of French society and its relations with the Third World, and also to press the case for reform or revolution. Speaking at a meeting in Moscow to commemorate the Revolution, Regis Debray – as the President's official representative – said that where he had once been a revolutionary he was now a reformist, and he spoke of the universal danger of revolution leading to a concentration of power, comparing Stalin to Bonaparte (D. L., 1989; '*Un retraite* . . .'). The revolutionary road was mapped out by groups who insisted on the bourgeois character of the Revolution and the ever-present revolutionary character of the proletariat. Against this, Noiriel (1989) has remarked that the manner in which 1848 and 1871 made reference to 1789 has created the myth of a revolutionary proletariat and produced the disillusion felt by writers such as André Gorz and Alain Touraine when they can no longer see such a potent force in modern politics. This revolutionary character was seen by some as having been betrayed by reformists 'who, on getting into power, smashed all opposition and all hope of change, demoralising and demobilising the working class' ('*Confédération* . . .'). The bourgeois character of the Revolution was seen to set inescapable limits to the rights that it established:

'[t]herefore, its "bourgeois democracy" allows a bourgeois equality (based on the great social inequality of the classes), "bourgeois justice" (equality before the law) and "bourgeois liberty" (freedom to get rich based on the freedom to buy and sell all goods, including human beings) – the whole lot resting upon one fundamental principle: that the rights of property owners come first' (Mouvement Révolutionnaire Internationaliste, 1989).

Within this classical model, the solution to current troubles is world proletarian revolution, which means that its absence explains oppression everywhere and establishes, for example, the capitalist character of Eastern Europe and China or, in one case, the feudal character of Communism (*La lettre* . . ., p. 3). Yet the developments in Eastern Europe seemed to be based on demands for some measure of market freedom, and what is more the problems of the planet have come to be seen as comprising an important ecological dimension. In trying to incorporate these matters within the Left's agenda, the events of 1789 proved far less useful than its principles. One radical-left group (Parti Socialiste Unifié), whilst writing of new 'bastilles to take' and of 'a new solidarity of grass-roots movements', named its target as 'capitalist, productivist, ecocidal development' (Labertit *et al.*, 1989).

In general, historical reflection on the events, personalities or even ideals of the Revolution was not used to explore new agendas on the Left. The symbols were used as a vocabulary to express positions defined largely in terms of the same class politics of the pre-revisionist histories: Juarès for the Communists, Guèrin for the Trotskyists. It did not have to be like this. One group which seems to have thought explicitly about the uses of history wrote:

'If one wants to play the game of historical parallels – which has at least a metaphorical value – one must speak of the "Third World", not as today's "Third Estate", but as a part of the "Fourth Estate". The Fourth Estate has always been excluded from history – from representation and power – even during the "radiant and terrible" days of the French Revolution. The Fourth Estate includes women, the proletariat exploited by the Third Estate; it includes the colonised, the angry people who no longer seem to be represented by any Cordeliers, or Hébertists' (IRIS, 1989).

The group went on to develop a very suggestive analysis of the relations between capitalist accumulation and the fall of the Soviet Empire:

'In this sense, the unreal Communism of real socialism is the philosopher's stone of capitalism: embodying the intensive, hyperreal repetition of primitive accumulation in a foreshortened space and time, it contributes to the extension of capitalism. . . . Clearly there is a risk that the gap created by the implosion of unreal Communism will encourage the globalisation and modernisation of capital: a process of reproduction extended from the individual to the communal level where money and the vote mean free markets and representative democracy. . . . But it is not impossible that the current crisis at natural, social and intellectual levels, and the contemporary recognition that we are all part of a global project, encouraging an immense unity in diversity which expresses a grouping of multiple autonomies, could take us towards a future which is communitarian, diversified, libertarian and ecological' (IRIS, 1989).

Even if this sort of imaginative repositioning of the present with respect to the global significance of 1789 was unusual, Mitterand's injunction to look at the ideals of the Revolution did have a number of positive consequences for the Left.

Using the Official Enthusiasm on the Left

The first positive consequence of this way of memorialising 1789 was that it placed the whole question of rights discourse before socialists. It is in this respect that Maurice Agulhon (1988) could speak of the 'principles of '89' as the 'best defence against barbarism'. Elsewhere he urged the Left to take possession of '89 as well as '93 so that it could have the flag of liberty. Only the far-Right still refuses the Revolution, he said, and isolating the fascists is the most important priority in French politics. A nationalism which bases itself on the 'Rights of Man' might help the French address what Agulhon (1987) saw as its greatest challenge: the integration of its immigrant workers. To their credit, the equation of the Third World with the 'Third Estate' led many on the Left to highlight the importance of the anti-racist struggle within France.

The second positive consequence was that it gave the Left resources with which to generalise local struggles, and to press them more urgently on the government. Two examples illustrate the power and limitations of this strategy. The 'Rights of Man' are presented in universal terms, so that individual cases where they are denied open the government to charges of hypocrisy: charges which have no legal status, but which can embarrass the government when it is placing itself in the spotlight as the advocate of human rights.

The first case concerned four leaders of Action Directe, an anarchist urban guerrilla group. They had been convicted of the murder of several business and military figures, including a French general, and were being interrogated over several other murders. This interrogation had been going on for three years, during which time they had been kept in solitary confinement. The anarchists claimed that this was an infringement of their human rights, calling themselves the '*sans cravattes*' (Conil, 1989a). On 20 April 1989 they went on hunger strike and by mid-July they were in a parlous state ('*Les quatre . . . mourir . . .*'). The investigating judge (*juge d'instruction*), Jean-Louis Bruguière, was opposed to any modification of their conditions of imprisonment (Logeart, 1989). The Magistrates Union in France accepted the anarchists' claim that prolonged isolation is akin (Conil, 1989b) to torture, but the anarchists' bid for the status of political prisoners under international law faltered on the fact that French law does not recognise that anybody is imprisoned in France for political reasons (Moruzzi, 1989). By 19 July one of the prisoners was being force-fed ('*Les quatre chefs . . .*'), and all four were in hospital ('*Action . . .*'). With the likelihood of a death on their hands, the Ministry of Justice increased its pressure on the investigating judge, and he lifted his insistence on solitary confinement. The four were to be allowed to communicate with each other. The two men were to be in neighbouring cells in one prison and allowed to meet, whilst the two women were to have the same rights in another. The hunger strikers initially insisted on being able to share cells (de Tanney, 1989) but soon dropped this demand, accepted their new conditions and ended their hunger strike ('*Les quatre . . . cessent . . .*'). There can be no doubt that the Bicentenary gave extra force to the human rights arguments in this case, and that this, together with the fear of a horrible death in prison on the anniversary of the storming of the Bastille, ensured some improvement in their conditions after three years of asking.

The second case concerned ten car-workers sacked by the management of Renault, the state-owned car company. They became known as the 'Renault Ten'. In the summer of 1986, as part of the restructuring of the state sector of the economy, Renault made 1,200 workers redundant from their factory at Billancourt, near Boulogne. In the protests over this action, there were violent clashes with the police on 30 July 1986, and on 1 August some workers broke into the factory's offices, allegedly injuring two office workers. Twenty workers were sacked, and charges were finally brought against ten individuals in Autumn 1987 for the theft and violence. In 1989 the National Assembly decided to mark the Bicentenary by granting amnesties to a series of people charged with minor crimes, and the 'Renault Ten' were amongst them. The National Assembly also said that they should be given their jobs back. The Constitutional Council argued that this was an unacceptable infringement of the right of the Renault management to manage, however, and annulled the amnesty on 9 July 1989 (Bouffin, 1989). From 16 May, the ten workers occupied their posts at work. The PCF and the PCF-dominated trades union organisation, the Confédération Générale du Travail (CGT), took up the case in order to protest against the redundancies in the state sector and to draw attention to how the law protects management. As the head of the rival blue-collar trades union group, more sympathetic to the PS, pointed out, 11,000 workers each year are sacked for issues relating to trades union activity (*'Jean . . .'*). One of the 'Renault Ten' was the son-in-law of George Marchais, the General Secretary of the PCF. The CGT were having little success in pressing the issue at Billancourt, where one-half of the workforce had been shed in the previous three years, leaving it 'an empty little Kremlin', increasingly peopled by immigrant workers reliant on the management for access to housing and unlikely to follow the CGT on to the streets (Herlich, 1989). The crucial battles were all in Paris. On 6 July the CGT demonstrated outside Mitterand's home in Paris, and called for a bus and metro strike for the 14 and 15 July in order to disrupt the Bicentenary (Thoraval, 1989; Taupin, 1989a). The President appointed a mediator, Jean Lavergne, to explore the common ground between workers and management. This took the heat out of the situation, and the ten car-workers ended their occupation. The march in Paris, of 12 July, passed off peacefully (Ortoli-Lanöe and Lecourt, 1989). By 18 July, with the threat of disrupting the Bicentenary now redundant, Lavergne revealed that the President had given him no instructions which might form the basis of a new initiative (Agudo, 1989), and – whilst Jacques Levy, the director of Renault, expressed a willingness to negotiate – it went no further than providing the workers with good references for jobs elsewhere (G. Lq., 1989). Indeed, by this stage an employers' organisation was protesting about what it saw as the special treatment being given to the 'Renault Ten' (Taupin, 1989b), and Nestlé had locked out seventeen workers whom they had been instructed to reinstate at Beauvais, a ruling that the company had initially followed (*Libération*, 22 July 1989, p. 44).

These two cases were certainly given greater significance for being presented as a test of the state's commitment to the principles of 1789. The pressure from the Ministry of Justice for an amelioration of the conditions of imprisonment of the anarchists, along with the appointment of Lavergne in the case of the 'Renault Ten', were clear concessions. Certainly, the possibility that the CGT might attempt to disrupt the Bicentenary could not be dismissed, although their strength in the crucial transport unions was probably not sufficient for them actually to bring Paris to a standstill. Nevertheless, the act of memorialising the 'Rights of Man' made the government vulnerable to these attempts at investing particular cases with much wider significance. In that sense, at least, a usable past had clearly been defined.

The Politics of Architecture

Mitterand's Political Project and the Appeal to History

I have described Mitterand's Bicentenary project – to colonise the centre-ground of French political culture from the Left and to displace the legitimacy conferred on the Right by the legacy of the patriotic general – and I have shown that his presuming to speak for the nation was contested from both Right and Left. I have also suggested the reason why the abstract principles of 1789 were made the focus of Mitterand's recourse to History. I have further documented the struggle which arose over what the implications of those principles were for contemporary France. The recourse of history would be pointless if it did not energise commentary in this way.

Staging the Bicentenary was justified, according to Mitterand, because the events had clearly 'touched a vital collective memory' (*Le Monde*, 21 July 1989, p. 7). As a Communist poet put it in July 1989: '[t]he memory of a people is not given once and for all, it is a struggle' (Ristat, 1989). Although the odds were stacked in favour of the stage-manager, the political content of memorialising 1789 was still negotiable and *Le Figaro*, for one, noted that Mitterand was playing with fire and predicted wishfully that, just as the Revolution had consumed its makers, so might its Bicentenary (*'Bonne ...'*). The attraction of directing the Bicentenary lay precisely in attempting to construct a consensus about something that still mattered, and at the same time marginalising some of one's political opponents. I have described the palimpsest of historical references which gave significance to the Bicentenary. The usable past that was recovered did not involve a long reach across two hundred years, but rather a set of multiple mappings of various events against each other. These events (the Liberation, May '68, the Centenary, 1789, and so on) raised a linked set of questions about political authority, patriotism, and the trajectory of French society. They also referred in quite specific ways to the City of Paris, to particular sites within the city and to a prevailing ideology about the city's custodianship of its past. In this connection it was important that *Le Monde* should conclude that the Bicentenary 'was expensive, but (that) the crowd did not seem to think the head of state had gone too far and they were able to appreciate for the first time the integrity of his contribution to the modernisation of Paris'. Indeed, the article went on to suggest that 'the success of the 14 July owed something to its setting, to the triumphal passage which, from the pyramid to the Arch, sets the jewels of yesterday in the light and materials of the year 2000' (Fontaine, 1989).

Here, perhaps, two notions of history interpenetrate: firstly, the history in landscapes, together with the uses of the past implicit in their conservation and transformation; and secondly, the commemoration of particular historical periods, together with the construction of a usable history. For these sorts of references to be credible, the cityscape must be readable in terms of a set of pertinent appeals to history. This means taking the symbolic order of the city seriously, and it means nurturing the practice of making symbolic appeals to history. We can briefly illustrate two important aspects of this form of appeal by looking at two examples from the Bicentenary celebrations. The two points I want to stress are that this nurturing is both deliberate and far from innocent.

Historical Symbols and the Challenge of Irony

The first example I want to take is Goude's 'Marseillaise' parade. I have already described the political context in which Goude received the commission, but what about the content? Watching on British television, Derek Jarman's overall assessment was caustic:

'The French government mounted a most inapposite procession to mark the 200th anniversary of the storming of the Bastille – designed by Grace Jones's hairdresser. Some young fashion models dressed in funeral black, like enormous tops, pirouetted unsteadily down the Champs-Élysées embracing embarrassed young boys dressed as Pierrots and Samurai. A dilapidated black conductor, looking like an illustration of a cocktail shaker in a twenties *Vogue*, hammered at a drum surrounded by girls flopping around in a pastiche tribal dance, black willies in classical tutus. . . . An event almost as big as Live Aid, said the commentator. I decided it was no coincidence that Mr. Goude's name resembled the dullest of Dutch cheese, the tedium was increased to hilarity by constant reference to his *avant garde* status' (Jarman, 1991, pp. 110–111).

Only against the pomp and circumstance of the military parade could Goude be considered nonconformist, yet this designer of television advertisements was regularly presented as such in the French press. The imagery that Goude peddles is generally based on an orientalist fascination with the grotesque, the incongruous. Yet while Goude was somewhat playful – his original intention of having a British postcard military tattoo marching in the rain was only scrapped when the dancers caught a cold during rehearsal – there were some genuinely revolutionary styles quoted in the parade. It is a pity that he did not realise his dream of having a live television link-up to James Brown in prison, but setting the American marching band to Brown's music did pay appropriate respect to the black contribution to American popular music, a revolution which is still too easily denied. Similarly, designing the Soviet contribution around 1920's constructivism highlighted a rigorous attempt to rethink the links between Art and Revolution; links with which the Soviet authorities still felt uncomfortable (it took six visits to get their approval). It is undoubtedly indefensible of Goude to take stylistic forms so seriously as to claim that the real revolution today is world music, and perhaps rather naïve of him to think that it signifies a real cultural marriage of any consequence; although Goude's show did bring forth a predictable charge of cosmopolitanism from Le Pen (Jacquemart and Monet, 1989). But, one might at least argue that if the translation of the ideals of 1789 into the medium of popular music and the language of style is acceptable, then Goude was trying to establish a set of relations in popular culture which echoed some of the principles he was to celebrate. Certainly, his worries about the flaccid costume drama that many wanted were probably not misplaced. For at least one Radio Canada commentator, Goude 'played with fire' (Biétry-Rivière *et al.*, 1989).

I find it interesting that *Le Monde* felt that in Goude's parade 'the golden age of cartoon comics was sent to the postmodern mill' (de Roux, 1989). It is interesting because this is precisely the sort of *avant garde* cachet that Goude and, more importantly, his sponsors would have welcomed. Without entering too far into polemics about the nature of postmodernism, I think it mistaken to term Goude's 'Marseillaise' postmodern. One commentator reported that Goude had intended his parade to be ironic, but that his political minders would not allow it (Lacan, 1989). Goude's ignorance of history would have been a serious handicap if any sort of costume drama had been planned, but the celebration of the principles of 1789 in an idiom with which he was familiar (pop-styles) posed fewer problems – in any case, Dupavillon made up scrapbooks of the story of the Revolution for Goude's education (Vincendon, 1989). If Jessye Norman singing the 'Marseillaise' alone at the obelisk in the Place de la Concorde recalled her performance of the same song at the centenary of the Statue of Liberty, it also brought the parade to the sort of dignified conclusion which certainly did not challenge the ordinary conventions of official events. Yet the real challenge to any strategy of dissembling before the principles of 1789 came in Tien An Men Square. The students who danced to house music, who built their own Statue of Liberty, also sang the 'Marseillaise' before they were crushed. Goude and his team made two visits to Peking. Before this demonstration of 'the universality, the reality and the force' of values of the Revolution, they dropped their intention of incorporating a Chinese float in the procession and sent it ahead with a minute's silent tribute.

This was very much an official enthusiasm. The Chinese complained that the French police had failed to prevent dissident Chinese students in exile breaking into their Parisian embassy. In fact a group, under the banner 'Peking People of all Countries Unite', occupied the offices of the Educational Service of the Chinese Embassy on 29 June 1989 and sent Deng Xiaoping the following (Maoist) fax: '[h]umanity will not be satisfied until the day when the last capitalist has been hanged with the guts of the last bureaucrat' ('*Pekins* . . .'). On 12 July Lang was at the Grande Halle of the Parc de la Villette where, in the company of two refugee leaders of the Chinese students' movement, he unveiled a replica of the copy of the Statue of Liberty which had been smashed in Peking (*Le Monde*, 14 July 1989, p. 3). The evening of 14 July saw windows at the embassy smashed ('*Les sept* . . .'). The 14th also saw two students executed in China for their part in the pro-democracy demonstrations, and the French government honouring a pre-arranged commitment to make a loan of FF830m to the Chinese Government (Marcovici, 1989). The 'Big Seven' condemned the repression, bringing the expected exasperation of the Chinese government, but – led by Japan as the largest foreign investor in China – they stopped short of sanctions. The French government was keen to harvest these new layers of significance added to both the Statue of Liberty and the 'Marseillaise' as symbols of French republican democracy. It chose to lay them out in their theatre of solidarity. Goude's political minders had no intention of allowing the parade to feign ignorance about the principles and events to which the symbols of the Bicentenary made reference. Irony was therefore *not* the order of the day.

The Use of Symbols: Censorship by Significance

The second example is more ambiguous since it shows both the power of the historical references of symbols and the shallowness of the debate about them. During the period of cohabitation, when Mitterand was president and Chirac was prime minister (1986–8), Chirac appointed François Léotard, leader of the Republican Party (RP), Minister of Culture. Little was done to prepare for the

Bicentenary during this period. One of the few projects Léotard did promote was a garden to be built at Versailles on the site of the Hôtel des Menus Plaisirs, the building in which the Estates General met in May 1789 and where many of the early, 'liberal' measures were passed, including 'The Declaration of the Rights of Man and Citizen' (26 August 1789). A Scots artist (poet, sculptor, publisher), Ian Hamilton Finlay, together with a landscape designer, Alexander Chemetoff, won the commission and promised a Revolutionary Garden. One quarter (FF800,000) of the cost was to be met by the Badoit mineral water company since it was celebrating its 150th anniversary in 1989 ('*Badoit . . .*').

Finlay rails against what he sees as the devaluation of art by a lazy, liberal consensus which is unwilling to engage with the heroic, and which lies about its past in attempting to exorcise extremism by insulating its heritage from Jacobin or Nazi contamination. He embraces classicism as an attempt to ground revolution and heroism in nature ('Neoclassicism needs you'). Throughout the 1980s he appropriated the symbols of the Terror and the mottoes of St. Just and Robespierre, and in 1985 he wrote 'in Revolution, politics become nature'. A garden is certainly an appropriate place to rethink these links between nature and culture, and history suits his purpose too because it is the phoney progressivism of changing fashions which sticks in his throat: '[r]everence is the Dada of the 1980s as irreverence was the Dada of 1918'. In a series of reworkings of classical themes from fine art paintings, Finlay has shown shepherds happening across SS relics. Edwin Morgan notes all of this and describes: '[t]hree prints of 1990 – *Two Scythes, Scythe/Lightning Flash*, and *Sickle/Lightning Flash* – instead of turning swords into ploughshares or pruning-hooks turn scythe and sickle, images of harvest and fecundity, into the SS lightning-flash, reminding us that politics and revolution are a part of Nature. The German connection is strong, and obviously problematic' (*Evening . . .*, p. 43). 'Problematic' is putting it lightly, and defuses the shock that Finlay intends. In 1982 he put together an exhibition entitled 'Third Reich Revisited', which tried to show that the intention of conveying the heroic which motivated Albert Speer's neo-classicism was rather general in Western countries at the time. To do this by adding swastikas to pictures of 1930s Scottish government buildings was, of course, provocative ('*Liberty . . .*').

In 1987 Catherine Millet, editor of the French cultural magazine *Art Press*, wrote about Finlay's use of Nazi imagery. No serious discussion of neo-classical architecture followed, and Finlay, guilty by association, became too hot to handle. On 23 March Léotard defended Finlay ('*Une commande . . .*'), but a week later, after someone had scrawled 'SS' on one of Finlay's sculptures, Léotard dropped the project altogether (*'Bicentenaire '89 . . .'*). The Nazi imagery retains its power, but, and paradoxically, it is difficult to think with it in public. Finlay issued (1987) a card with the motto 'Myriam Salomon [Assistant Editor, *Art Press*] owns the Second World War and you are not allowed to mention it' (*Evening . . .*, p. 43). He also consigned Millet's guillotined head to a basket in his installation of 1987, *4 Heads* (*'Glasgow's . . .'*). Finlay had fought (1942–1945) the Nazis in the War, but neither this nor the fact that he was genuinely keen to rekindle an important part of the Revolutionary symbolic heritage counted against the 'fascist' smear. The power of symbols is clear, as is the fact that this power places some of them off-limits in certain contexts; in this case an official, patriotic one.

It is significant that architectural taste was the sign which betrayed Finlay to his critics. Writing in 1978, Albert Speer avowed:

'Karl Arndt is right to detect Hitler's desire for power and the submission of others in my buildings. The main character of my architecture expressed this urge. In those years, I believed in the claim to absolute domination and my architecture represented an intimidating display of power, as Georg Friedrich Koch describes in an essay. The National Socialist movement represented more to me than the mere incarnation of political power. It was the fulfilment of a claim to dominance over a nation. Every person, if they wanted to survive, had to submit to it' (Speer, 1985, p. 213).

Speer went on to claim:

'That a monument's value resides in its size is a belief basic to mankind. Only the boldest and the most spectacular buildings seemed suited to represent the noble image of human reason under the French Revolution. All my plans appear conventional when compared with the French revolutionary architects' (Speer, 1985, p. 213).

In taking up classical architectural forms, the architects of the Revolutionary period appeared to be engaged in a form of revivalism. It certainly disappoints Charles Jencks (1988, p. 106): 'I think that the architects Ledoux, Lequeu and Boullée were very reactionary. Their projects consisted essentially in a return to Roman archetypes.' Jencks claims that it was not republican Rome but the fascist period of the Caesars which attracted the architects of the First Republic. The inhuman monumentality of the architecture reveals its fascist pretensions. The only way to indulge people's taste for historical rootedness without yielding to repressive monumentality would be to deploy a sense of irony, so Jencks suggested, and he commended Ricardo Bofill's monument 'to nothing' in his Centre de la Place des Arcades du Lac at Saint Quentin. Jencks's characterisation of the stylistic attractions of Antiquity for the Revolutionary architects is certainly contestable – one might argue that they were interested in what they saw as a return to natural forms in classical architecture rather than in its monumentalism (Béret, 1988) – but it does point to some of the hazards of historical quotation in architecture, and to some of the difficulties that there would be in recovering the Revolutionary heritage for commemoration.

Revivalism Versus Modernism: The Political Construction of Architectural Difference

In one very striking sense, the ideological implications of architecture for politics have remained constant; or at least the terms of the discussion have. As Bruno Zevi summarised an unofficial competition for the redesigning of Les Halles in 1980, the possibilities on offer were those of 'retro', 'backward-looking neo-classicists' or the 'Modern Movement' (Barré *et al.*, p. 76). Indeed, the old market halls at Les Halles that were demolished in 1972 illustrated some of the ways in which this opposition between looking forward and looking backward has been valorised. The pavilions had to be designed by Baltard in cast-iron and glass, since Haussmann had sold this style to Napoléon III. Haussmann's aesthetic has been termed 'eclectic' (Choay, 1983, p. 232). In fact, Haussmann promoted neo-classical facades for residential developments, to ensure uniformity and a greater spread of land values along a street than individual buildings would allow (Roncayolo, 1983, p. 112), and he favoured functionalism for commercial buildings to express the dynamism of industry and commerce. Les Halles was Haussmann's first (1854) commercial building, but 'once Les Halles had been built the model spread: about forty more [in Paris] between 1860 and 1880' (Marrey, 1989, p. 40).

In planning the replacement for Les Halles, politics took up architectural forms in terms of very similar oppositions, although now the iron canopies were part of the tradition. Giscard d'Estaing, on becoming President in 1974 following the death of Pompidou, was clearly looking backwards in insisting that the empty space receive 'a large garden *à la française*' (Rosner, 1980, p. 13). When Chirac became Mayor of Paris and made it clear that he wanted to interfere, Giscard withdrew. Chirac announced: '[t]he architect of Les Halles? That will be me, and no bones about it' (Muret, 1980, p. 77). In fact, faced by the challenge of respecting the 'historic' buildings on site (the Bourse and the church of Saint Eustache) and of not upstaging the nearby monument to his political mentor (the Pompidou Centre), Chirac opted for an architecture of self-effacement. Rosner's summary is just:

'A politician's project, but without any policy, its chief characteristic was banality and inadequacy. A project that was anti-monument, a project of created emptiness, which itself became monumental. An administrator's project which would probably sell well, offering Commercial Space, offices and a luxury hotel . . . Chirac wanted a dull plan for ordinary people, a plan which would blend into the landscape he was magnifying at the same time' (Rosner, 1980, p. 15).

In return for ceding his interest in Les Halles, Giscard was given free play on another site by Chirac: La Villette (Chaslin, 1985, pp. 64–65). On this site an enormous and unused meat market was to be redesigned as a Museum of Science and Industry, and a musical centre was to be created out of other buildings there. For the rest of the area Giscard wanted a park, 'sober and classical', laid out à la française with regimented rows of trees, 'in the Tuileries tradition'; and with terraces to hide the Parisian circular route, the peripherique (Chaslin, 1985, p. 66). One element of the winning design which Giscard cancelled was an enormous shiny globe (the Géode) which was to house a futuristic cinema. The President considered it 'an inelegant excrescence counter to French taste' (Ellis, 1986, p. 16). On becoming President in 1981 Mitterand reinstated the Géode but cancelled the park, and he held a competition to plan 'an urban park for the twenty-first century'. Already in this competition brief Mitterand marked a clear separation from the grosser forms of revivalism. In fact, he presented his *grands projets* as attempts to forge a new future for the city in the face of its dislocation by economic and demographic changes. His sense of historical mission is almost palpable in this statement of January 1982: '[w]e will have achieved nothing if in the next ten years we have not created the basis for an urban civilisation' (*Architectures*, p. 11). In 1987 he wrote: '[m]y wish is that the major projects help us to understand our roots and our history; that they will permit us to foresee the future and to conquer it' (*ibid.*, p. 8). Such a programme defines in broad terms the relationship between politics and history in Mitterand's major projects.

Mitterand's Modernisation of Paris and the Refusal of Pastiche

Laloux's railway station on the Quai d'Orsay had been threatened with the wrecker's ball in the early 1970s, but the destruction of Les Halles and conservative disquiet at the modernism of the Pompidou Centre (Beaubourg) brought a reprieve. Giscard proposed in 1977 that it be a Museum of Impressionism which, he argued, would celebrate the traditional architectural values of the nineteenth century in presenting the French art which had cornered the greatest international acclaim in the twentieth. The President's 'retro' tastes, his wish to create an 'anti-Beaubourg', and his continual modification of the artistic terms of reference of the museum created an unsettled period of planning (Chaslin, 1985, p. 47). Under Mitterand the chronological sweep of the museum was extended (1848–1914), its artistic terms of reference broadened from fine art to include journalism and photography, and the artistic movements on display were to be set in the context of the social and economic history of France. The reorientation of the purpose of the museum was directed by the Marxist whom Mitterand placed in charge, Madeleine Rebérioux, and the primacy of political interpretations of art was to be underlined by beginning the collection with the year of revolutions, 1848. The old building was not simply left alone. New elements were introduced to define the sequence of museum spaces. In speaking of the 'insertion of a new architecture which would respond to the original', the museum's director, Jean Jenger, referred to an attitude to the past which is almost a leitmotif of the major projects – 'any idea of creating a pastiche was rejected from the start' (*Architectures* . . ., p. 56).

The rejection of pastiche was nearly ritual. Speaking of his designs for the Opéra-Bastille, another of Mitterand's major projects, the architect Carlos Ott claimed that they were 'opposed to the eclectic reaction which associates – with a mixture of naïvety and nostalgia – the notion of structured urban space (avenues, streets, places and squares) with architectural pastiche employing formal organisation devoid of content'. Instead, he argued that he wished to be 'respectful of both human and urban history' (*Architectures* . . ., p. 109). In terms of the look of buildings, pastiche is very commonly taken as the badge of postmodernism. Christopher Norris, for example, suggests that 'Post-Modernism makes a lot of sense in connection with architecture because what you have there are various collages and quotations, citations, the mixture of styles' ('*Jacques* . . .', p. 76). This corresponds to what Hal Foster (1984, p. 145) has termed 'neoconservative' postmodernism which is 'defined mostly in terms of style, it depends on modernism which, reduced to its own worst formalist image, is countered with a return to narrative, ornament, and the figure'. Such a (mis)understanding of the intentions of 'new-classical' and post-modern architecture clearly lies behind the polemical resistance to pastiche in these Parisian works.

Given Mitterand's avowed historical mission, this is not surprising. The Opéra-Bastille, for example, was planned as an attempt to bring opera to the people. The project began in the 1960s, and its explicit antithesis was the elitist clientele and ornamental architecture of the Palais-Garnier, criticised in the Vilar report (1968), the Bloch-Lainé report (1976) and the Conseil Économique report (1980). The new building was to be the largest opera-house in the world (Duponchelle, 1989). It was because he would not put on more and better-known operas that Daniel Barenboïm was relieved of artistic direction of the opera in 1988 (Soulier, 1989). Ott repeatedly stressed that his architecture was 'anti-elitist' (Steinbach, 1989), and his attacks on pastiche are intended to convey his respect for the significance of the site, abutting the Place de la Bastille. How these views translate into particular design principles is not clear in this case, and the criticism that it is simply a 'steel-and-glass camembert' is perhaps fair. Indeed, the main design principles seem to flow from what Ott conceived to be his anti-elitism. Since his building is for everyone, he said, 'it must be simple and easy to understand' (Rancé, 1989).

A more thoroughly worked-out critique of pastiche is present in Ieoh Ming Pei's pyramid at the Louvre. The polemical force of denying postmodernism in historically significant contexts is apparent. Critics of Pei such as Andre Fermigier dismiss the pyramid as a 'post-modernist whatnot' (quoted in Heck, 1985, p. 49). On the other hand, Pei himself claims a different heritage:

'Modern architects, Mies van der Rohe, Le Corbusier, posed the foundations of a movement which is still in its infancy and is far from having explored all the possibilities offered by modern technology. I consider myself part of this movement and feel that much remains to be accomplished. Very nearly all of current architectural production is pastiche, the latest style, and will not last. It has no future' (*Architectures* . . ., p. 42).

Pei is hence sure that it is his modernism which causes offence, suggesting that: '[i]f this project is controversial, it is perhaps because it is the first time that such a modern statement has been made in a historic context' (*Architectures* . . .). The idea of conserving the Louvre in aspic would be to bring an end to a process of continuous change which has reshaped the Louvre many times before: 'modernisation is the norm not the exception' (Stein, 1985, p. 46). The pyramid reaches back to the harmonies of geometry and balance which informed earlier classicism, but which are not defined by it:

'The modernist pyramid, as Pei sees it, is inherently classical and therefore comple-ments the classical façade of the Louvre. Much earlier Pei considered both explicitly classical and postmodern schemes and rejected both approaches, feeling that they could not measure up to the Louvre' (Stein, 1985, p. 25).

But Pei went further in noting that, as a result of its repeated modernisations, much of the 'classical' façade of the Louvre's buildings is already pastiche so that 'to copy them would be a pastiche of a pastiche' (*Architectures* . . ., p. 42). He also rejected any self-conscious display of technical mastery by wanting a pyramid that drew as little attention as possible to the difficulties of its construction:

'Concerning the design of the elements, Pei and his team have chosen the approach quite the opposite to "high-tech". All of the bolts and articulations associated with metal construction have been banished. The smooth forms offer no resistance to the wind – or to the view . . .' (Allain-Dupré, 1987, p. 69).

Susan Stein remarks that the 'controversy goes beyond the architectural issues and has become a political struggle emblematic of conservative resistance to the Socialist government' (Stein, 1985, p. 25). This is only partly true, since Chirac took his stand elsewhere and moreover defended Pei's vision to the press at the height of the battle over the pyramid – in February 1984 – with a 'conviction and clarity' (Chaslin, 1985, p. 130) which impressed the architect himself. However, one can see the ideological attraction of portraying the Socialist President (and, alone among the major projects, Pei's commission was given personally by the President without a public competition) defacing France's great architectural heritage. In this regard, Pei's attack on what Foster would call 'neo-conservative' postmodernism was certainly apposite.

Like La Villette, Musée d'Orsay and the reorganisation of the Louvre, and unlike the Opera-Bastille, the major project at La Défense was a development which was already in train when Mitterand came to power, but one which he also revised and accelerated. La Défense is the only major project sited outside the city boundary, and it lies on an axis going north-west from the Louvre through the Arc de Triomphe. With restrictions on high-rise office building in the city itself, this district was planned as an overspill growth pole of finance and banking services. It was poorly conceived. Competitions came and went, but with the state only half-heartedly developing the region – coupled to some pessimism about the area's true potential for office growth – business equipped the site with a rather cheap and nondescript set of steel-and-glass slabs. To mark the district's organic connection with the centre, secured in practical terms by a high-speed rail link, there was a recognition that the Louvre–Arc de Triomphe–Défense axis needed some sort of terminus at the heart of the forest of office blocks. Giscard would not contemplate anything taller than the Arc de Triomphe, for fear the grandeur of the established icon would be shaded; and this defined the terms of the competition of 1979. Robert Lion, president of the public–private group set up to manage the commercial district, recalled that:

'In February 1981, I was compelled to write an article denouncing the lack of ambition manifested by the projects and their radical inadaptibility to both La Défense and Paris's history. One hundred years earlier, France had produced the Eiffel Tower – a legendary demonstration of technical prowess for the capital – was it imaginable a century later to consider a modest, shrinking proposal reflecting a lack of self-confidence and a fear of one's shadow? Was French architecture incapable of producing monuments, condemned to adapt itself to the omnipresent worship of surburban form? In 1982, the President of the Republic requested that a new challenge to architects the world over be launched: architectural imagination was to be freed; no constraints were to be imposed . . .' (*Architectures* . . ., p. 16).

Johan Otto von Spreckelsen's Cube won. From von Spreckelsen's brief and poetic entry we can see what led Mitterand to praise its purity and power:

'An open cube
A window to the world
As a temporary Grand Finale to the avenue
With a view into the future
It is a modern "Arc de Triomphe" . . .' (*Architectures* . . ., p. 19).

Soon renamed the Grande Arche de la Défense to make its kinship even more obvious, it has been universally praised. Not being located among historic buildings that it might qualify maybe made it easier for the Cube, but it also has an audacity and a symbolism which are difficult to resist. As von Spreckelsen explained: '[t]he façades of the open cube appear with a bright and smooth surface, symbolising a micro-chip, showing the lines of communication – an abstract graphic work inspired by the most brilliant invention of modern electronics' (*Architectures* . . ., p. 22). The point that nothing could be further from Giscard's 'retro' tastes hardly needs labouring.

The Challenge of Poststructuralist Postmodernism

The notion of postmodernism against which these repeated refusals of pastiche and celebrations of contemporary context seem to rail would not be recognised by many students of the subject. For my purposes it is only necessary to distinguish Mitterand from Giscard, and to underline the reference to history at stake in these polemics against pastiche. Certainly a form of postmodernism such as Charles Jencks (1986, p. 7) identifies as a double-coding – making a point of its possibilities of choice from a past repertoire of styles and combining this choice with a statement about the limits of modern practices, an 'eclectic mixture of any tradition with that of the immediate past' – could rescue much of this Parisian architecture (as well as a great deal of all architectural practice anywhere at anytime) for the postmodern camp. Furthermore, Foster's second brand of postmodernism (in addition to the 'neoconservative' strand) is one that he terms 'poststructuralist', and this is rather different from the ghostly figure rejected in the Parisian debates. Where the 'neoconservatives' are concerned with 'pastiche', the poststructuralists are concerned with 'textuality':

'"Poststructuralist" postmodernism . . . assumes "the death of the author" not only as originary creator, but also as privileged subject of representation and history. This postmodernism, as opposed to the neoconservative, is profoundly anti-humanist; rather than a return to representation, it launches a critique in which reality is shown to be constituted by our representation of it' (Foster, 1984, p. 145).

This sort of radical 'dismantling of stable conceptions of meaning, subjectivity and identity' (Dews, 1987, p. xi) would have serious implications for the nurturing and deployment of historical significance that I have described so far, were it to be adopted as an architectural strategy. Moreover, to move from the poststructuralist analysis of discourse to the postmodernist analysis of society, the account of the redundancy of historical appeals in a society which is too pluralist and fragmented for such appeals to have anything but the most local and short-lived significance is in flat contradiction to the seriousness present in Mitterand's cultivation of these appeals. This latter point, I would suggest, is an empirical rather than a primarily theoretical question: can such appeals realistically be made in modern societies?

The winning entry to build the 'urban park for the 21st century' at La Villette raises these questions in a particularly acute way because the architect, Bernard Tschumi, adopts a self-consciously poststructuralist tone when writing or talking about his work, and also because Jacques Derrida was among the people directly involved in elaborating Tschumi's architectural programme. Whilst recognising the pertinence of Tschumi's account of his work, I want to suggest that it is heavily qualified by the political context in which the work was executed, and I am unsympathetic to the claim that it is primarily in terms of his texts about architecture that Tschumi's architecture should be considered. Something like this claim is present in Andrew Benjamin's remarks on how we should read the architecture of Tschumi's garden:

'Do you want to go to the Parc de la Villette and say: "there is an example of Deconstructivist architecture"? If that is the sort of relation for which you're looking, you're not going to find it. A bridge is provided by Derrida and Eisenman's own writing on architecture. ... It is precisely these writings that provide the way of understanding how architectural thinking could be exemplified. We must allow the question of exemplification to be heterogeneous and recognise that exemplification within architecture is not purely a question of applied theory in practice. There is a different link between a philosophical dwelling on architecture and an architecture that takes up and sustains that philosophical thinking' ('Jacques . . .', p. 134).

Tschumi wrote in his competition entry:

'Developments in architecture are generally related to cultural developments motivated by new functions, social relations or technological advances. We have taken this as axiomatic for our scheme, which aims to constitute itself as image, as structural model and as a paradigm of architectural organisation. Proper for a period that has seen the rise of mass production, serial repetition and disjunction, this concept for the Park consists of a series of related neutral objects whose very similarity allows them to be "qualified" by function. Thus in its basic structure each folly is bare, undifferentiated and "industrial" in character. In the specialisation of its programme, it is complex, and weighted with meaning. Each folly constitutes an autonomous sign that indicates its independent programmatic concerns and possibilities while permitting maximum programmatic flexibility and invention. ... In contrast to the Renaissance or nineteenth-century spatial organisation, the Parc de la Villette presents a variation on a canonical modern spatial scheme, the open plan' (*Architectures . . .*, p. 134).

At one level this project *is* recoverable within the same sort of reading that I have presented of the other major projects. Pure revivalism is rejected, and a rethinking of appropriate architectural forms is proposed for articulating a forward-looking vision of the capital. Looked at in this way it would be sufficient that Tschumi's proposal not sully any obvious historical references for its *avant-garde* status to give Mitterand the sort of progressive image that modernism elsewhere had provided for him. But Tschumi does more than this. He presents a rigorous rethinking of the park in precisely the terms which the competition called for, and he defends his design principles in terms of 'deconstructionism'.

There are three elements to Tschumi's garden. The first is a grid of structures, the follies, which all take a red cube as their point of departure before hybridising into a series of buildings to be appropriated for particular purposes as the park develops. Several of the first follies to be completed were by Tschumi himself, and they quoted 1920s Soviet constructivist designs. Many of the others will be by other architects, and one, for example, is an Eisenman–Derrida collaboration. The second element to the park is an elevated serpentine walkway which tracks a series of separate gardens in a manner akin to how the soundtrack on a film strip is placed alongside the discrete, but related, frames of its images. The final element is a series of surfaces between the follies and the walkway which will be devoted to particular sets of leisure activities.

The competition brief asked those submitting proposals to think about the relations between nature and culture in the modern city, and it explicitly dismissed the nineteenth-century Parisian park as 'exclusive and irrelevant to the city, and city life' (Cann, 1987, p. 52). Tschumi (1987a, p. 3) responded:

'The fact that Paris concentrates tertiary or professional employment argues against passive "aesthetic" parks of repose in favour of new urban parks based on cultural invention, education and entertainment. The inadequacy of the civilisation versus nature polarity under modern city conditions has invalidated the time-honoured prototype of the park as an image of nature.'

In this context the replicability of the grid and the genetics of the follies express an image of a machine age which neutralises space. It responds precisely to the concerns of the competition brief. However, exactly the same features take on a different significance when Tschumi elaborates his design principles in terms of deconstructionism. The three elements of the park's design are in conflict, so he claims, and the fact that they cannot be resolved into a single hierarchical form is a critique of the architectural ideology of a 'unified, coherent architectural form' (Tschumi, 1987b, p. 111). Similarly, the extendability and lack of focus of the grid offers 'a potentially infinite field of points of intensity: an incomplete, infinite extension, lacking centre or hierarchy' (Tschumi, 1987a, p. vi); and the fact that the functions of the follies are not fixed, and that other architects with different design principles might develop their own strains from the red cube for placing on the site, is proof to Tschumi that the park's authorship and identity are not fixed and will be qualified by practice. At one point Tschumi wants to call this questioning of authorship and reference a 'posthumanist' architecture (Tschumi, 1987b, p. 111), and he clearly has in mind something like the two opposed schools of postmodernism identified by Foster (1984):

'The Parc de la Villette project thus can be seen to encourage conflict over synthesis, fragmentation over unity, madness and play over careful management. It subverts a number of ideals that were sacrosanct to the Modern period and, in this manner, it can be allied to a specific vision of postmodernity. But the project takes issue with a particular premise of architecture, namely, its obsession with presence, with the idea of a meaning immanent in architectural structures and forms which directs its signifying capacity. The latest resurgence of this myth has been the recuperation by architects of meaning, symbol, coding and "double coding", in an eclectic movement reminiscent of the long tradition of "revivalisms" and "symbolisms" appearing throughout history. . . . La Villette [instead] looks out on new social and historical circumstances: a dispersed and differentiated reality that marks an end to the utopia of unity' (Tschumi, 1987a, pp. vii–viii).

His canonical structure (open-plan office/grid) may resist the hierarchical coherence of more focused designs, and his provision for collaboration on site may qualify the authority of the author, *but* the symbolism of his references to the machine age remains, reinforced by the constructivist quotations; and his remarks on the place of nature in the modern city show in a very direct way that his park *represents* his understanding of people–environment relations in modern urban society. I make these remarks not to question his poststructuralist/postmodernist/posthumanist credentials, but merely to show how easily recoverable his project is in terms of both the original competition brief and the historical mission which animated all of the major projects. Tschumi's awareness of this political dimension is actually evident in the third rationale he offers for his design principles:

'The numerous unknowns governing the general economic and ideological context suggested that much of the chief architect's role would depend on a strategy of substitution. . . . Hence the concern, reinforced by recent developments in philosophy, art and literature, that the park propose a strong framework while simultaneously suggesting multiple combinations and substitutions' (Tschumi, 1987a, p. iv).

Gerry Kearns

The Politics of Space

Mitterand Cultivates the Tourist Gaze

Mitterand's new Paris was on show in July 1989. Hoping to add the pyramid and the Cube to the 'absolutely distinct objects' (Urry, 1990, p. 12) which tourists must personally have gazed on, Mitterand was also reminding punters of impressionist paintings at the Musée d'Orsay, the Mona Lisa and the Venus de Milo at the Louvre, as well as of the Eiffel Tower and the Arc de Triomphe. Yet there is more to the 'tourist gaze' than photographing the great wonders of the world. Urry (1990, p. 140) notes that: '[p]hotographic images organise our anticipation or daydreaming about the places we might gaze upon'. This organisation places objects in context, and these contexts may be represented through very slim clues. As Culler has written: '[a]ll over the world the unsung armies of semioticians, the tourists, are fanning out in search of the signs of Frenchness, typical Italian behaviour, exemplary Oriental scenes, typical American Thruways, traditional English pubs' (Culler, 1981, p. 127; quoted in Urry, 1990, p. 3). Urry (1990, p. 140) says much the same thing: '[t]he gaze is constructed through signs, and tourism involves the collection of signs. When tourists see two people kissing in Paris what they capture in the gaze is "timeless romantic Paris".' Mitterand's new Paris gives another set of signs, signifying that '[i]t is a city of imagination, ideas, youth . . .' (Mitterand in *Architectures* . . ., p. 8). This is not to deny history and tradition, of course, and it is precisely the idea of making modern statements which are respectful of historical context which locates the 'Mitterand-style' away from revivalism and pastiche. For his domestic audience, this type of modernity serves as a feel-good politics alert to the desire for rootedness which the French esteem as their love for history. (It is striking that the 1980s also saw Thatcher and Reagan engaging in national boosterism primarily for domestic political reasons.)

Nostalgia for Haussmann

The respect for history is a particularly fraught question in the case of the Parisian urban fabric; it has never been innocent. And it is primarily a question of fabric, to be sure, for so many other things inevitably change: the social and economic content of the urban form, its integration through transport technologies and the ideological significance of particular forms. Take, for example, the Paris of the boulevards planned by Haussmann between 1853 and 1870. It is true, as Sennett (1990, p. 62) has recently commented, that there are clear differences between the American grid and the Parisian boulevards in their operation as ideologically significant planning instruments. The grid neutralises the previous enviroment, denying nature and history to extend urban space without beginning, without end: in contrast, the boulevards emphasise centrality and a hierarchical system of significant foci. Monuments were placed in these new foci to take the gaze from one to another. This gaze also helped surveillance, but this was not its only function. At least as significant, particularly in the west and north-west of the city, was the speculative development of new land values. Here there is a great difference between the French and American cases: the grid institutionalises the competitive development of fragmented and privatised space, whereas in Second Empire Paris a more corporatist solution was devised with the use of compulsory-purchase orders and even the freezing of the land market to spread improved rents over as wide a space as possible. This aim was also served by Haussmann's emphasis on improving circulation to extend accessibility. The goal was to maximise the global sum of rental value, on which taxes were levied, in contrast to individual financing of development on private sites which centralises values at the monopoly locations of the most prestigious activities (Roncayolo, 1983, p. 112).

The creation of massive credit, through mortgaging the improvement in taxable income that urban development would bring, defined the political economy of Haussmannisation. Controlling elements of the land market, raising credit through the semi-public Crédit Foncier, sustaining the speculative process across the blips of the economic cycle to wait for an upturn that would realise the investment and justify the enterprise: such strategies required a strong and unassailable commitment from the state. Long-termism had clear political conditions of existence. Second Empire Paris meets Savitch's (1988, p. 16) criterion for corporatism: 'does the state orchestrate and direct a limited number of select organisations?'

Spatial considerations were central to Haussmann's vision of an efficient, regulated and capitalist city, but this was less the utopian solution of a city architect and more the corrective surgery of an engineer. This, in part, defined Haussmann's treatment of history. He recognised the potential of the royal sites in Paris to bring grandeur to the space of central Paris, and he was willing to press almost any monument into service so as to anchor the localities of the city in public spaces knitted into an urban hierarchy of such places. Sennett (1990, pp. 91–92) comments that Parisian squares bring people together, whereas those of contemporary London were reservoirs of nature denying urban civilisation: Haussmann thus subordinated greenery to the geometry of his streets (Choay, 1983,

p. 204). In fact, he tried to subordinate almost everything to his spatial system:

'He was fond of "system", a design that allows faster circulation, along thoroughfares, across bridges, around squares, opening up communications between all the sensitive parts of the city: he connected the two banks of the river, the centre and the periphery with its major access routes from the provinces, the railway stations, residential and business and recreational quarters. The removal of the barriers and obstacles which characterised Balzac's Paris giving a distinct juxtaposition of distinct neighbourhoods, having their own societies, morals and proper language, gave way to the unified metropolis of Zola' (Choay, 1983, p. 169).

Pierre Pinon is convinced that '[n]othing of great value was lost though a few churches on the Left Bank were demolished' (Champenois, 1991), but this was not how it was seen at the time. Haussmann disliked and distrusted the congeries of workshops and tenements which made up artisanal, medieval Paris, the centre which had sustained the Revolution of 1848, and Gaillard (1976, p. 3) believes that in creating an integrated metropolis,

'Haussmann deprived citizens of their *milieu*, lifting them out of a culture more important to them than the church or school – the workshop, the street, meeting halls, the wine market. Nostalgia for this lost way of life contributed to the fall of the Empire but taking back a city centre from which work had expelled them was not their primary objective, rather they wished to reclaim a personality which had died during the Empire and for which the old quartiers of historic Paris were both a symbol and a locale.'

The creation of a new urban space for the metropolitan play of capital thereby involved the suppression of other spaces, and it is revealing that Ross (1988) sees the Commune of 1871 as in part a reassertion of this earlier, local, less hierarchically-ordered set of spaces.

Haussmann's spatial strategy had a clear cultural politics:

'The baron badgered his architects for imagery, for scale, for points of focus. He disliked the neutrality of the Place de la Concorde with its indecipherable obelisk, and dreamt of replacing it with something stronger; he did not seem to realise that there are places in every city which disqualify themselves from the symbolic order – by the very density of different histories that have claimed the place and spilt blood. He was not above cutting down the Liberty Tree in the Jardin du Luxembourg, the last survivor from the Great Revolution, and putting Jean-Baptiste Carpeaux's *Quatre Parties du Monde* in its place – imperial messages shouldering out republican ones' (Clark, 1984, p. 50).

This cultural politics was of a piece with the economic dimension of the spatial system in his new Paris. The centre was sanitised, industry expelled, the working class following, to the north-east of the city, and monuments were used as perspectival points to integrate the city and thus to support the spread of land values. In doing this the local significance of places was denied. Monumentality was seen to reside in the decontextualised appreciation of individual buildings. For the rest of the urban fabric, uniform façades, straight streets and height restrictions were necessary to set off the monumentality of the decorative points triangulating the urban system. As Gaillard points out, the reaction to the social, cultural and political programmes of the Second Empire partly took the form of a nostalgia for old Paris. This nostalgia increasingly animated an ideology of preservation which soon came to receive quite different social and political contents. It is this transformed nostalgia which provides the context of Mitterand's spatial politics.

Sutcliffe has described some of the reasons why nostalgia became official policy. Firstly, after 1914:

'The city stopped changing, and conservation was transformed from a pipe dream to the natural state of affairs. The preservationist movement lost some of its momentum now that there was little to protest about, but its philosophy became more acceptable to those who governed the city as they renounced their hopes of being able to pursue ambitious modernisation policies in the foreseeable future' (Sutcliffe, 1970, p. 330).

Secondly, after the Second World War, '[p]artly out of respect for the historic areas, and partly because of the high cost and limited return of carrying out improvements in the centre, the city developed a policy for the historic core which increasingly acquired a character of out-and-out conservation' (Sutcliffe, 1970, pp. 330–331). Sutcliffe sees this as a refusal to modernise central Paris on the grounds of cost, but there is also a crucial political geography behind the treatment of the 'historic core'.

The Geopolitics of Nostalgia

The problem of public order, with which Haussmann was concerned, had not really gone away, and Paris has continued to be a densely populated workers' city. Given the extreme centralisation of French politics, the rivalry between city and state in Paris assumes very great significance. After the Second World War there was a conscious policy of gerrymandering French cities to weaken their left-wing cores – segmenting these cores amongst middle-class suburbs. In the case of Paris this involved continuing to deny the city a formal democratic status, distributing political power instead among a number of different *départements* (Seine, Seine-et-Oise and Seine-et-Marne). When Paul Delouvrier took up the question of giving Greater Paris some sort of development plan in 1961, he left this political impotence in place and sought to disinfect further the left-wing city by placing new industry in satellite towns that were connected to the city by train and road but remained politically quite discrete. In 1964 the three *départements* were dissolved into eight: 'Leftist strongholds in the communes around Paris were diluted within the more heterogenous *départements*' (Savitch, 1988, p. 104). By the mid-1970s central Paris, the twenty *arrondisements* of the old *octroi* tax, had been sufficiently deindustrialised (un-modernised) to entrust them to their electorate, and in 1977 Paris got its first mayor, Jacques Chirac. The political geography of Paris is now so 'safe' that in 1989 not one *arrondisement* was controlled by the Left. As Garcias and Meade (1986, p. 24) remark: '[t]he principle function of French town planners being to forge social euphemisms, they contrived urban categories which have made it possible to retain the upper stratum of the middle classes within the city, using the alibi of history and culture'. What this means is that failure to arrest the deindustrialisation of the city, along with the failure to provide substantial social housing (in the name of the historic character of particular districts), is in itself a socio-political spatial strategy. In this respect Lojkine (1972), when speaking of the urban strategy as the instrument for creating unity among the dominant classes, is simply referring to a different aspect of urban policy than Savitch (1988): a policy may be elitist in intention and corporatist in implementation.

Mitterand and his advisers claim publicly that they have broken with the elitist patterns of the postwar period. Certainly, they have returned to the geopolitical questions concerning the identity of Paris and its relations with the state. Having a strong and confident right-wing administration in the capital did not suit the Socialist government, and they proposed yet another reform of the administration of the city:

'Ironically, Chirac's standing as mayor was bolstered by the Socialists' rise to power in 1981. After the Left won control of the national government Chirac dramatised the Gaullist opposition from his office at the Hôtel de Ville. When, in 1984, the Socialists tried to undercut Chirac's authority by proposing that Paris be divided into twenty self-governing neighbourhoods, the Mayor took up the cudgels to protect a united city. He handily won the contest and, in the process, gained the admiration of many Parisians' (Savitch, 1988, p. 107).

But, having failed to cut Paris down to size, the Socialists have also blanched at the prospect of allowing it to swell into the potentially Greater Paris of its journey-to-work region: such an entity would too easily challenge presidential authority.

The second way in which the Socialists claim to have addressed the elitism of the Parisian urban strategy is in their promotion of public uses for their new spaces; they charge the Right with abdicating the city to private capital. There is certainly some justice in both claim and charge. This has been one of the most contentious issues between Left and Right. For example, during the cohabitation Alain Juppé (as Chirac's Minister of Finances) cancelled the funding for the International Communications Centre which Mitterand wanted to place in the Cube at La Défense. This government-directed initiative was to position France at the centre of media developments in a future Europe. It was also to respond to the universal ideals that Mitterand planned to celebrate in 1989. In 1986 Chaslin (1986, p. 30) pointed out the significance of this move: '[t]he arch at Defense will be completed, somewhat lacking in meaning now offices and commercial exhibition spaces have taken the place of the great public monument first intended'. Instead, Faure saved the symbolic status of the Cube by founding a Centre for Human Rights there out of Bicentenary funds (Bonnet, 1989b). Robert Maxwell made up some of the rest by taking a large part of the exhibition space and offices, but failure to get private sector clients meant that on their return to power in 1988 the Socialists were still faced with taking twenty of the thirty-two floors for government use (Duponchelle, 1989b). Elsewhere, as at Bercy, the Right's reluctance to let the state use prime sites has been assuaged by the recognition that government office development will lead private sector interest, and that in locating thousands of civil service jobs in the district the government is also providing much of the infrastructure needed for commercial development. Indeed, nothing that the Socialists have done is counter to the general programme to 'tertiarise' (Lojkine, 1972) the city and to produce a functional differentiation of urban space progressively marginalising the work and homes of the working class.

Yet the Socialists speak of a spatial strategy which is quite *different* to the model that I have just described. They talk in terms of reintegrating the neglected east end into the modern city, and of promoting a new relationship between centre and suburb with the push to the north-east through La Villette and to the west through La Défense. Politis (1988, p. 187) talks in terms of a new generation of planners 'refusing to abandon the centre', suggesting that:

'This new momentum was amplified by the Plan for East Paris, which was approved in 1983 and which aims to improve the quality of life in the most deprived districts. . . . [T]he Plan for East Paris is designed to create modern, multifunctional *quartiers* and so ensure the continuing development of the city, the restoration of a number of urban continuities, the respect for existing configurations, the revalorisation of sites, and a concern for questions of human scale.'

However, the upgrading of the east end lowers housing densities, raises rents and, providing little in the way of new jobs for the working class, contributes to its gentrification. This is the real way in which a new social space is emerging:

'Today the old dichotomy between west and east Paris has begun to fade. The new middle classes search for space in all parts of the city and they are not inhibited by imaginary lines. To be sure, eastern Paris is still a far cry from the elegance of the Champs-Élysées and grand boulevards near the Avenue Foch. The Nineteenth and Twentieth *arrondisements* still belong to *les class populaire* and their overcrowded streets are filled with Arab-speaking workers and cheap shops. But much of eastern Paris has been transformed for the middle class and young professionais' (Savitch, 1988, p. 115).

The new dynamic of Parisian development lies not in the attention paid to the poor *people* in their neglected *quartiers*, nor in a centrifugal push towards the working-class satellites. Instead, a new east–west spine of office development stretches from Bercy in the east, taking in Les Halles at the centre, and pushing along the axis of the Champs-Élysées to La Défense in the west. Strict zoning regulations through the 1970s drove new office development to La Défense, and Mitterand's new Paris is now allowing the 'tertiarisation' of the east. Everywhere that has had the hand of culture laid upon it, the process of 'revalorisation' signals the arrival of the professional functionaries of a world-city of service employment: '[a]nd the Opéra-Bastille, according to its critics, has only served to further destabilise an area which was already being renovated' (Bonnet, 1989a).

In this regard it is not surprising that 'behind the scenes' there has been considerable co-operation between the City of Paris and the Socialist government. Mitterand had no intention of picking an unnecessary fight as Giscard had done before him: '[w]ith Giscard's failures in mind, Mitterand was careful to negotiate the projects with the municipal authority . . .' (Chaslin, 1988, p. 48). Chirac was concerned to make a point of the Socialists' reckless spending, to contest Mitterand's right to host the Bicentenary and to win the presidential election of 1988. Yet no right-wing politician could complain about the continuity of spatial strategy with the 1970s, nor would any political figure want to gain a reputation for wrecking any of Paris's new embellishments. They may have been jealous that Mitterand took up and ensured the completion of a suite of projects which exploited spaces left empty by the deindustrialisation of the city, a landfall in central Paris equalled in size only by Haussmann's *extension* of the city. They may have worried about the financial wisdom of creating the massive credits necessary to see the projects through, and again the only appropriate comparison here is with Second Empire Paris (Lortie, 1989, p. 39). The opportunity will not recur. Mitterand has said that 'our only grandeur lies in what we leave behind' (Bonnet, 1989a). Similarly, Haussmann (1890, p. 7) contrasted the 'subjective and temporary' value of political administration with the 'visible and permanent' record of 'monuments'.

In this sense the Bicentenary and the *grands projets* reinforce one another. They imply that the Socialist accession of 1981 was a new beginning which revivified the principles of Republican France. They also say that the Socialist party refuses the mantle of revolution. In other words, they are a continuation in words and stone of the project begun with that first visit to the Panthéon in 1981. History is a contested signifier in all of this, and it represents something different in each of these media. In the politics of spectacle, it refers to a positive evaluation of the Revolution which allows it to serve as the nation's founding myth. In the politics of memory it constitutes a commitment to certain principles and a willingness to be evaluated on that basis. In the politics of architecture it represents a respect for context and a refusal of irony. In the politics of space it becomes the fig-leaf for a redrawing of social classes in space which sits uneasily with the principles being memorialised.

Notes

1 I want to thank Hannah Moore, Anne Moore, Peter Compton, Mike Heffernan, Stuart Corbridge and Gerard Toal for their help. I must thank too the staff of the Communications Library at the Pompidou Centre for their assistance in using their newspaper-cuttings collection. Thanks also to the University of Liverpool Research Support fund for the air fares. Much of this research was done during field classes in Paris, and I want to acknowledge the co-operation and encouragement of two groups of students as well as of the graduate students who worked with me, Naomi Williams, Peter Compton, and Sally Sheard. Much of this research is based on French newspapers. The main ones used were the following: *Libération*, practically the official newspaper of Mitterand's Socialist Party; *Le Figaro*, a strongly anti-Mitterand newspaper favouring free-market liberalism; *Le Quotidien de Paris*, a paper which has consistently opposed Mitterand's politics of spectacle; *Le Monde*, a centre-right newspaper less stridently critical of Mitterand than the previous two; *L'Humanité*, the official paper of the French Communist Party which persistently attacks Mitterand for selling out on any Socialist principles he may once have had.
2 Proposed by Jack Lang, French Minister of Culture, to the EC Commission, 13 June 1985 (*Bulletin, EC*, Vol. 18, No. 5, 1985, point 2.1.61).
3 Against a cost of FF500m for the Bicentenary, the Minister for Tourism, Olivier Stirn, insisted that the current account surplus on tourism would be up by FF5,000m over the 1988 figure, an increase of 20% (*Le Figaro. Fig-Eco*, 14 July 1989, p. iv). He further argued that the Bicentenary had occasioned 800,000 extra tourist nights spent in Paris alone (Faujus, 1989). In terms of the current account balance of the French economy, then, the Bicentenary spectacle might be defended as a prudent investment. In July 1989, in fact, GATT decided for the first time to take up the question of tourism in recognition of its status as the world's largest employer, providing 22–27% international receipts in services for countries like the US and Great Britain but 70% for Spain and Portugal, and 82% in Indonesia ('*Le GATT* . . .'). In 1985 the Council of the OECD had looked at the question of tourism: '[t]he adoption of the International Tourism Policy Decision-Recommendation constitutes a significant advance in the concept that tourism is a service industry in its own right to be treated on a par with banking, insurance, transportation or communications' ('*Tourism* . . .'). **Figure 1** shows the developing contribution of tourism to the French current account balance, and it can be seen that the balance on services has become increasingly important in view of a deteriorating balance on trade of goods. Furthermore, it is apparent that the balance of trade on tourism

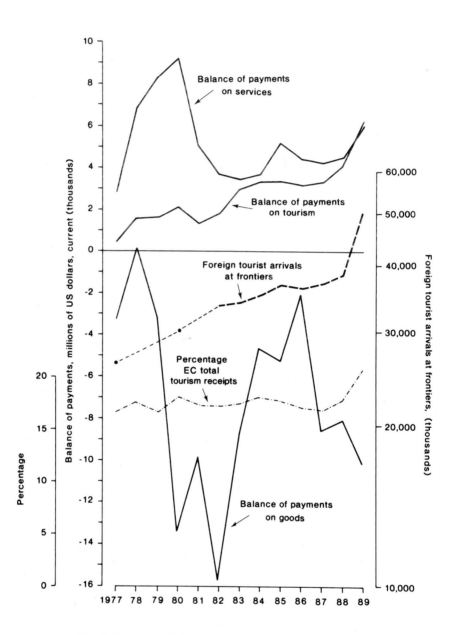

FIG. 1. The changing performance of tourism in France.

TABLE 1

Tourism Receipts (a), Tourism Expenditure (b), & Current Account Balance (c) for a selection of countries (in 1,000 million current US dollars) (n.a. = not available)

		1985	1986	1987	1988	1989
Canada	a	3103	3860	3961	4655	n.a.
	b	4130	4294	5304	6316	n.a.
	c	−1470	−7600	−7060	−8330	−16593
France	a	7942	9724	11870	13783	16500
	b	4557	6513	8493	9713	10292
	c	−35	2426	−4446	−3549	−4299
West Germany	a	4748	6294	7678	8449	8658
	b	12809	18000	23341	24945	24129
	c	16998	40093	46120	50466	55477
Italy	a	8758	9855	12174	12403	11984
	b	1880	2758	4536	6053	6773
	c	−3540	2912	−1663	−5446	n.a.
Japan	a	1137	1463	2097	2893	n.a.
	b	4814	7229	10760	18682	n.a.
	c	49177	85831	87020	79610	56783
United Kingdom	a	7120	8163	10225	11008	11182
	b	6369	8942	11939	14614	15195
	c	4756	158	−7373	−26733	−34065
United States	a	17937	20454	23505	29202	n.a.
	b	24517	26000	29215	32112	n.a.
	c	−112750	−133230	−143700	−126580	−105900
Spain	a	8151	12058	14760	16686	16174
	b	1010	1513	1938	2457	3080
	c	2851	3965	−233	−3784	−10932

is now of the same order as the total current account balance on services. The steady growth in tourist arrivals during the 1980s is clear, as is the sharp upturn in 1989. In 1989, France's receipts from tourism (see **Table 1**) exceeded those of Spain for the first time since 1983 and, as **Figure 1** shows, gave France a record share of EC tourism receipts. French tourist arrivals were up 31% over 1988; in the rest of the EC the figure was only up 5% (12% in West Germany, 10% in the UK, but 1% in Italy and 1% in Spain). The improvement in receipts was 20%, exactly what Stirn predicted, against a growth of only 1% in the rest of the EC (2% in West Germany, 2% in the UK, 3% in Italy and 3% in Spain). In terms of 'spectacle as investment' this was more than satisfactory. (The data on balance of trade in goods and services is taken from *International Financial Statistics Yearbook, 1990*, 'Trade balance', pp. 140–141; 'Current account balance', pp. 142–143; data on tourism is taken from *European Marketing Data and Statistics, 1992*, 'Trends in tourism receipts', pp. 430–431; 'Trends in tourism expenditure', pp. 432–433; 'Foreign tourist arrivals at frontiers', p. 434.)

4

Trading History, Reclaiming the Past: The Crystal Palace as Icon[1]

MARK BILLINGE

'St Pancras was a fourteen year old Christian boy who was martyred in Rome in AD 304 by the Emperor Diocletian. In England he is better known as a railway station' (Sir John Betjeman).

Preamble

In the magazine supplement to its edition of 22 October 1988 *The Independent* newspaper featured a scheme then under consideration to rebuild the Crystal Palace. The reconstruction – or, more properly, the *resurrection* as it was described at one point – was to take place on the Sydenham site occupied by the modified original from 1852 until its destruction in 1931. Memorable as much for its tone of offended righteousness as for its analysis, the discussion of the scheme protested the inadequacy of the imposter ('little more than a travesty') whilst reaffirming the sanctity of the original ('that great symbol of the Victorians' faith in progress and enlightenment'). The planners and their stormtroopers, the builders, were encroaching on tricky and even sacred territory. To change the metaphor, a nation's birthright was under the knife. Was it safe in their hands? The author, Gavin Stamp, thought not.

Anybody could see that the plan was wrongheaded. But only a savant knew that the wrongheadedness lay in the practice not the principle. The premise was simple. A national symbol of 'almost mythical' status, the Crystal Palace was arguably due a reappearance. It would cheer people up; rekindle the Greater Britain that went down with the original on that bleak night in late November 1931. In other words, it was ripe for forgery. But – and this was the problem – as forgeries go, it would be a particularly inept specimen. Inaccurate in its dimensions, it was also inauthentic in its feel. A 'proper' forgery (both senses intended) would be true not only to the nuts and bolts of the original but to its spirit as well. It would aspire to 'reflect what the Crystal Palace stood for'.

In negotiating the can of worms which is the latter – seldom has a building 'stood for' so much – the author was more conventional and at the same time less adroit. In a gesture which speaks volumes for 1988, he grabbed for the nearest sentimental encomium and found the one which had been attached to the Palace at its burial and rehearsed almost universally ever since. As a result, dewy-eyed nostalgia for what had arguably never been got the better of hard-nosed consideration of what might yet transpire. Percipience was the first casualty.

Concentrating on what the Crystal Palace has 'come to mean' – important as that is – involves a serious diminution, as this essay will try to demonstrate. Not the least problem is the general laundering, even sanitising, of 'the Victorian' which the politics of the last dozen years or so have sought (with some success) to effect and which symbols like the Crystal Palace have been dragooned into legitimating. To take it at its apparent face value is to capitulate to the rules of a particular political game. It is also to forget that to its contemporaries of 1851, the old Palace was an *issue* and one which forced them to confront raw aspects of their own society. The way you lined up over the Palace said much about the way you lined up in general. In 1851 the Palace was catalytic and dangerous. By 1988 it had been made, and *The Independent* article echoed this, comfortable, revivalist, neuter, conformist. My concern is to put these two moments in context and for two parallel reasons: to rescue the old Palace from its anaesthetised image, and to recognise the damage which partial and unscrupulous historical borrowing can do to any society foolish enough to accept it uncritically. The plan for the resurrection was part of a blueprint for something bigger, and in this something bigger past and present were deliberately conflated. Separated again they tell a different story.

We can begin perhaps by allowing that the timing of the resurrection scheme was not accidental. But if we are to make sense of it, then we must add two important recognitions: firstly, that over and above its commercial viability, it was the Palace's capacity to generate *political* capital which guaranteed the support of the Conservative-dominated London Borough of Bromley; and secondly, that the Palace's susceptibility to what would amount to political exploitation was rooted in a 52-year absence during which time the general affection for it had so strengthened that the public was no longer capable of appropriate critical response. 'Today there is every reason to fantasise about the Crystal Palace, as the reality no longer exists', *The Independent* opined, failing to realise that this was a weakness as well as a strength. Having begun life as a building, it had become an apotheosis: its power to transcend time and reason finally guaranteed by its spectacular immolation of 1931.

But such recognitions are at once too specific and too general. If it can be agreed that the Palace had acquired a particular and accepted meaning, then it must also be admitted that this meaning was capable of wilful manipulation. By 1988 the invitation to view its reincarnation would be an invitation to adopt a particular attitude toward the achievements, not just of our Victorian forebears, but more seriously of Mrs. Thatcher's Britain too. We would do so furthermore in a context that Mrs. Thatcher had virtually invented: the rediscovery of our Victorian ancestors' secret of success; their unstinting (the word was 'staunch') devotion to wholesome values with all the practical benefits that such a litany implied. In this sense the invitation was an invitation to compare, but it was also an invitation to conflate: to conjoin past and present; to see the merits of the past reborn in ourselves; and to see the bad in ourselves consigned to the past. It was history rearranged: Victorian with nobs on.

For this reason if no other the Crystal Palace had already been appropriated unofficially as the symbol of Mrs. Thatcher's Britain, and – like a particularly bad case of chickenpox – it had spread through the body politic, incubated in a thousand festering precincts. For almost a dozen years miniature versions of Paxton's masterpiece – or at least its most prominent feature: the barrelled arch of the great transept – had been appearing throughout the land. Versatile to the last, it graced shopping arcades, office blocks, science parks, the more progressive churches, fashion emporia, cinema complexes, health clubs, executive massage parlours, laundromats and the smarter city centres. If you wanted to get ahead, you got yourself a Paxton hat.

The Independent noticed this too:

'The Crystal Palace has become a symbol and a totem. One of the most remarkable architectural phenomena of recent years is the literal homage paid to Paxton by architects, especially those of the high tech persuasion, who see the Crystal Palace as pioneer and paradigm. Like the grin of the Cheshire cat, that round-arched glazed transept . . . has become ubiquitous.'

For *The Independent*, such associations were wistful and at the same time purposive. What the original Palace had pioneered was progressive engineering, Victorian architectural 'high tech', and it was this which was now being transferred by association to those who were high tech in our own terms. Though described as a totem and a paradigm, it was in this analysis more akin to a baton passed from one era to another: a seal of approval, a badge of honour worn by members of a very exclusive club racing towards a technological nirvana. But, as an interpretation this is again too narrow, too technocratic and too literal. Deeper associations are missed if we stop here. Certainly the mini-crystal palaces – by no means confined only to the obviously high tech – announced their users' identification with the spirit of inventiveness, but they screamed more loudly their capitulation to the political and social assumptions of the enterprise culture as a whole. Architectural historians will doubtless find post-modern impulses in the spawning that went on – curvy lines, recycled ideas, exotic materials – and romantics will discover the influence of nostalgia just as much as will the 'anti-carbuncle' brigade. But chroniclers of *society* will surely find a whole political manifesto entombed amongst the ironwork. At heart the buildings were but physical manifestations of the bigger idea; pawns in an altogether larger game.

Here as elsewhere the political analysts were not slow to miss the point. If Mrs. Thatcher was ambitious to emulate former glories and burnish her image in the light of another's achievement, then the role model was not, as many then alleged, the Churchill of World War II. Rather, and more logically, it was the last British female to enjoy even remotely comparable power: namely, Victoria. Rhetoric about the Victorians and what they could tell us – as well as show us – poured out in a torrent of political ballyhoo which reached its crescendo in the mid-1980s. No ribbon was cut, no magnum shattered, no mega-store mega-launched without the solemn drone of instruc-tive Victorian counterpoint. If all went according to plan, the late-twentieth century would be remembered not as the second Elizabethan Age but as the second Victorian. For Victorian meant indisputably good things, and a mere glance at history proved it. Politically, then, the secret was to accentuate the positive. The Crystal Palace, emblem of the Victorian upside, was the perfect image for the 1980s upswing. The downside was expunged; the downturn would come later.

From this perspective, the only way forward was back: and it was no coincidence that just as this theory was undergoing its most detailed elaboration, Bromley – guardian of the sacred site – decided to produce the rabbit from the hat: the *real* Crystal Palace rising from the ashes of the old. The crowning glory, it would be mission accomplished; prophesy witnessed; destiny fulfilled. At least it seemed so for a time.

In the end it all came to nothing. Objectively, there is a great deal to be made of this failure and any number of reasons to be grateful, but there may still be one sneaking cause for regret. Had Mark II made it from the drawing board a series of – surely unintended – signifiers as to the real state of Britain in the 1980s would have been preserved for future analysis. For Mark II differed in several respects from Mark I, and the differences are instructive. To begin with, it was a veritable dwarf: two-fifths the height, one-half the length, and one-quarter the volume. Leaner and doubtless fitter, like the Midlands indigenous car industry, you could see it all in an hour. Victorian with nobs on was Grandeur on a shoestring. Worse, the Palace was to use public land for private purposes and private capital for private reasons. Though the original had also been a private venture, public access and the public interest had been maintained through The Crystal Palace Act, an act that Bromley Council was now in the process of trying to repeal. An exclusive shopping mall for the well-upholstered, and a hotel and leisure complex for the seriously rich, the 'People's Palace' Mark II was largely dedicated to the titillation of the jaded. It would relieve them of their cares about money, largely by relieving them of their money. Formerly a temple of achievement for the achieving millions – at least that was the official line – it was now the cathedral of conspicuous consumption for the conspicuously few: flexible accumulation in a riot of glassware. To round out the absurdity, this most British of symbols, built at a time when Britain was Great and it was great to be British, was to be financed largely by foreign capital. To be specific, by the Al Houda Hotel and Tourism Company of Kuwait. Like its predecessor, it was more symbolic than it knew.

Interlude

According to her private diary, 'it was a day to live forever'. She had not seen its equal and did not expect to do so again. It was an extraordinary day; 'magical, glorious . . . touching'. Victoria was thirty-two years old; Victorian England a mere fourteen. Fifty years remained to them: years which would give shape and meaning to the concept 'Victorian' as to perhaps no other. Yet here, the Queen insisted, almost at its beginning was its apogee. If we are to make sense of what follows, we must remember this above all.

Come the official high water-mark, the Jubilee of 1897, neither the Queen nor the world – the greater part of which she then ruled – would be the same. By then she would be seventy-eight and a widow of twenty-six years: a portly figure, publicly stern, privately uncompromising; no longer given to the enthusiasm of youth. Tempered by the times to which she had given substance, embalmed in a mood and manner of her own creation, she was by then the archetype of *late-Victorianism* – an animal only tenuously connected to its former

107

promethean self. In the world too the political, social and economic shadows had lengthened. The zeal had to an extent been tempered; the sense of purpose blunted; the certainty of vision occluded. Come the Jubilee, ceremonials excepted, high optimism was not the mood of that day.

But, if there was ever a day in which all things were truly bright and beautiful then the day in question – the one which would live forever – was indisputably it. It was the first day of May 1851. The world over it would become known as Labour Day, but in 1851 labour was decidedly not the issue. The issue, and it had been *the* issue for more than a year, was the opening of an exhibition of manufactured goods in a temporary building in Hyde Park. On the face of it, there was nothing very remarkable about it; little out of which even the most fevered political brain could make a *real* issue. Exhibitions had happened before, royalty had opened them, and 'issues' had never come into the equation. On a European scale, *expositions* had been growing in both scope and frequency ever since the French had identified them as a novel way of clearing the state warehouses of unsold (and *de facto* unwanted) goods during the English-inspired blockades of the 1790s. The principle, that of selling shoddy goods to a credulous (later internationally credulous) public, had found general favour and had even been refined. But it had never become an issue. And nor would financial propriety be the issue here, even though almost everything else would. The list of principles, philosophies, creeds, sensibilities, sensitivities and plain good-senses against which the Great Exhibition in general and the Crystal Palace in particular were held to offend would be legion. Every society or group in the country, however fly-blown and ephemeral, was either for it or against it, and each had its uncontestable axe to grind: its pet contention sharpened by the spectre of the Palace; its soap-box line to defend. But at the opening ceremony itself – 'with the eyes of the world upon us' – the issues momentarily took a back seat. Something remarkable was happening, and intuitively the Queen grasped the fact. It was not just another exhibition and this was not just another building. History was being made, so she felt, and she was right. A hundred and thirty years later, the history banked today would be taken out of deposit and used to counter a run on credibility seriously threatening the current account.

In her uncharacteristically upbeat assessment of the Exhibition's opening, the Queen was not alone. *The Times*, sixty-six years old and no more circumspect, concluded that such a day 'has never happened before and . . . in the nature of things can never be repeated'. It cited the public as witness: 'they who were so fortunate as to see it hardly knew what most to admire, or in what form to clothe the sense of wonder and even mystery which struggled within them'. The naïve bystander acknowledged the spectacle, could do no other; but those with a wider grasp sensed the extraordinary: the birth of an image from which a decisive metaphor subtended. To such onlookers, *The Times* continued:

'It was felt to be more than what was seen or what had been intended. Some saw in it the second and more glorious inauguration of their sovereign; some a solemn dedication of art and its stores; some were most reminded of that day when all ages and climes shall be gathered around the Throne of their Maker; there was so much that seemed accidental, and yet had meaning; that no one could be content simply with what he saw.'

To the Queen's sense of 'magic', *The Times* had now added the altogether more orthodox and comforting sense of religion. Not just a High day, 'it was a holy day', the newspaper sermonised. In its admission to the Pantheon – the Supreme Head of the Anglican Church and its Chief Executive, the Archbishop, were both present at the obsequies – the Crystal Palace was personified, sanctified and canonised in the time it took to buy a ticket. Thrice blessed, it could properly anticipate transcendence. Later, there would indeed even be talk of resurrection.

Unaffectedly to begin with, many of the early enthusiasts were in the simple though vital business of quelling fears and allaying suspicions; deflecting the criticisms, negating the critics. Given the controversies which had surrounded the building as building, its promoters as people and the Exhibition as an event, these deliberate attempts by committed supporters to counteract a hostile press are hardly surprising. But what was begun in good faith on that day was added to more knowingly thereafter, until those attached to the Palace found themselves riding a machine with a momentum all of its own. Many would remain astonished at the success that they had generated. By the Exhibition's close the Palace itself had achieved genuine though unexpected popularity: its first and longest step towards achieving immortality. Thereafter, following its move to Sydenham amidst almost universal acclaim, meanings accreted with startling speed and little reason. Fragments of transient popular currency glued themselves to the fabric until a Crystal Palace undreamed of even in the mind of the most rampant ideologue began to emerge: a rich palimpsest of triumphant associations and sedimented patriotic-images. It was a festival of Britain in itself; the best of British decanted into Britain's best; a hope for the future and a comfort from the past.

Narrative

The Event: Balloons, Cannons and Magic

Though intermittent bad weather had dogged the previous month and raised all kinds of alarm amongst the orchestrating committee, as well as somewhat shameful approval from the critics, the skies cleared overnight on 30 April 1851 and the morning of 1 May was bright. By eleven o'clock the breeze was light and the sun making an appearance. Sir Charles Spencer, the 'great aeronaut', reported conditions ideal for his balloon ascent, timed to the minute to coincide with a gun salute from a full artillery company stationed at the Tower, which – in a dizzying chain of command – would be signalled in turn from the Palace by semaphored news of the last chord of the fanfare which marked the official moment of opening. At eleven thirty, the crew of a model frigate afloat on the Serpentine made final preparation for its own salute (a full round of cannons), despite a warning from *The Times* that the concussion would 'shiver the glass roof of the Palace' with the result that 'thousands of ladies will be cut into mincemeat'. Inside the Palace, the as yet unminced meat, who had presumably yet to catch up with their copy of *The Times*, watched two hundred and one musicians (the two hundred and first seated at the Grand Organ) and six hundred choristers (the combined forces of the Chapel Royal, St Paul's, Westminster Abbey, and St George's Chapel, Windsor) fiddle nervously with scores of the Hallelujah Chorus. The younger boys, slowly poaching under the glass, fanned themselves unrhythmically until, admonished by the choirmaster, they too stood stock-still waiting for the most important down-beat of their lives. A less than serene Archbishop settled in his place, rehearsed his blessings and then set to counting them.

Outside 500,000 people gathered in expectation, though in expectation of what is not entirely clear. Unable to see anything except the Palace itself, they were 'drawn as by magnetic device', and like so many iron-filings stayed stuck. By eleven forty-five a traffic jam stretched immovably from Hyde Park to the Strand, unable to decant its contents to join the 1,000 carriages of state, 1,500 cabs, 800 broughams, 600 post carriages, 300 clarences and 300 assorted vehicles that had already made it to the Park. Neither seven battalions of infantry nor 6,000 extra policemen could do anything to accelerate their progress.

At eleven fifty, a Chinaman – the possessor of both a junk on the Thames and a keen eye for publicity – prepared to push his restaurant and his luck by standing ready to join the Queen on her journey to the dais. Thinking him an invited dignitary, no-one would challenge him or their assumption, and the little man would see it all from the Queen's side, bowing every now and again in authentic oriental style. Amid so much exotica, he was easy to miss.

At twelve o'clock, a cursory tour of the transepts completed, the speeches made and the ceremonial blessings dispatched, the declaration of opening was made. As the echo of the chorus died away –

the noise in the nave was tumultuous, but according to one commentator parked in the extreme west end of the Palace it registered only as 'similar to that of a musical snuff box' – the fanfare started. As the fanfare ended, the semaphorist began and so on down the line, until Sir Charles was aloft and, in a stiffening wind, last seen heading for Dover.

'The glimpse of the transept through the iron gates, the waving palms, flowers, statues, myriads of people filling the galleries and seats around, with the flourishes of trumpets gave us a sensation which I can never forget and I felt much moved. The site as we came into the middle where the steps and a chair (which I did not sit on) were placed, with the beautiful crystal fountain just in front of it were magical – so vast, so glorious, so touching. One felt as did so many whom I have spoken to – filled with devotion. More so than with any other service. . . . And there too my beloved husband – author of this peace festival which united the industry of all nations of the earth – all this was moving indeed and it was and is a day to live forever' (Queen Victoria).

The Context: Changing Politics and Moralities

The world into which the Crystal Palace was born was a world pregnant with events; so much so that it was hard for any one of them to seem especially significant. 'Pivotal moments' are easier to spot in retrospect, and – though in the course of the nineteenth century any number might plausibly suggest themselves – most of the better candidates seem to have chosen 1851 in which to happen. What appeared then to the Duke of Wellington to be just the usual headlong dash into the future, now looks like a turning point, and not only because it falls exactly mid-way in a hyperactive century.

In Europe, the dust of the revolutions of 1848 was settling, and the political landscape which emerged through the miasma seems to have had a different and distinctive feel to it. In Germany, the failure of the middle class to achieve their demand for wide-ranging reform set the seal on the future by promoting the interest of the militaristic Junker aristocracy of the Prussian heartland. This in turn determined the course and character of German Unification; but less parochially, it had more than a passing influence on future political relations and relativities between the European states. In France the short-lived Second Republic fell into the hands of the charismatic Louis Napoleon following the *coup d'état* of December 1851, and by the following year France had an Emperor and set about finding him an empire. (He and the Prussians would test out their respective European theories in 1871: odds on a Prussian victory would shorten dramatically in the interim, giving further evidence that 1851 heralded significant alteration.) In France itself, the Belle Epoque had begun: an era which still has the power to resonate. During it the economic, social and political being that was France would be transformed almost as much as its self-conscious symbol, the Paris of Haussmann's plan. Meanwhile in Italy – the rubber duck that Napoleon III played with in his political bath – the promotion of Cavour to ministerial office in 1851 proved one of the fruits of the otherwise abortive risings of 1848/9. Hollow as that might seem, there is a plausible case to be made for regarding unification through stealth as beginning from here. Nothing as complex as the *Risorgimento* can ever be pinned to one man, but Verdi – a shrewd reader of his times and himself a hewer at the coal face of the movement – saw Cavour as a united Italy's best hope: thus a great deal of what was achieved over the next ten years, in philosophical liberalism as well as in territorial rationalisation, had its origins in that unpromising promotion.

Finally in Britain itself, the European uncertainties of the late-1840s had been mirrored in more sporadic though equally diagnostic challenges. Chartists and the Anti-Corn Law Leaguers, alongside a hundred un-christened groups protesting bread prices, enclosure, urban conditions and the rights of labour, climaxed their demands in the decade before 1851 and then fell silent – though some only temporarily – thereafter. In a country whose leaders have always turned paranoid at the mere sight of dissent, it is hard to be certain just how significant such challenges were. For some commentators, the 'defeat' of many of these causes left the world unchanged; for others it is possible to detect in their 'success' the origins of a major rethink in government policy (as well as in the character of government) from 1851 onwards. In the latter view political hegemony itself shifted, though not always in predictable ways. The aristocracy, challenged but not eclipsed in 1832, gave way to the middle class; coercion gave way to consensus; moral economy gave way to political economy; and the country gave way to the city. The working class, meanwhile, simply gave way.

Setting aside the *fons et origo* of all this, and ignoring a thousand significant disputes about it, there is still a case to be made for regarding 1851 or thereabouts as the point at which one set of assumptions finally gained ascendancy over another, though it may be claiming too much to say that it was only now that the Regency (with all of its assumptions) truly gave way to the Victorian (with assumptions all of its own). Variously though misleadingly labelled: urbanisation, secularisation, modernisation – the terms do not really matter – these underlying re-orientations, planted like time-bombs in the latter half of the eighteenth century if not before, began systematically to explode from 1851 onwards. Naturally these were read as the hallmarks of a certain kind of progress, though contrary to the image the authorities were trying to portray, and for some the Crystal Palace would be the portrait's centre-piece: Britain would feel less not more secure as a result.

If the political climate was evidencing major squalls, then the moral climate – joined to it at the hip – was held, by many contemporaries at least, to be exceeding the statistical range allowed for by the Beaufort scale. In the year in which more than half the population of the country became urban, and urban life became in consequence a scourge not a privilege, the religious condition of the people was probed for the first and only time. As poverty rose and conditions worsened – the largest sustained cholera epidemic was only two years past – their accompanying shadow of crime began to attract attention. Authority, eager to make forensic connection quickly attributed criminality – as evidence of depravity – to the absence of religious observance. A heathen population was *de facto* a criminal one, an allusion made the more plausible when zeal and insensitivity, certainty and intolerence went side by side. The fact that a significant proportion of the urban destitute were Irish Roman Catholics added fuel to the fire; the image of a Jewish Fagin torched it. When one body of opinion denounced the Great Exhibition as a papist plot, and another detected in it an attempt at world destabilisation by Jewish marketeers, their paranoia was typical not extraordinary. For the evangelists and xenophobes, it was all grist to the mill; and the urban campaigns, which were intended to return the sheep to the fold but which in practice scattered them to the farthest points of the mountainside, gained their legitimacy through reference to the official statistics of 1851 and the implied susceptibility of the weak-minded to religious guerrilla warfare of the type allegedly perpetrated by covert international exhibitors.

Spiritually and materially these were indeed confusing and even troubled times. The image of the Exhibition portrayed by its protagonists admitted none of it, though, and the incongruity would be a major focus for the critics. Even in practical terms intention and reality could be starkly at variance, as a Swiss visitor who had travelled to see the world's greatest exhibition in the world's acknowledged capital discovered. As confusion gave way to disappointment, his sense of humour evaporated in the heat. On one occasion he could get neither refreshment nor a cab because it was Sunday, could not settle his daily account at his hotel for the same reason, and felt disinclined to walk to Hyde Park for the air (the Exhibition itself would be closed to appease the Sabbatarians) since a substantial part of the walk was more a wade – through the accumulated effluent generated by the capital's human resources – and through the same thoroughfare in which he had been robbed the previous Tuesday. He could not attend Mass, since no-one was able or inclined to tell him where the Roman Church was. Falling in desperation to conversation, he addressed a young lady at his hotel on the subjects of his distress only to find himself soundly rebuffed. The proprietor explained later: the lady and he had not been introduced.

This then was the world which would be home to the Exhibition: a world made up of the confused and the confusing, a world in which political reaction was of the knee-jerk variety and the idea of a policy was still confined to the offices of fire-insurers. In the same month in 1851 in which internationalism was signalled, as the first submarine cable from London to Paris was laid, Parliament passed a Bill declaring all titles bestowed by the Pope to be illegal and by implication all holders of the same to be ideologically suspect if not traitorous. In the same month that the building of King's Cross Station began and several hundred destitute families were evicted in consequence, Livingstone reached the Zambezi, the investors calculated their margins and the scramble for Africa began. As Teddington, Sir John Hawsley's 3 to 1 favourite crossed the line to win the Derby at a breeze and Cambridge beat Oxford by an innings and four runs, Holloway Prison, the Army and Navy Club and Colney Hatch Lunatic Asylum were completed: homes for a triumvirate of Victorian types: the criminal, the military and the insane. In such a world it was hard to tell which was which.

The Architects of the Vision: the Prince Consort and Paxton

Benjamin Disraeli – a man given to speaking in the Upper-Case and apt to find followers of Machiavelli everywhere – considered the timing of the Exhibition to be as suspect as the motives of those behind it. He thought it '[a] Godsend to the Government of this Nation, diverting Public attention from their continuing Blunders'. But for once he could not have been more wrong – in the motivation department at least – because the government actually had scandalously *little* to do with the Exhibition, the consequences of which would only reveal themselves fully at a later date.

The timing too was a matter of dynamics internal to the Royal Society of Arts. In 1847 the Society had found itself a new President. He was a well-connected German, recently naturalised and with a strong interest in the sciences. Albert – for such was he – was Victoria's husband, the Prince Consort, and a man who had recently made two important discoveries: that there was no role for him running the country (a failed attempt to intervene in a parliamentary debate had rather driven this message home) and that there was nothing much else for him to do either. Finding himself supremo only of a Society, he hence set about trying to transform it into a country, honing his statecraft by developing his Presidency. His first aim was to try to put his principality on the map. This he achieved by staging a series of increasingly spectacular events which emphasised his subjects' achievements in the arts and sciences. Domestic displays took place in 1847 through to 1849 under the administrative care of Henry Cole, Assistant Keeper at the Records Office. Cole was the type of Victorian that the Victorians liked to think of as a type. Amongst other lifetime achievements he was responsible for the implementation of Hill's penny-post and for standardising railway gauges, thereby increasing the chances of getting from **a** to **b** without a wad of tickets heavy enough to induce a hernia. He was a painter, author of childrens' books and publisher of the first commercial Christmas card.

He was also the man – history would naturally record his superior's name – who conceived the idea of the Great Exhibition as a result of a visit to the Paris Exposition of 1849. Though the Paris event was still a domestic affair, thus preserving London's claim to have held the first *international* event, it was bigger than anything London had seen. Cole, first minister in the principality's 'war cabinet', considered expansion: in scope – the Society should invite the world to London; and in size – the Exhibition would abandon Somerset House Gardens and take over Hyde Park. It would commission a building (opposite Knightsbridge Barracks) to house the exhibits and the building itself would be as exquisitely fashioned as it was mechanically inventive, so representing that marriage of art and science which was the Society's *raison d'être*. To some members of the government of the 'real' state, the combination of Prince Albert, a radical scientific creed and an uncompromising public display of strength was a dangerously powerful cocktail; each element worthy of opposition in itself. If the threat – as symbolic as it was real and as political as it was non-partisan – was to be diminished, then they would need to act quickly. They would need to mobilise the natural forces of opposition and encourage the expression of any other appropriable discontent along the way. If the Committee of the Royal Society of Arts had chosen almost any other building, the fight might have been decided on points, so relentlessly were the forces of opposition deployed: but armed with the Crystal Palace, though the battle was bloody, the outcome was never in serious doubt. Credited as the architect of the exhibition hall, Joseph Paxton was in truth architect of the greater victory as well.

Paxton was born in Milton Bryant near Woburn in 1801, and by the age of twenty-three was a gardener in the employ of the Royal Horticultural Society. Talent-spotted by the Duke of Devonshire, in that year he was set to remodelling the Duke's Chiswick estate, and having proved himself as a miniaturist he was let loose on the real thing – Capability Brown's Chatsworth – a year later. He was a man out of whom enterprise shone like the proverbial sun, and ideas leaked like a drain. The range of his responsibilities expanded wildly. Organiser of the Duke's North American Botanical Exhibition; designer of the new village at Edensor on the Chatsworth Estate as well as the Duke's Buxton properties; organising social supremo of Czar Nicholas of Russia's Chatsworth visit – for which he received the Grand Cross of the Order of the Knights of St Vladimir – he was no mean achiever in a society well supplied with mean achievers. A self-made man steeped in the spirit of self-help he rapidly learned to eschew the selfless, and through self-promotion helped himself to almost anything that was on offer: business directorships, newspaper proprietorships (he was Dickens's boss at *The Daily News*), railway company stewardships (the Crystal Palace would be first sketched on the Midlands Railway Company's blotting paper at one of its interminable board meetings) and parliamentary opportunities (MP for Coventry in 1854, he could have stood for the monarchy and been elected). Begun as a doodle, his plan for the Crystal Palace turned an exhibition into a public statement, an event into a symbol, and an ephemeral moment into an enduring and charismatic summation of an epoch's achievement. It was Albert's Palace of Westminster and Canterbury Cathedral rolled into one: the secular and the sacred brought into a bliss so consummate it was supercharged.

Given the mood of the critics, had it been proposed to build the Sistine Chapel the design would have been found wanting, the decorations uncouth. As it was, Paxton handed them controversy (but not victory) on a plate. To begin with, the design was submitted after the closing date for the competition, indeed after the winning entry by Isambard Kingdom Brunel (in a style dubbed 'St Paul's Cathedral out of Euston Station') had already been selected. Paxton's scheme was forced into prominence through deliberate leakage to the press, and hotly debated for its outrageous use of materials hitherto untried on the scale proposed. To the charge of impropriety was added that of structural improvidence and aesthetic ineptitude. More recent debates about the Palace's stature have been defined by arguments between the modernists (Henry-Russell Hitchcock's claim in 1937 that it was 'the most prophetic monument of the mid-nineteenth century; a monument often hailed with pardonable exaggeration as the first modern building') and the revisionists (Sir James Richards's belief that 'historically [it] is more accurately described as the last of the great engineering feats of the early-nineteenth century'), but at the time it was more a debate about whether it would stand up, and if so for how long? Paxton later confessed that the design – an expanded version of the great greenhouse at Chatsworth – was based on the structural properties that he discovered in the radiating leaf-ribs of an exotic Guinean water-lily cultivated in the paddle-driven hot water tanks of the Chatsworth grounds: wisely he declined to vouchsafe this until practice made the theory less newsworthy. The aesthetic charges – as perhaps befitted them – were met with confident silence rather than with the usual dead-weight of statistical fact, the design's sheer grandeur and exhuberance being left to speak for themselves. Next to Brunel's conception, which according to Brougham threatened to make an 'exhibition' of both itself and the country in the wrong sense of the word, it was an object lesson in restraint. Douglas Jerrold writing in *Punch* christened it the Crystal Palace and, capturing its spirit to perfection, the name stuck. Beyond its self-announcing splendour, the Palace's single most important asset was its evident impermanence and thus its suitability for a temporary exhibition in a public park. Who after all expected a greenhouse to last a lifetime? The pessimists gave it a fortnight.

It is always claimed that the detailed design took Paxton nine days, though Paxton himself considered that an exaggeration: he had spent two of the nine watching the opening of Stevenson's Menai Straits Bridge, so it was seven at most. Construction took 185 days between 30 July 1850 and 31 December. From January 1851 attention switched to the interior as over 2,000 workers applied the finishing touches, decorating the structure as well as erecting the 991,857 square feet of exhibition stalls required by the 13,937 exhibitors. It was finally completed on 28 April, two days prior to the opening. The structure was tested by a company of Sappers who exercised as vigorously as they could whilst trained experts – ears to the glass – waited for signs of oscillation. They detected nothing, but left, it would be wonderful to record, vibrating like tuning-forks. During construction, an initially hostile press – which had published Paxton's plans only for the mischief that they might cause – began to realise that it was backing the wrong side and, with that sense of principle for which the metropolitan press was renowned, quietly deserted the post of critic and turned instead to proselytising with the uncritical evangelism which is the stock-in-trade of the recently converted. Now the newspapers recruited the Palace and its promoters in the age-old battle for circulation. Much of the hysteria surrounding the project – as well as the foundation of its popular success – was generated by a press eager to use news of the Palace to trump the hand of its nearest rival. Hardly a pane of glass found its place in the jigsaw without being sketched, engraved, eulogised and turned into a scoop.

Ironically, as it rose above eye-line the Palace had decreasing need of such support, and the newspapers' more outrageous claims on its behalf became counter-productive. Even reduced to statistics, it was impressive enough to establish its own case. Four times the volume of St Peter's in Rome, six times the volume of St Paul's, it was the largest covered space ever conceived let alone spanned:

Length of main building	1,848 ft
Breadth of main building	408 ft
Floorspace	772,824 sq ft
Height of nave	64 ft
Height of transept	108 ft
Weight of iron	4,500 tons (700 wrought; 3,800 cast)
Columns	2,300
Timber	600,000 cu ft
Panes of glass	293,655 (900,000 sq ft)
Guttering	24 miles

The cost was to be £150,000 or 1d per cubic foot, though the contractors Fox and Henderson offered to build it for £79,800 if, when the Exhibition was over, they could salvage the materials. In the event public enthusiasm would favour preservation and new deals would have to be struck. In its appropriate context, such enthusiasm is easy to understand. But what of the opposition?

The Forgotten Discourses of Opposition

It was inevitable that an event like the Great Exhibition, knowingly promoted to draw the eyes of the world towards it, would find itself embroiled in whatever political fashions happened to rage at the time. Equally, as we have seen, there were private and public interests standing ready if necessary to manufacture pretexts for a full-frontal assault. Necessarily, then, we have to disentangle two sometimes contradictory strands which anchored the case against: the one made up of issues which were attached to the Palace for no reason other than its prominence, and the other which related directly to the Exhibition and what its promoters were allegedly trying to achieve. The first would require another and different chapter. Suffice it to say that the Crystal Palace provided a front-page opportunity to draw out into the open many of the festering disaffections otherwise hidden in the larger politics of 1851. Zealots, xenophobes (particularly the anti-Irish), anti-moderns, revivalists (sacred and profane), bigots (sacred and profane), spiritualists (sane and insane) and even proto-environmentalists posted notice of their existence and threatened the coming of their kingdoms to expunge the evil epitomised in Hyde Park. The environmentalists, for example, had protested the destruction of a number of Hyde Park trees which – in a classic exercise in political inflation – would become known as John Bull's Trees of

Liberty, the hopes of all England reputedly pinned to them. They stood strategically in the centre of the Palace's site, and had the potential to bring everything to a halt. In the event the Palace would again triumph over temporary adversity. Sir Charles Barry modified the original plan to add the arched barrel vault of the transept which could accommodate even the tallest tree, and he thereby created the most notable feature of the Palace's design: the feature which would stay resolutely in the mind and make the whole so instantly recognisable, whilst reinforcing still further the overall greenhouse effect (nineteenth-century sense, of course).

Two groups should be added to this preliminary list of incidental opponents: the constitutionalists and the republicans. Since both were opposed to the Prince Consort and therefore to his project, it was often difficult to tell them apart, though a significant difference existed. The constitutionalists wanted to foil the Exhibition in order to keep Prince Albert firmly in his place; the republicans added it to the panoply of arguments designed to remove him from any such place altogether.

Groups such as these were joined in their opposition by particular individuals with motivations all their own. Some remained obscure in their reasoning: Lord Brougham, for example, once a committed internationalist, whilst others were only too transparent. Amongst the most famous – and a man who typifies the range of injustices for which the wilful could hold the Palace responsible – was Colonel Charles de Laet Waldo Sibthorpe. MP for Lincoln, he was a well-known eccentric; ultra-conservative and ultra-protestant, but above all ultra-skilled in the art of issue-making. As his biographer says:

'A man of eccentric dress and manner, he successfully opposed the granting of £50,000 a year to Prince Albert because he, the Colonel, did not like foreigners. He was against Catholic emancipation through his private dislike of Catholicism and he had always opposed railways, because he personally opposed railways. He was a sturdy adversary of Parliamentary reform and free trade and early in 1850 had fought the Public Libraries Act because, as he said in the House, he did not much care for reading at all.'

Such a man would naturally find any number of plots lurking in the Exhibition halls. Having detected Popery, an international plot to import foreign prostitutes and the makings of Irish nationalism, he predicted epidemic diseases and even climatic change (all that glass) before finally getting to the point:

'It is the greatest trash, the greatest fraud, the greatest imposition ever attempted to be palmed off on the people of this country. The object of its promoters is to introduce foreign stuff of every description, live and dead stock without regard to quantity or quality. It is meant to bring down prices in this country and to pave the way for a

cheap and nasty trumpery system. . . . All the bad characters at present scattered over
the country will be attracted to Hyde Park. . . . That being so I would advise persons
living near the Park to keep a sharp lookout for their silver forks and spoons as well
as for their serving maids. Let the English beware of mantraps and spring-guns. They
will have all their food robbed and a pie-bald generation half black and half white
for an inheritance' (Sibthorpe).

For individuals like Sibthorpe, then, as for many of the organised
groups, the Palace was merely a pretext: a political football. By
pretending to kick the ball, you could cripple the player; even better
for some, you had a sporting chance of taking out the captain.

Amidst these more or less notorious opponents stood the critics of
the Exhibition *per se*. Again they were varied in substance and in style,
but a technique common to all was to highlight an argument advanced
by the Exhibition's promoters and then to turn it on its head. Three
issues in particular dominated these critiques: the role of the Exhibition
as a measure of internal progress, more broadly 'the condition of
England' question; the role of the Exhibition as external audit, more
broadly its *de facto* calibration of the development 'league tables'; and
the role of the Exhibition in promoting international co-operation,
amity and comity in a turbulent world.

For exhibitors the world over, the display of a nation's commercial
and industrial wealth was a significant measure of its progress and
standing in the league of nations. Like so many bits of litmus-paper,
these events – they would be more or less decennial after 1851 –
dipped into an economy, tested its material well-being. Much could
be read into the findings, particularly if easy formulae were applied
to transform measures of material wealth into symptoms of social
well-being. Massage the image and Britain was not only powerful,
rich, and technologically advanced, but stable, equitable and secure.
Its people were enterprising, hard-working and productive, *ergo* loyal,
solvent and content. For the critics the matter was necessarily more
contentious, the conflations more sinister. Behind the goods on
display lay a system of production, and the backbone of that system
was the people whose conditions mocked the achievement that their
employers claimed. The Exhibition took place only three years after
Engels had undertaken his study of the working class of Manchester
and in the same year as a census which would at last go some way
towards exposing the true conditions endured by the labouring and
non-labouring population of the land. Dickens, amongst others, felt
uneasy with the conjunction: 'I have always had an instinctive feeling
against the Exhibition, of an inexorable sort. When I view the
predicament of so many of our people and the injustices with which
everyday life confronts them, I cannot in conscience tolerate so empty
a display.' *The Eclectic Review* sounded warnings too: '[t]here are hard

facts weltering beneath the rose-pink surface. This great farce will one day prove the opening of a great tragedy.'

For many of the critics there was nothing accidental in this emphasis on goods rather than on well-being, on productivity rather than on commonwealth; on value rather than on values: for them the Exhibition was a public charade with an express purpose, to divert attention from the real issue – the true condition of England; the gulf between the Two Nations – and to disguise a national insensitivity towards the plight of the 'other England' comprising the industrial north, the metropolitan homeless, the urban poor. Even Sibthorpe, amidst the idiosyncratic poison, could sometimes hit this particular nail on the head:

'Are the elms to be sacrificed for one of the greatest absurdities ever known? For myself I will do nothing to encourage foreigners – nothing to give secret service money to them in the shape of premiums paid to strangers when they are going to spend £26,000 on this building *whilst the Irish are starving*' (Sibthorpe: emphasis added).

For some, therefore, the Exhibition's emphasis on the fruits of production together with the Palace's embodiment of blatant over-consumption served to throw into yet stronger relief the contrasting plight of those bound into that production yet denied an appropriate share when the social goods were carved up and distributed. One of the ironies – amidst many – thrown up by such opposition is that the government actually got the message and began to worry at the bone of the issue, eventually distancing itself from the Exhibition and announcing this symbolically by refusing to commit any state resources to the project. But if this got them off one particular hook, it was later to skewer them on another. By making the enterprise entirely self-financing, the government determined that the interests of private profit would fix entrance charges at a level which precluded (even on the cheapest days) the participation of many ordinary workers. Not only could the workers whose labour was tacitly acknowledged in the Crystal Palace not share in the consumption of the products on display, in 1851 they could not even get in to see them.

If one problem was that exhibitions failed to stimulate the kind of *internal* auditing which the socially sensitive felt necessary, a second was that in the *external* or international arena technical auditing worsened an already acute problem. A nation's performance at an exhibition mattered a good deal; so much in fact that no effort was spared to improve it. The downside was that the corners cut in the process exacerbated the already horrendous conditions of labour as the spotlight shone increasingly on production at all costs. 1851 gave immense impetus to these tendencies by emphasising which nations had proven themselves important enough to be invited, and thereafter by allocating space to them on a strictly hierachical basis. At the Crystal Palace over half of the 19-acre site was reserved for Britain and its Empire, with the rest arranged in a pecking order which trumpeted inferiority with no blushes spared. Like the geography of reserved pews in an unreformed Anglican church, it quickly and effectively established to whom the forelock should be tugged and, alternatively, who it was safe to spit on. In 1851 the results were intriguing: France, Belgium, Germany (Steuerverein, two Mecklenburgs and the Hanse towns), Russia, Austria, America, Holland, Italy, Denmark, Norway/Sweden, Switzerland, Turkey, Spain, Portugal and Greece. With the league table thus established, it became the priority of each country to advance its position towards the top. The nightmare was to loose ground between shows. Even in Britain the paranoia would become intense: winning in 1851 by a considerable margin, it was now a case of running to stand still, and as the field closed over the next few decades so the inquests began.

In practice, there were two ways to accelerate progress. One – denied by all but attested by example – was to take careful note of what everybody was up to and then copy the most advanced innovations. The second was to throw national resources at the problem. The critics were quick to point out the consequences of the latter, even prospectively: technical improvement rather than social progress or rational development (conservationists, had there been such, would have joined in the despair), and a penchant for the flashily impressive rather than the genuinely useful. It was this sort of mentality which gave Victorian Britain sixty-four different patent walking sticks and eighty-nine types of collar crimper, but no serious weapon against infectious diseases or a large-scale urban sanitation system which worked.

A perfect example is provided by France. It was generally agreed that of all the 'foreign' exhibits at the Crystal Palace, the French had been the most tasteful. In a world obsessed by meccano on a manic scale, tastefulness was not the acme of distinction. Rococo scrolls and floriated capitals, fine on a statue, cut less ice on a steam engine. Worse, the French won few prizes and then only (or so they claimed) when the British had no exhibit in that category. Their dolls were admired and even gained commendations along the lines of 'best design of a non-useful mechanical object', but it was a category which British hauteur prevented the hosts from entering and thus a rare opportunity for a foreign victory.

'The French porcelain tapestries and delicate textiles were widely admired, as was some of the Algerian fancy cabinet work. In the field of electricity, long a French speciality, there were important exhibits such as Volta's electrical apparatus and a double-magneto current electrical device. There were also a number of domestic novelties such as a patent spring mattress, patent concentrated milk and an alarm bed . . . which at the selected hour sprang forward throwing the occupant (without distress) into first an upright position and then (if desired) a tub of cold water for instant rousal. . . . There was in addition an interesting departure in the direction of naturalism in a marble statue of a billiard player. . . .'

Interesting exhibits, then, but hardly enough to impress the world or to establish the French as a nation to be reckoned with in industrial or trading affairs. In fact as one cynic more pithily remarked in *The Times*: '[m]ore interest was aroused by a single item in the American display – Mr Colt's improved and concealable revolvers – the precise significance of which every visitor can deduce at once'. The impact was as decisive as it was immediate. Anxious to consolidate his new government, Louis Napoleon ordered an inquiry, and a ludicrously disproportionate number of French civil servants was commandeered to 'investigate, report and recommend lessons to be learned in the light of the recent Exhibition in London'. By 1873 – twenty-two years after the Exhibition, eighteen years after the Paris Exposition and three years after the Franco-Prussian War – an admittedly much smaller phalanx of state employees was still working on the report and its implications. Meanwhile, French efforts at industrialisation were proceeding apace, spurred by the sting of 1851 and aiming at the international model that it provided.

A similar inquiry in England headed by Lyon Playfair and Charles Babbage was to have a significant impact on the Education Act, in that it laid responsibility for any failures in the English performance squarely at the door of a technically deficient education system. Eric Hobsbawm argues that education in late-nineteenth-century Britain was far from being a cultivation of the intellect and in consequence 'not a high road to anywhere'. Instead it was a utilitarian litany of useful facts designed to produce factory workers from factory schools and a labour pool unskilled in the dangerous art of inquiry. In this prescription the Exhibition bore and indeed continues still to bear much of the responsibility. From Dickens, in summary of the prevailing ethos:

'Herein lay the spring of the mechanical art and mystery of educating the reason without stooping to the cultivation of the sentiments and affections. Never wonder. By means of addition, subtraction, multiplication and division, settle everything and never wonder. Bring me yonder baby just able to walk, and I shall engage that it shall never wonder.'

Finally, there was a third source of anxiety amongst the critics and one which was to prove explosively pertinent. It had long been the contention of Exhibition promoters – those of 1851 made the contention a faith – that such events were the harbingers of world harmony and co-operative progress. Like all the best theories this one embodied a simple principle: brought together in a spirit of friendly rivalry, nations would learn first to respect, then to like and later to love each other. Co-operation, knowing no boundaries, would redraw the map of Europe. The augury here was not however the Olympic torch, but the brighter flame of science: that most internationalising and non-partisan of nineteenth-century European movements, the sheer intellectual force of which guaranteed rational progress along the empirical Appian Way toward Priestley's Empire of Reason. Religion divided, science – the 'book of nature' revealed – united. Many of the architects of the movement, intellectual radicals to a man (and it was of course a 'man') and bound by silent oath to an international code of collective progress, saw the Exhibition as the first concrete manifestation of European unity: a tide of Apollonian energy carrying poor, disfigured Dionysian society in its wake.

But if this was the hope, it was a pious one; for the reality – as the critics prophesised and at the same time feared – was very different. The juxtaposition of inequity in national wealth and national achievement brought with it an intensification rather than a diminution of rivalry, together with a deepening chauvinism and a suspicion of others which verged on the certifiable. At the same time, an exhibition which emphasised material goods necessarily sanctioned the acquisition of the same by whatever means. Sadly, science was but one of these means, and Livingstone's arrival at the Zambezi presaged another. Since from 1851 onwards exhibition rules allowed colonial goods to be displayed under the banner of the occupying power, the extent of a state's territorial domination over other parts of the world now stood graphically revealed for the world – and occasionally his wife – to comprehend. The Indian stand at the Crystal Palace, which was particularly admired, started a trend which between 1851 and 1900 focused world attention not on the morality nor even the political consequences of the scramble for Empire, but upon its success. Germany, France, Belgium, Holland, Italy and above all England ransacked the known world for their own economic purposes and displayed the results at countless exhibitions, simultaneously spurring each other to yet greater rampages as the system of colonial exploitation intensified and as the penalties of having no new conquest to parade at the next international jamboree increased. Colonies became just so many trophies 'bagged' for display.

Even on the European stage such friendliness and co-operation as was evident in 1851 was quickly shattered as the participant states systematically fought each other for power and nationhood. Like the jolly pun leavening the Archbishop of Canterbury's beatifying speech ('peace is within our walls and plentiousness within our palaces'), Queen Victoria's ambitious sentiment at the opening ('[t]his wonderful hall of achievement inspires us all to carry the banner of co-operation abroad') looked faintly ridiculous when only three years later the Crimean War erupted and the Light Brigade charged. The Universal Exposition of 1867 in Paris was followed – again within three years – by the Franco-Prussian War and recipes for rat soup, and on this evidence alone it is hard to believe that the critics were wrong in insisting that exhibiting generated quite as much heat as light, as much resentment as back-slapping. In an already imbalanced continent of unstable administrations, developmental and political differences were more not less evident in the harsh light of Paxton's greenhouse, and the result was a friction which no amount of pomp or special pleading could disguise. Even the carriages of the royal houses were entered into competition: Britain won.

So, the many and varied voices expressed their outrage, disappointment and forebodings as the plans for the Exhibition matured. Systematic as some analyses became, and sustained as their campaigns were, the critics enjoyed little success in diminishing the momentum of a populist pro-Palace campaign that had the force of a train and the subtlety of a steam-roller. Amid the mania the controversy which the Palace created and for some even embodied – the questions which it had asked of its society – were slowly engulfed and, over a decade, gradually forgotten. Drowned amid the clamour, the darker voices of the sceptics fell silent; crushed, some alleged, under the massive weight of misrepresentation. All they had on their side, wrote one, 'was righteousness'. But in Victorian Britain, stacked to the rafters with the self-confessed self-righteous, the real thing was not enough. By the Paris Exhibition of 1861, though many of the dire forebodings were no longer unhappy projections but even more unhappy facts, exhibiting was also a fact: a fact of international life. By then too the Crystal Palace was a national treasure and, more crucially, a profitable enterprise. Most could not even remember what all the fuss had been about.

The Ambiguities of What Could be Seen and of Who Could See

Though the significance of the Palace was ultimately to transcend that of the Exhibition it housed, for a time the two were inseparable to the extent that we need to pause briefly to consider precisely what the nation – or at least a part of it – came to see. Charitably, provided they could pay, visitors saw a lot for their money; on the other hand – and even allowing for a jaded twentieth-century palette – what they saw was a bizarre and tasteless jumble of the world's most extravagant kitsch, the cumulative effect of which would have made the crowded walls of an average middle-class Victorian sitting room look like a study in understatement. It was bric-a-brac on a pestilential scale; the village bring-and-buy as Albert Speer might have imagined it; less white elephant than white elephantiasis.

The somewhat ambiguous tone was almost inevitably set by Britain, whose stands combined the best (in other words, the largest) feats of British engineering – Mr Naysmith's collosal steam hammer constructed to forge the paddle-shafts of Brunel's S.S. *Great Britain* – with the most delicate exotica of the Indian stand. The result was a collision of cultural artifacts deeply resonant in both scale and intent of the vassal/servant relationship that they unwittingly attested. Such juxtapositions were immediately in evidence. Even in the entrance avenue the exquisite and the tasteless vied for simultaneous attention: a single 24-ton block of Welsh coal confronted a statue of the madonna and child; whilst a doric column carved from Cornish granite (European high art from English base material) outstared a pointless and quite random collection of stuffed fauna too large for exhibition inside.

Once within the Palace's walls, the contrary themes of the refined and the tasteless, the bizarre and the commonplace, were taken up with even greater vigour and renewed purposelessness. Most symbolic of all perhaps was the ornate rococo-style papier mâché chair of the French stand which doubled as a commode and would appear amidst its naughty inventiveness to constitute the world's first dissolving toilet. Or perhaps the floating church of the American stand, designed to minister to godless sailors and to convert them before their taint reached the shoreline and more particularly the women of the New World. Audacity married to impracticality, glitz without purpose, a show without Punch, they were metaphors for the whole.

From the British section, the visitor could choose to assault his or her sensibilities with four different categories of wonder: *Raw Materials*, including 'novelty guano – a superior fertilizer' and rhubarb champagne (exempted from the otherwise stringent alcohol ban on the grounds of ingenuity: the French, refused a permit for brandy, were not well pleased); *Machinery*, including 'a window cleaner for the protection of female servants from accidents and public exposure' and a lifeboat with a decorative figurehead in which could be boiled eight quarts of coffee; *Manufactures*, a huge category which included a corset which opened automatically 'in case of emergency' (nature unspecified), artificial legs, an artificial nose in sterling silver, an artificial jaw fitted with 'a compensating swivel which allows the wearer to yawn without dislocating both upper and lower sets', elastic chest expanders, a patent 'ventilating hat' which opened to allow heat to escape once a critical internal temperature had been reached, a set of bleeding irons with decorative bowl and lid to permit the continuance of social intercourse during treatment, a pair of rifles described as 'a twelve shilling gun for the purpose of barter with African natives together with a different model for shooting him down', and a safety lock which should anyone attempt to pick it fired a revolver straight at them; and finally, *Sculpture and the Fine Arts*, dominated by the two stereotypes of mildly titillating erotica on the one hand (naked Greek sirens and the like: on removal to Sydenham, where they would be ogled by rude mechanicals, they would have discrete fig leaves added) and the patently sycophantic on the other (the future Edward VII as a shepherd of his flock, Victoria as goddess, Albert as Hector, Albert as Hercules and – a stab at the critics this one – Albert as Daniel).

William Morris described the whole as 'wonderfully ugly', whilst *The Times*, reserving as always its most severe criticisms for foreigners, nevertheless admitted to the rather tacky aesthetic underlying it all: '[t]he Americans have everything to learn in matters of design. They have inherited our ignorance on that subject and increased it with vulgarities all their own.' But if the overall impression was one of ludicrous jumble, there were also ominous undercurrents which revealed – and not only to the hypersensitive – a different and altogether less innocent spirit of endeavour. Two stand out: a Birmingham firm exhibited 'a splendid array of shackles, manacles, pinions, handcuffs and fetters for export to the American slave states' to remind us that colonial wealth and even the international balance of trade was founded on the transportation of items other than cocoa, and often on a labour process even more bestial than that on which most of the displays were dependent; and on the German stand appeared an exhibit which in retrospect was wholly ominous in its portent, being described in the catalogue as 'Krupp, F. (Essen near the Ruhr) Steel Gun. Six pounder. Complete'. The Krupp munitions factories were to be in overdrive for a significant proportion of the next hundred years, of course, and unlike this particular item the majority of its manufactures were for use rather than display.

Ironically, as far as the Exhibition itself is concerned, the contents – collectively or individually – did not really matter; were not even finally the point. The Exhibition had happened; furthermore, it had happened in London, it had been a great occasion and, because it was repeatedly proclaimed as such, it had been a triumph. It had given birth to the Palace. The medium was the message.

By the Exhibition's close on 11 October no less than 6,039,205 people had visited. At two o'clock on the afternoon of 27 October

92,000 were doing so simultaneously. The breakdown by ticket type is highly suggestive, and not only of the sedulous character of the turnstile officials:

Season Ticket Holders	773,776
£1 Days [2]	1,042
5s Days [28]	245,389
2s 6d Days [30]	579,579
1s Days [80]	4,439,419

[Figures in brackets denote the number of such days]

Faced with the statistics – of both building and Exhibition – it is easy to fall into the obvious trap: that of admiring the achievement and of thereby missing the point. The officials of the committee were keen to publish the statistics however absurd (they even included the number of ladies using the toilet facilities over the four and a half month period) for precisely this reason. At a time when big was beautiful, small was failure and failure was inadmissible. And yet to take a single example – that of the visitors – failure is evidenced quite as much as success. Though stories abound of excursions by all manner of people from all parts of the country, and though these stories contain the seeds of a certain truth, there is still no doubt that those most systematically excluded were the ordinary workers, particularly those of the industrial North and Midlands. Their absence is attributable as much to a conspiracy of circumstance as to one of policy. Though both committee and government made tactical errors, the simple fact is that in the end a combination of high admissions charges (one shilling on the cheapest days), sporadic excursion services (some of the regional railway companies proved more enterprising than others) and non-Sunday opening (the Sabbatarians at it again) effectively precluded all but the most determined and relatively prosperous provincial workers from making the pilgrimage. That too was symbolic: out of sight, they were to all intents and purposes out of mind. When the odd stray appeared in the Palace, they appeared as emissaries from an alien world. As Lord Briggs has written:

'The Exhibition was designed to honour the working bees of the world's hive, but served instead to cast tacit reflection on the drones. This was the lesson – the workers of all types stand forth as really great men. But it was the workers who were excluded from the hive and the drones who, in Europe at least, created the pattern of their own decline.'

It was insensitivity compounded by ignorance; the unattractive obverse of the caring, prosperous and secure image which the

Exhibition sought to portray. As the correspondent of the *Manchester Guardian* reported from the midst of the British section:

'The viewing public displayed as much sly mirth from the display of clogs as worn by Lancashire and Yorkshire operatives as they did from the exotic designs of the Indian stalls.'

To be sure, nineteenth-century capital could produce its material wonders drawn from every habitable and even uninhabitable part of the globe and could, at the click of its fingers, assemble them on a given meridian in a palace of splendour: but it flourished in a society where the rewards remained ludicrously disproportionate in their distribution, the lot of the average citizen inconceivably hard and the whole accomplishment fatally tainted by the very circumstance of its achievement.

The Eclectic Review had suggested that the great farce might one day prove the opening of a great tragedy, and a trawl of the later-nineteenth century would produce any number of contenders for the title of prophesy fulfilled. A graphic if somewhat parochial harbinger was almost immediately to hand. Eighteen months after the Exhibition closed, a London newspaper was preoccupied by the trial of Thomas Bare. He and his family had been evicted, like so many others, as a consequence of the demolitions associated with the construction of the King's Cross railway terminus. He was an unskilled, casually employed labourer who, a witness argued, had become increasingly unbalanced as a result of watching helplessly the poverty and stress his family endured. It was the prosecution case that on the night of 15 October 1851 (three days after the Exhibition closed) he had taken a triangular saw file and stabbed to death his entire family. He had done it, counsel argued, 'for want of future, want of hope'.

If the Crystal Palace – later the 'People's Palace' – came to represent the best, the unchallenged and undisputed summit of High Victorianism, the symbol of achievement, 'the chalice of hope', it carried a significant burden of responsibility, heavier perhaps than any other. These weighty yet simple, austere yet assimilable qualities made it, to later generations, a symbol of awe. It had certainly not started out that way.

Afterword

Narrative is easy, evaluation contestable. So what in the end was it all about? More importantly, what is its real legacy to us beyond its availability for instantaneous tactical deployment at moments of national rediscovery? Simplistically, we can point to the spin-offs: the physical legacy of the Exhibition, its profits (£185,000 plus £100,000 worth of exhibits presented by their manufacturers) and point too to the institutions created out of them and still clustered around 'Exhibition Road' (the Victoria and Albert Museum, the Science Museum and Library, the Natural History Museum, the Imperial Institute, the Royal College of Science, the Royal School of Mines, the City and Guilds College, the Royal College of Art, the Royal College of Music). Of the Crystal Palace more specifically, we can speak too of the removal to Sydenham and its exciting though somewhat chequered career thereafter as the world's greatest festival site: the late-nineteenth century's Disneyland, home of circus, fireworks, sport and music.

But the political and cultural legacy is harder to pin down, though to confront it brings us full circle; back to where we began. Two issues still stand taller than the rest: the systematic impoverishment of the Palace's associations and their typically collaborationist thrust. When the Crystal Palace was fixed in the aspic of afterglow, a portrait of Victorian England was fixed with it, and unlike that of Dorian Gray the portrait remained beguilingly fresh – fixed in its Sunday best – whilst the society that it portrayed, never as sheerly beautiful as the representation implied, aged, multiplied, divided and developed. One issue is thus concern for the icon and for its society; for the subsequent eradication of the diversity which made them interesting, made them complex, made them tick. The second is concern for ourselves and for our own society, disturbingly revealed in the sad contrast to be drawn between the attitudes toward the building and rebuilding of the Palace evident in the two periods. In 1851 the Crystal Palace initiated a vigorous debate – best summarised as the 'condition of England' question. It does not begin to matter that many of the objections were quirky and wrong-headed, based upon false oppositions and unfounded, even manufactured, fears: what does matter is that there were critics, there was opposition, there was debate. In 1988 plans to resurrect it sparked no such opposition, stimulated no equivalent internal inventory, demanded no such political circumspection. What the proposal denuded of the past, it revealed as starkly denuded in *the present*. Architecturally, as *The Independent* would doubtless agree, the mercy is that the rebuilding never happened. Politically, the tragedy is that the debate did not either.

Happy Ever Afterword

This chapter was finally penned in late July and early August of 1991. As it neared completion, an article appeared in *The Times* of 7

August 1991 under the byline of 'Leisure-age Crystal Palace gets a Tudor pub and a bowling alley'. Bromley was at it again. Since no words of mine can do full justice either to the sweep of the scheme, or to the historical grasp displayed by the chair of Bromley Leisure Services Committee, I quote verbatim from the article:

'The Crystal Palace is to rise from the ashes, not so much a phoenix, more a popinjay. Paxton's 1851 glass exhibition hall is to be re-created in south London with a modern glass edifice featuring a 36-lane bowling alley, a Tudor theme pub and a three-star hotel.'

'Bromley Council, which owns the 12 acre site, has approved a £25 million plan by the Chester firm THI to develop the area left derelict since the original palace burnt down in 1936. But the building Prince Albert saw as a showcase of all that was great about Britain at the height of the Industrial Revolution is now aimed squarely at the entertainment market.'

'John Wykes, chairman of Bromley Council's leisure services committee, said: "The site has been derelict for 55 years; it has been defunct and gone for all that time. We are not putting on an exhibition for Britain now. That was Prince Albert's idea then, when we led the world with great engineers like Brunel whom we were proud of. We feel now that some use should be made of the site and the most suitable is a hotel."'

Paxton vobiscum.

Note

1 I am grateful to Christopher Greenwood for drawing my attention to *The Times* article of 7 August 1991 and grateful also to the Royal Geographical Society for ensuring that an earlier and even less satisfactory version of this paper was heard by so few people. Please note that this is an unreferenced paper, written very much as an interpretative essay.

5

Revisioning Place:
De- and Re-constructing the
Image of the Industrial City

BRIAVEL HOLCOMB

The Rise of the Place Marketing Industry

In the last dozen years the marketing of places, which has a long and colourful history in the United States, has been transformed from an activity which was essentially amateur (in the original sense of the word which derives from the Latin 'to love', meaning that places were extolled by local topophiles) to an increasingly profession-alised and costly manifestation of interplace competition. Place marketing in the United States today is a multi-billion dollar industry as a growing number of consultants and public relations firms specialise on the packaging, advertising and selling of cities, states, retirement communities and tourist resorts (Holcomb, 1990). Places are now commodities to be consumed. Though Laurence Alexander, President of Downtown Research and Development Institute in New York City (an urban economic marketing organisation), recently declared that cities are 'the most undermarketed product in America today', the city is obviously seen as a product to be even more aggressively sold in the future (Ashworth and Voogd, 1990).

The primary goal of the place marketer is to construct a new image of the place to replace either vague or negative images previously held by current or potential residents, investors and visitors. Place marketers tend to define their task in rather more idealistic terms as, for example, 'the awareness, the concepts and the tools that make us understand marketing as a dynamic process of society through which business enterprise is integrated productively with society's purposes and human values', or '[m]arketing is thus the process through which economy is integrated into society to serve human needs' (Drucker and Parlin, quoted in Bailey, 1989, p. 1). The profession distinguishes between marketing and mere selling. Marketing

'. . . is a philosophy, an ideology, and an orientation to business that seeks to distinguish itself from old-fashioned selling. Business leaders speak of the development of the "marketing revolution" or the "marketing concept" as a matter of transcendent significance. In this view, selling is a practice millennia-old but marketing is the opposite of selling. Selling, according to one leading text, is "trying to get the consumer to buy what you have". Marketing is "trying to have what the consumer wants. . . . Selling focuses on the needs of the seller, marketing on the needs of the buyer"' (Schudson, 1984, p. 29).

One approach of significance is that place marketers do not see their task as purely promoting and advertising, but also as adapting the 'product' (that is, the place) to be more desirable to the 'market'.

A category of places which perhaps best exemplifies both the construction of new images to replace the old, and the efforts to recreate places to be consonant with the preferred images, are the previously industrial cities now being restructured for the post-industrial, service economy. Glasgow epitomises these transformations from heavy industry to silicon glen and from the notorious Gorbals of my youth to the culture capital of the 1990s, its images equally transmuted through the agency of the 'Glasgow's Miles Better' campaign. In what a London *Sunday Times Magazine* article referred to as 'Glasgow's very American public relations campaign' (Jack, 1984, p. 37), Glasgow is celebrated as Britain's first major post-industrial success and bathes in a radiant glow of civic pride. Though some criticise the campaign as a 'grandiose exercise in self-delusion, a placebo which offers no cure for a terminally-ruined economy and the wretchedness of mass unemployment' (Jack, 1984, p. 38), what is undeniable is that the image of Glasgow, both within and outside the city, has been radically reconstructed.

A similar story can be told of some North American cities whose economies have been restructured and their images reconstructed with varying degrees of efficacy. Such cities include those as large as Baltimore, Detroit and Chicago, as small as New Brunswick (New Jersey), Columbus (Ohio) and New Britain (Connecticut), and as distressed as Newark (New Jersey) which had the highest indicators of poverty, unemployment and population loss of any of the largest fifty cities in the US in 1980, yet has recently attracted £6 billion in new investment and last year was ranked by *Money* magazine as the twenty-fourth best place to live in America. Two cities which perhaps best exemplify successful marketing campaigns are Pittsburgh and Cleveland. After briefly describing and illustrating their associated place marketing programmes, some of the issues that they raise are discussed.

Selling the Transformed Pittsburgh

In the late-nineteenth century Pittsburgh's moniker was the 'Forge of the Universe'. The Sheffield of America, Pittsburgh was described in a quotation variously attributed to James Parton, Lincoln Steffens and Charles Dickens (in three different newspaper articles circulated by the Greater Pittsburgh Office of Promotion!) as 'hell with the lid taken off'. In a description rendolent of Engels's Manchester, H. L. Mencken wrote as follows of Pittsburgh in the 1920s:

'Here was the very heart of industrial America, the centre of its most lucrative and characteristic activity, the boast and pride of the richest and grandest nation ever seen on earth – and here was a scene so dreadfully hideous, so intolerably bleak and forlorn that it reduced the whole aspiration of man to a macabre and depressing joke. Here was wealth beyond computation, almost beyond imagination – and here were human habitations so abominable that they would have disgraced a race of alley cats. I am not speaking of mere filth. One expects steel towns to be dirty. What I allude to is the unbroken and agonizing ugliness, the sheer revolting monstrousness, of every house in sight' (quoted in Gill, 1989, p. 74).

It might be said that industrial Pittsburgh had a bit of an image problem! In the 1940s Frank Lloyd Wright, when asked by the city's leadership how the city could be improved, replied: 'abandon it'. As recently as twenty years ago, Pittsburgh was known as a city where the smog was so bad that street lights stayed on all day and homemakers washed the curtains once a week. But today the headlines of media stories about the city read:

'Pittsburgh cleans up its act: for decades it was a grim, smoke-belching city. Now, minus steel, it's hailed as prime living space' (*New York Times*).

'A Pittsburgh visit clarifies demise of old grimy image' (*Allentown Morning Call*).

'Blooming in the rust: new uses transforming old mill towns' (*Los Angeles Times*).

'Transformed Pittsburgh goes from smokestacks to software' (*The Portland Oregonian*).

This transformed image is a product both of economic changes and of an energetic marketing campaign. In the period between 1979 and 1987 more than 67,000 jobs in the steel industry and 63,000 jobs in heavy manufacturing were lost in the Pittsburgh area as the domestic steel industry lost competitiveness with overseas sources. Companies closed overnight, leaving silent mills and virtual ghost towns in the Monongahela Valley. As people left in search of new employment, the city's population dropped below 400,000, from the 700,000 in 1950. However, Pittsburgh has a tradition of co-operation between wealthy Republican industrialists and a powerful Democratic city government. This coalition had instigated a major rebuilding of downtown, accompanied by programmes designed to clean the air and the rivers of the city, prior to the collapse of the local steel industry in the late-1970s. By 1982 the public–private partnership of the Allegheny Conference convened a commission of over two hundred local leaders to develop a strategy to move from a reliance on heavy industry toward a more diversified economic base.

The strategy was successful by several measures. A marketing agency was established to attract foreign and out-of-state companies; venture capital for the establishment of new enterprises was raised, a programme to assist local businesses in procuring Federal contracts was launched, and various funding and support programmes to attract R&D (Research and Development) and high technology enterprises were instituted. Between 1975 and 1987 over 100,000 jobs were created, mostly in education, health care and research. 'High tech' is now the metropolitan area's fastest growing industry. In 1988 only 22,000 were still employed in Pittsburgh's metropolitan area in the steel industry. These are included in the 134,000 total in manufacturing, compared to 786,000 in services. In that year the biggest steel company, USX, employed fewer people than the University of Pittsburgh (Pittsburgh Facts, 1989).

Part of the success of this economic recovery is attributable to the aggressive marketing campaigns of such organisations as The Greater Pittsburgh Office of Promotion, the Urban Redevelopment Authority of Pittsburgh, the Neighborhoods for Living Center (which markets the residential neighbourhoods of the city) and Penn's Southwest Association (a non profit-making regional economic development organisation). These organisations produce attractive brochures, information packages and videos which depict the assets of the city for business and for residence (see **Figure 1**). The campaign received a tremendous boost in 1985 when the *Rand McNally Places Rated Almanac* listed Pittsburgh as the 'most livable city in the United States'. Today, although Pittsburgh has slipped from first position in the most recent edition of the Almanac, one is still greeted at Pittsburgh Airport with huge banners and assorted gift shop paraphernalia (T-shirts, coffee mugs, key rings) proclaiming 'we're #1 !'. (In fact, Pittsburgh may be the only airport in America whose gift shop theme is not 'I ♥ X' as in the original 'I love New York' slogan). A typical piece of publicity is a glossy and attractively illustrated list of 'The Facts of Life in Pittsburgh, or 101 reasons why the grass is greener and the skies are bluer'. These include the claim that 'the quality of air in our fair city is about the same as that over the green, rolling hills of Louisville, Kentucky', that Pittsburgh ranks among the best in the nation for safety from personal crime, that the average cost of a new home in Pittsburgh is about 16% less than the national average, and that they have more golf courses *per capita* than any other city in

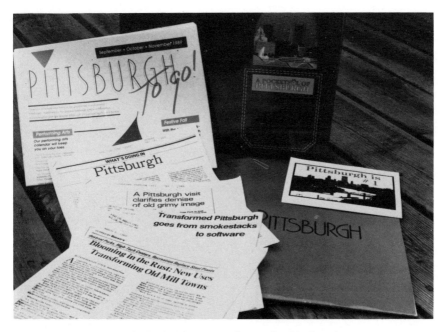

FIG. 1. Promoting a new image for Pittsburgh

the United States. It quotes the architectural critic of the *New York Times*, Paul Goldberger, as saying: 'This is the only city in America with an entrance. Pittsburgh is entered with glory and drama', and Jean-Jacques Servan-Schreiber, founding editor of *L'Express*, calling Pittsburgh 'the cradle of a world renaissance . . . the center of a new revolution, the knowledge revolution'. A frequently quoted *New Yorker* article claimed that if 'Pittsburgh were situated somewhere in the heart of Europe, tourists could eagerly journey hundreds of miles out of their way to visit it' (Gill, 1989, p. 74).

Marketing the Recreated Cleveland

Cleveland's 'rebirth' has been slightly less spectacular than that of Pittsburgh, but the transformation in its image is certainly as radical. During the 1970s Cleveland experienced a more profound psychic decline accompanying economic distress than did other ailing industrial cities. It became the butt of jokes even on national television. Cleveland was called 'the mistake on the lake' and 'the armpit of the nation'. Johnny Carson asked 'What's the closest you can come to a week in hell? A day in Cleveland.' Bad things kept happening to Cleveland. Not only did it have to contend with the conventional

FIG. 2. Promoting a new image for Cleveland

indicators of distress such as loss of employment (manufacturing employment in the city declined 59% between 1947 and 1982) and population loss (the city's population fell by 177,000 or 24% during the 1970s: most of those leaving were white, often middle-class taxpayers), but one summer the Cayahoga River caught fire! (Actually it was an oil slick which burned, thus reinforcing the image of a polluted river.) A few years later the mayor accidentally set his own hair on fire while wielding a blow torch at an opening ceremony of a factory (Lewis, 1988). The nadir of Cleveland's image was probably in 1978 when it became the first American city since the Great Depression to default on its loans.

The default marked the start of an increasingly sophisticated campaign to reverse the city's image. The Greater Cleveland Growth Association (alias the Chamber of Commerce) initiated the effort with conventional tactics and a local scope. They produced the usual bumper stickers and coffee mugs with 'I ♥ Cleveland' and a version from Cleveland of Steinberg's famous 'New Yorker's view of the world'. For a time Clevelanders bathed in purple as Cleveland took a leaf from New York's Big Apple slogan and declared 'Cleveland's a Plum'. Soon after the default, however, the publisher of the city's main newspaper, the *Plain Dealer*, began to raise funds for an organisation known as The New Cleveland Campaign whose mission was (and still is) to market the city and to improve its image. The public relations firm of Edward Howard and Company, a Hill and Knowlton affiliate and one of the most respected PR firms, was hired for the campaign. Glossy, sophisticated promotional materials have been produced and distributed nationwide and internationally (see **Figure 2**). One such package, entitled 'The Cleveland Perspective', contains two royal blue and purple brochures sheathed in a silver case, one asking rhetorically 'Are There Any Civilised Cities Left? A View from Cleveland' (which elaborates on the cultural and recreational amenities of the city) and the other asking 'Will the United States Economy still lead the World in 1990? An Emphatic Yes from Cleveland' (which describes Cleveland's economic climate and business advantages).

An innovative aspect of the New Cleveland Campaign is its segmentation of the market and its appeals to particular groups that it wishes to attract to the city. There is a special yuppie brochure, 'Cleveland: A City Guide for Young Professionals', which extolls the city as a place to work and play with photos of waterskiing on Lake Erie, sailplaning over the countryside, and TGIF (Thank God Its Friday) parties in the financial district. Another booklet is directed towards women, both spouses of relocated executives and professional women themselves. Since Merrill Lynch estimated that by 1990 75% of corporate relocations would involve dual career households, Cleveland's image as a place where professional and business women can find jobs and an attractive lifestyle is critical. The New Cleveland Campaign has been highly successful at 'placing' laudatory stories about Cleveland's rebirth in the national media, and in 1984 the Campaign won the highest award of the Public Relations Society of America, 'The Silver Anvil'. A telephone survey by the Citizen's League of Greater Cleveland in 1988 found that 82% of the 750 respondents thought that Cleveland's national image had improved in the last five years, and 76% were very or somewhat optimistic about the city's future.

This improvement in the city's image and self-confidence has been accompanied by redevelopment in Cleveland's downtown. New office, commercial and entertainment facilities include Playhouse Square, the largest theatre restoration project in the United States, a new Rock 'n Roll Hall of Fame, with plans for a project similar to Baltimore's Inner Harbor and a new domed stadium. The Flats industrial warehouse district has been quiched and ferned, and the city is gradually undergoing transformation from a gritty city of heavy industry to a corporate headquarters centre with banking, legal and professional support services.

Image Versus Reality

When one looks beneath the patina of the new images of Pittsburgh and Cleveland, however, a rather similar palimpsest of industrial economic decay and associated human suffering emerges. In Pittsburgh many of the young have left, and Allegheny County has a higher percentage of people over sixty-five years old than any county in the nation except the retirement haven of Dade County, Florida. Blacks form a quarter of Pittsburgh's population, but have benefited little from the revitalisation of the city. Black infant mortality rates in Pittsburgh in 1982 were the highest of any American city and nearly two and a half times greater than white infant mortality rates (Sbagria, 1989, p. 111). In 1984 Pittsburgh ranked second of the ten metropolitan areas with the highest black unemployment: '[t]he Pittsburgh metropolitan area had 26.8% black unemployment compared with 7.6% white unemployment . . . those for the Cleveland metropolitan area [another of the 'top ten'] were 18.8% and 7.2%' (Sbagria, 1989, p. 112). A report to the mayor in 1985 concluded that 'what we have here is a tale of two cities: one ranked number one for quality of life and the other number two for rate of black unemployment' (quoted in Sbagria, 1989, p. 112).

Similarly in Cleveland, despite the vastly improved image and the very real newly exposed or rebuilt bricks and mortar downtown, the city continues to lose jobs in basic industries which provide solid wages, unions, benefits and pensions. The downtown building boom is leaving older office space vacant. Crime and public school quality are major problems. Nearly 40% of Clevelanders live below the poverty line and Cleveland lives up (or down) to its reputation as the second most segregated city in America. More than 65% of the population of three black neighbourhoods on the East Side is in poverty (Keating, Krumholz and Metzger, 1989).

The rise of the place packaging industry has coincided with that of postmodernism as the prevailing theoretical fashion. Perhaps this is no mere coincidence. Postmodernism is a perspective, a way of seeing, a way of constructing an understanding of the world by deconstructing our experience of it. Postmodernism is eclectic. It juxtaposes, blends, splices, copies, combines, repeats ideas, attitudes, aesthetics and forms:

'It neither embraces nor criticises, but beholds the world blankly, with a knowingness that dissolves feeling and commitment into irony. It pulls the rug out from under itself, displaying an acute self-consciousness about the work's constructed nature. It takes pleasure in the play of surfaces, and derides the search for depth as mere nostalgia' (Gitlin, 1989, p. 52).

At first blush it might be assumed that place marketing, with its enthusiastic embrace of place, its appeal to the supposedly unique attractions of particular locations, and its passionate text, is anything but postmodern. Yet ultimately, the deconstructed discourses of the packaged newly post-industrial cities replicate the same images, amenities, and potentials and contain the same silences with respect to poverty, race and blight. The pastiche of upscale places is contextless: presumably intentionally so, since the fashionable fern bars are often in not-yet-completely gentrified neighbourhoods. The time of places marketed is present and future. The only past that matters is the packaged past of the heritage industry. The Homestead steel works, Andrew Carnegie's plant which once employed 15,000 people and supplied massive amounts of armour plate in World War Two, is now a waterslide amusement park and a light industrial park with street names like 'Open Hearth Avenue'. One building may be preserved as a historic landmark, the site of a bloody labour dispute in 1892, which is said to have set back the unionisation of steel workers for fifty years.

Characteristically postmodern in its indifference to consistency and continuity, the most striking new building in Pittsburgh is PPG Place (PPG standing for Pittsburgh Plate Glass, although the present owners are a pharmaceutical firm). Designed by John Burgee, its largely decorative towers are intended to evoke the Victorian Gothic of the Houses of Parliament. As Brendan Gill (1989, p. 74) remarked, the PPG building is 'a bizarre echoing . . . a pastiche of a building that is itself, after all, only a pastiche'. In Cleveland, the heavy machinery of the working waterfront is judiciously posed in proximity to jazz clubs and restaurants as the landscape of downscale production is replaced by one of upscale consumption.

Cleveland's marketing materials focus almost exclusively on the downtown and Flats area, with the only residential neighbourhood usually mentioned being that of the exclusive Shaker Heights. Pittsburgh, in contrast, is marketed as a mosaic of neighbourhoods, each of which is both unique but the same. The Neighborhoods for Living Center offers advice to newcomers in selecting a residence in 'a community which reflects some of your own interests . . . where the scenery satisfies your particular version of pleasant surroundings. Happily Pittsburgh has more than seventy neighbourhoods to choose from, *each* with unique qualities and *all* with a reputation for welcoming neighbours' ('A Pocketful of Pittsburgh', 1989, p. 21: emphases added). Ethnic tastes reflected in architecture or local restaurants are celebrated, especially if they are English, Irish, Welsh, Scottish or German. Race is not. A perusal of the panoply of neighbourhood pamphlets revealed no depiction of an African American adult. The 'blurb' on one predominantly Black neighbourhood (not so identified) notes that 'the chaotic sixties were a time of decline for inner city neighbourhoods, with many residents fleeing to the suburbs, [but] the sixties are long past and today Homewood-Brushton is experiencing far better times' ('A Pocketful of Pittsburgh', 1989, p. 29).

The Commodification of Place

A tautological relationship exists between the product and the packaging of post-industrial American cities. The packaged image reflects the aesthetic tastes of postmodern society, with its eclectic conformity, its fragmented palimpsest of past times and distant spaces, its commodified ethnic culture and sanitised classlessness. The city is rebuilt to conform to this increasingly international aesthetic so that, although the beer is better in Glasgow, the chablis and the spider plants are indistinguishable from those in both Cleveland and Pittsburgh. As Harvey wrote, 'aesthetics has triumphed over ethics as a prime focus of social and intellectual concern' (Harvey, 1989b, p. 328). Meanwhile, a laid-off steelworker watching the demolition of the Homestead steel mill has a plea. 'There are two Pittsburghs. . . .

There's the upscale Pittsburgh of the renaissance downtown, and there's this Pittsburgh that most people don't see. I'm not bitter about all this promotion of the new corporate image – that's good, I think they should do it. But they shouldn't turn their backs on the people who built the town' (quoted in Rause, 1989, p. 59).

6

The City as Commodity: The Contested Spaces of Urban Development

MARK GOODWIN

Introduction

The selling of cities is a process as old as commodification itself, beginning in earnest, according to Harvey (1989a, p. 15) with the civic boosterism of the Hanseatic League and the Italian City States in Medieval Europe. Such urban entrepreneurialism has gained in intensity and sophistication ever since, but it has constantly provided a key site where cultural and economic speculation intersect. This chapter will explore this intersection in contemporary London and Sheffield, and I will stress the local political practices that help to shape its specific unfolding in any particular place. Although the chapter concentrates on these two cases, it initially looks ▪t Los Angeles – a city which has been the subject of more cultural speculation than most. Partly because of this, it has recently been the focus of attention for geographers and cultural theorists alike (Baudrillard, 1988a; Davis, 1990; Soja, 1989), and hence offers us a chance to indicate some broad lines of inquiry at the outset.

Los Angeles and 'City Myths'

Los Angeles was the creation of real-estate capitalism, a place literally built by speculation at the beginning of this century, but this burst of economic speculation was underpinned by the production of a fictional landscape which portrayed Southern California as the 'promised land' (Davis, 1990, p. 20). According to Mayo:

'Los Angeles, it should be understood, is not a mere city. On the contrary, it is, and always has been since 1888, a *commodity*; something to be advertised and sold to the people of the United States like automobiles, cigarettes and mouth wash' (Mayo, quoted in Davis, 1990, p. 17).

The early growth of the city in fact 'required the continuous interaction of myth-making and literary invention with the crude promotion of land values' (Davis, 1990, p. 26), an interaction which was orchestrated by a set of developers, bankers and transport magnates, operating through the city's Chamber of Commerce.

Throughout the century which has followed the creation of this mythical cultural landscape, Los Angeles has been the subject of a continuous 'revisioning' (see Holcomb, this volume) through novels, films and canvas, as well as in the brochures of property developers. The most recent vision involves the promotion of the city as 'the arts mecca of the coming century' (Burnham, quoted in Davis, p. 70), a vision which relies on the incorporation of museums, libraries and arts centres into downtown redevelopment schemes. According to Davies (1990, p. 71):

'The large scale-developers and their financial allies have been the driving force behind the public–private coalition to build a cultural superstructure for Los Angeles's emergence as a "world city". . . . They have become so integrally involved in the organisation of high culture, not because of old-fashioned philanthropy, but because "culture" has become an important component of the land development process, as well as a crucial moment in the competition between different elites and regional centres. Old-fashioned material interest, in other words, drives the mega-developers to support the general cultural revalorisation of Los Angeles.'

In other words, Los Angeles has been the focus of a constantly changing '*city myth*' (Davis, 1990, p. 23), which has helped to fuel the material transformation of the landscape, and the images produced in and about the city have been vital components in shaping its transformation from small town to metropolis. It provides an excellent example of the ways in which cultural and economic speculation are inexorably intertwined in large-scale urban development. This intertwining needs to be kept firmly in mind, and it is important to

'avoid the idea that Los Angeles is ultimately just the mirror of Narcissus. . . . Beyond its myriad rhetorics and mirages, it can be presumed that the city actually exists' (Davis, 1990, p. 23).

And, moreover, it can be presumed that something is actually being sold. The selling of places is ultimately just that – and urban cultural capital is in the end valorised like any other form of capital. The rest of this chapter explores two recent examples of such development in Britain, those in London's Docklands and in Sheffield's Don Valley, and highlights the current processes of 'cultural revalorisation' which are central to the redevelopment of each. It is not so much concerned with those urban regeneration strategies based on cultural events or centred around spectacular projects *per se*, but seeks instead to identify the ways in which urban images or 'city myths' are produced in order to promote investment potential more generally. In looking at this intersection between the cultural and the economic, the chapter also highlights the political agencies which help to draw these twin aspects of speculation together.

The Selling of Cities

If Los Angeles is exemplary, it is by no means alone in being marketed and packaged like 'automobiles, cigarettes and mouth wash'. Several factors have recently combined to heighten the significance of urban 'place-marketing', and the production of the 'city myth'. Firstly, from the early-1970s many Western cities have had to cope with the triple problems of deindustrialisation, a falling tax base and declining public expenditure. Across Europe and North America urban governments have been faced with the same difficulties of reviving declining economies, regenerating derelict land and controlling an increase in social and political tension. The key contextual question, in the words of Harvey (1985, p. 213), was 'how could urban regions blessed largely with a demand-side heritage adapt to a supply-side world?' Areas that had flourished in the post-war Keynesian boom, based on demand-led growth, now went into a frenzy of competition to improve their respective positions within the spatial division of both production and consumption. The 'speculative construction of place' (Harvey, 1989b, p. 8) moved centre-stage in this intense competition.

Secondly, however, it did so against the background of widespread cultural and aesthetic change. Economic change has weakened what might be termed traditional or mass industrial culture. According to Stuart Hall (1988, p. 24), society has entered a 'Brave New World' consisting of

'greater fragmentation and pluralism, the weakening of older collective solidarities and block identities and the emergence of new identities associated with greater work flexibility (and) the maximisation of individual choice through personal consumption'.

In this new world, the emergence of style as identity (Hebdidge, 1979, 1988) and consumption as a form of self-definition (Baudrillard, 1988b) have gone hand-in-hand with an increasingly reflexive and aesthetic appreciation hastened by the proliferation of visual images and electronic media. The selling of an urban lifestyle thus becomes part and parcel of an increasingly sophisticated commodification of everyday life, in which images and myths are relentlessly packaged and presented until they become 'hyperreal' (Clarke, 1991; Eco, 1986), whereby any distinction between the 'real' and the 'representation' is effaced.

Thirdly, the last decade has witnessed the re-emergence of political structures and ideologies which are based around the notions of privatisation and deregulation, twin processes which supposedly promote the unfettered operation of so-called 'market forces' (see Cloke, 1992). The infrastructure of urban government is becoming increasingly privatised, along with the ideologies and discourses of regeneration and revitalisation. Where public agencies were once seen as an essential part of the solution to any urban crisis, they are now viewed as part of the problem itself. Hence, market forces are promoted as 'the only possible way' of reviving urban economies (Tom King, Minister for Local Government, quoted in Duncan and Goodwin, 1987, p. 127), and new political agencies have taken the place of local authorities in many urban areas to ensure that such rhetoric is put into practice.

Changing Urban Cultures and the Restructuring of Urban Space

The economic, cultural and political processes which have combined to promote the conscious marketing of urban space are only a part of the myriad processes which help constantly to transform and to remake the urban landscape. In the words of Neil Smith (1986, p. 21):

'A given built environment expresses specific patterns of production and reproduction, consumption and circulation, and as these patterns change, so does the geographical patterning of the built environment.'

The social and spatial landscape of any city is constantly changing as urban geographies are continually developed, abandoned and restructured (see Harvey, 1982, 1985; and see Smith, 1984, on the geographical implications of uneven capitalist accumulation).

Often these urban environments will be formed into particular territories through what Harvey (1985, p. 146) calls the 'structured coherence' of production and consumption. Processes and infrastructures, both physical and spatial, combine to produce a particular space suitable for a specific route in producing surplus value. Forms and technologies of production, inter-industry linkages, markets of labour supply and demand, patterns of consumption, standards of living, social hierarchies: all of these seemingly interact to produce such coherence. Hence the appearance, so Harvey argues, of distinct urban regions with distinctive local labour markets in which labour power with particular specialisms can be substituted on a daily basis. Both the London Docks and the Don Valley in Sheffield were once at the centre of such regions of 'structured coherence'.

This coherence is often cemented, and reinforced, through the creation of fixed and relatively immobile spatial configurations. But these in turn can act as a barrier to further accumulation. In Harvey's words (1985, p. 150):

'Capitalism builds a physical and social landscape in its own image, appropriate to its own condition at a particular moment in time, only to have to revolutionise that landscape, usually in the course of crises of creative destruction at a subsequent point in time.'

Again, both the London Docks and Sheffield's Don Valley present prime examples of this 'creative destruction', and both areas are in the midst of complete physical and social change. But such change does not just happen as the automatic imperative of some technically-determined economic process. It may be a truism that capital is a social relation, and not a 'thing', but the conceptualisation is none the less important for this. It is of course people who create and destroy these urban geographies, even if they do not always do so in freedom from others or in conditions of their own choosing. People are able to monitor and learn from their experiences, and they build economic, social and political organisations to protect and to further their interests. The urban region is thus more than a simple coherence of production and consumption (and even this is never guaranteed). It is a complex collection of individuals and communities, which in certain instances develop particular regional and local cultures, formed by social relations and practices outside of capital's narrow logic. Together these movements and cultures can be important in helping to sustain or to destroy the coherence of a particular place. The 'building' and 'revolutionising' of an urban landscape is thus never just physical and economic: it is also social, cultural and political, and changes in these processes can play a vital role in easing economic transformation and helping to form a new round of coherence (and maybe a new patchwork of local coherences).

Urban change, then, is never pre-given, or guaranteed, but instead is actively shaped by competing social forces. It is in this competition that the intersection of cultural and economic speculation plays such a crucial role. The promotion of new urban images, of new lifestyles and of new 'city myths', is often a necessary prelude to the establishment of new urban economies. Importantly, however, the formation of these new images themselves is an issue of challenge and contestation, an issue which is often fought through particular political agencies and institutions. As my two examples will show, 'places do not have single, unique "identities"' (Massey, 1991, p. 29). They can have multiple meanings, and indeed do. A closed steel works, or an abandoned dockyard, means something completely different to a redundant worker than it does to a property developer or to a local politician. As will be shown, these meanings can be a potent force in the material transformation of the city, especially when they are produced and marketed through powerful development agencies. As the London Docklands Urban Development Corporation noted, 'the poor . . . image of the area as the backyard of London keeps land values depressed' (LDDC, 1982, p. 4). Depressed land values of course prevent economic speculation and regeneration, and hence new urban images are crucial in promoting such regeneration. I will now explore this intersection of the cultural and the economic in more detail, by examining the latest attempts to transform the 'structured coherences' of the London Docks and Sheffield's Don Valley, beginning with the latter.

The Restructuring of Sheffield's Don Valley

The Don Valley was for over a century the centre of Sheffield's metal and heavy engineering industries. The growth of the heavy steel industry around the turn of this century was accompanied by the development of a distinct social and political community in the city's east end, huddled along the slopes of the Don Valley around the blazing steel works and furnaces. These workers and their families were to earn Sheffield the soubriquet 'Steel City', a title reflected in the fact that by 1971, out of a total city workforce of 300,000, over 80,000 worked in metal goods, engineering and manufacturing. In the mid-1970s 45,000 people were employed in the Don Valley alone, three-quarters in metal manufacture and mechanical engineering. But between January 1980 and January 1983 41,000 redundancies were announced in the city, 35,000 of them in manufacturing. This swift and dramatic decline left only 16% of the workforce in mining and metal goods manufacture by 1988 (Goodwin, 1986a; Lawless, 1991; Lawless and Ramsden, 1990). The contraction in the steel and engineering industries, which employed the majority of skilled male workers in Sheffield, had important political and cultural ramifications. These male workers had been able to dominate local politics via their industrial strength, thereby excluding other groups and other politics. In the evocative words of Campbell (1989):

'Sheffield's labour movement was its fifth estate – it was a proletarian city with an amicable division of labour between the Labour Party, the Communist Party, and the trade unions. Labourism has defined this city's political ambience, securely and seamlessly, for decades – a monument to traditionalism.'

The secure and seamless ambience of politics described by Campbell can be interpreted as part of a localised coherence of economy, society, politics and culture which held for much of the post-war period. With the collapse of its economic foundations, however, that coherence became unstable. Local councillors who had grown up and learnt their politics in a labour movement founded upon the dominance of Sheffield's skilled workers in the steel and engineering industries, now had to formulate a response as this dominance quite literally crumbled to the ground. Their first move was a predictable one, as they sought to regenerate the local economy through protecting the interests of the skilled male working class by fighting factory closures in Sheffield's traditional industries. In the Don Valley the local authority tried to preserve Sheffield's 'landscape for labour' (Campbell, 1989) by re-affirming earlier plans which had zoned the area for continued industrial development. In successive Draft District Plans of 1979 and 1981 all suitable land was allocated for general industrial use, and the 1981 plan identified an additional 460 acres of land as having industrial potential. The 1979 plan did accept that manufacturing employment in the Valley may fall, but stressed that a probable loss of 2,500 jobs by the mid-1980s 'was not sufficient to alter the basic assumption that all suitable land would be allocated for industrial development' (Lawless and Ramsden, 1990, p. 39).

These land use plans were given a political twist in the early 1980s when the ruling Labour Party on Sheffield City Council underwent a marked radicalisation (Duncan and Goodwin, 1985a, 1985b; Goodwin, 1986a, 1989; Seyd, 1990). Among its policy initiatives was the setting up of a totally new Employment Committee, serviced by its own Employment Department, committed to exploring 'different aspects of the employment problem, and to try out innovative solutions' (The Labour Party, 1982, p. 20). The new Department had an explicit brief to create employment, but was also set up to help 'build a confident, local working-class movement' (Sheffield City Council, 1981). Hence the initial stress on regenerating industrial activity, as a key route to maintaining the strength and confidence of the city's labour movement. We see here the idea of economic regeneration being allied to political recomposition, resulting in cultural speculation of a particular hue. As David Blunkett, then Leader of Sheffield City Council, put it (1982, p. 13),

'councillors . . . can . . . be involved in changing people's perceptions of the reasons for economic policies, and with it put back some of the fire and direction which raises people's aspirations for a socialist society.'

The Department's overall approach was based around the idea of 'restructuring for labour' (Goodwin, 1989), whereby economic intervention resourced by the City Council would seek to improve the position of labour – its working conditions, pay and organisation – and enlarge public accountability at the same time as it created jobs. These ideas fed into an alternative strategy for the Don Valley, one that was

'based on three principles which contrast sharply with those underlying Enterprise Zones: (i) planned growth led by the public sector – in contrast to any further removal of democratic controls over private development and industry; (ii) regeneration of the local economy under local determination and control, with wide consultation from the bottom up, releasing the resources and skills of the community; (iii) an emphasis on direct intervention to preserve existing employment and create new jobs, in contrast to an indirect approach through the property market' (Sheffield City Council, 1982, p. 11).

But such plans proved easier to put on paper than into practice, especially given the amount of redundancies and unemployment experienced by the city in the early 1980s. As one officer of the Council remarked:

'the new scheme was supposedly built around specific employment-led initiatives, but no-one was able to find these, and the more we tried to beef-up the employment aspects the more the scheme became property-led. . . . The lesson of the Lower Don is that rhetoric is no replacement for the hard facts of economic life' (interview with author, 1985).

The irony is that rhetoric *was* soon to be a major part of economic life, as the Council's attempts to regenerate the Don Valley took a markedly different turn. Increased central government pressure on the city's finances which reduced the Council's room for independent manoeuvre, coupled with continued local economic decline, caused a reconsideration of policy. The Conservative's second successive General Election victory in 1983 led, in the words of an employment officer, to a 'dramatic rethink' (interview with author, 1985). It was recognised that private sector resources were essential to any future regeneration, and hence the Council's relationship with the private sector 'flipped over from confrontation to partnership' (interview with author, 1985; see also Duncan and Goodwin, 1985a, 1985b; Lawless, 1990, 1991; Seyd, 1990). In 1986 the Council helped to set up Sheffield Economic Regeneration Committee (SERC), comprising thirty members drawn from the Council, Chamber of Commerce, central government departments, the Council for Racial Equality, the city's University and Polytechnic, the local trade unions, the Church, Members of Parliament and representatives from local industry (Cochrane, 1991; Lawless, 1991). This new agency was to be of crucial importance in promoting a new type of urban development. In 1987 SERC funded a consultants' report on the regeneration of the Don Valley which proposed implementing regeneration via six flagship developments costing some £320m, of which £250m would come from the private sector. 'Hence, just three years on from the 1984 Environment and Employment Plan, with its emphasis on public sector intervention and control, a new approach was to emerge based much more on market initiatives' (Lawless and Ramsden, 1990, p. 40). The following year central government agreed to fund regeneration, but only to the sum of £50m over seven years, and only if this was controlled by a centrally imposed and unelected Urban Development Corporation (UDC) (see below for further details on the initial UDC legislation; see also Sadler, this volume).

The Council is now attempting to make its influence felt as strongly as possible with the UDC, and has signed an agreement setting out a joint code of practice for intervention (*Local Government Chronicle*, 18 August 1989). But such intervention seems increasingly property-led, with less and less emphasis on the nature and amount of employment created. The original notions of restructuring for labour, so prevalent when the Employment Department was set up in 1981, have been firmly abandoned. Where the steel works of the Don once produced most of the country's special steels, derelict sites are now being converted into 'domed palaces of leisure and consumerism' (Halsall, 1989). In 1990 the Meadowhall Retail Centre was opened on the site of a former steel works, at a cost of more than £235m. Described variously as 'the largest retail and leisure complex in Europe' (Lawless, 1991, p. 6) and as 'the hottest retail centre on both sides of the Atlantic' (Gardner and Sheppard, 1989, p. 112), the size and cost of the Centre has the following rationale:

'The traditional mall no longer has a competitive edge. As North American experience shows, the retail environment must provide more than shopping. It must be a dramatic celebratory space, a three-dimensional fantasy for everyone' (Ron McCarthy, project consultant, quoted in Gardner and Sheppard, 1989, p. 113).

Although the employment potential of such schemes is open to question (Lowe, 1991), the development is a vital component in the attempt by SERC to promote a new image for the city. Indeed, three leaders of the city's planning team, commenting on the Council's rigorous assessment of the scheme, stated that '[t]he final decision was that . . . this flagship development would boost the city's image and regeneration' (Bradley, Mercer and Sturch, 1990, p. 17). Image and regeneration intersect here in the same breath, the former seen as crucial to the success of the latter. The 'celebratory space' of the 'three-dimensional fantasy' is thus seen by the Council and by SERC as more than employment and shops. Its completion, on the site of a redundant steel works, was necessary to serve as a flagship for the city's new image, replacing both that of 'smoke, scrapyards and steel' and that of a derelict 'industrial moonscape' (Bradley, 1990a, p. 6).

The other 'flagship' and 'catalyst' of the 'economic regeneration strategy' (Darke, 1991, p. 6) is the World Student Games. In 1986 the Council had actively sought, and then accepted, an invitation to stage the 16th Universiade. This was a deliberate strategy of leisure-led regeneration, working not through the employment potential of the Games themselves, but via the enhanced image of the city that they would present. As the Director of the Games Administration has said:

'We had a choice, garden festival or leisure facilities. We needed something to demonstrate that Sheffield's future was secure, that there were people who believed in the continuing life of a great city' (Willis, quoted in Darke, 1991, p. 4).

It was decided that the physical presence of the new facilities, as well as the event itself, could help this demonstration of belief, and the new cultural script was accordingly written around the Games. The Head of Planning Strategy in the Council has remarked that '[a]n exciting new international sports stadium rising on the site of a former steel works in the emerging East End Park is a symbol of Sheffield's strategy to harness the growth in leisure to combat industrial decline and its social and physical side effects' (Bradley, 1990b, p. 8). Whether the Games will be anything other than a symbol, by bringing sustained benefit to the city, is open to question. In terms of type, number and cost, the direct employment created by the Games seems disappointing (see Darke, 1991: Foley, 1991; Lawless, 1991). In addition to the central controls over Council policy and finance must now be added cutbacks necessitated by debt charges on the £150m that the city borrowed to build the Games stadia (Darke, 1991; Lawless, 1991). But, if the direct benefits are still uncertain, the role of symbol and catalyst remains crucial. The Council may realise that the privatised provision of leisure and shopping will never replace the employment lost when the industrial sector collapsed, but it is also conscious of the need to promote the city through prime-time television slots marketed around international sporting events. In this way the Games can help throw off the city's continuing industrial 'clothcap' image. Thus, 'the 1991 Universiade provides a multi-project programme that would contribute to the city's new image, new economy and environmental renewal' (Bradley, 1990b, p. 8). Image and economy are placed together again. As Foley puts it, '[t]he legacy of excellent sporting facilities after the Games together with Meadowhall should enable Sheffield to stand out from the clamour of towns and cities claiming to be different' (1991, p. 13). Representatives of the Regeneration Committee certainly stress the importance of the images presented by the old and new Don Valley, comparing the former corridors of blazing steel with the lights of the new athletics stadium and shopping centre. 'Each of these spectacles presents an image' (Adsetts, 1991). The newer images are seen as a vital component in the city's strategy to attract external investment and rebuild internal confidence.

According to Peter Price, Deputy Leader of Sheffield City Council, the new images are necessary because the collapse of the steel industry, together with the political confrontations of the 1980s, 'left an image which frightened any potential investor' (Price, 1991). The flagship projects were needed to 'raise the image of the city', 'build confidence', 'boost civic pride', 'promote the city' and 'gain prime-time television exposure' (Price, 1990). In this manner they contribute to changing not just the economy of the area, but also its politics and culture. It is not only the economy of the previous coherence that has to be altered. The legacy of the radical politics of confrontation with central government, and the working-class culture of steel workers and their families, also have to be replaced. But for this to happen, specific local political changes have had to take place as well. The initial plans for the Don Valley at the beginning of the 1980s stressed industrial provision, and continued to view manufacturing industry as the bedrock of any regeneration strategy for good political reasons. That this has now changed is not just due to the severity of economic decline. It is also because of the changing nature of local politics and of local political agencies. Thus, where a Council strategy initially saw publicly-controlled economic change as part of a platform from which to rebuild a bruised and battered labour movement, an appointed Urban Development Corporation now works alongside a regeneration committee to promote property-led private investment. The increased importance given to the conscious marketing of the city's new image must be seen in the context of these political changes, whereby 'the anti-capitalist ethos of the early 1980s was replaced by a new strategy of collaboration with local capital' (Seyd, 1990, p. 339). These new images of Sheffield are hence themselves the result of political contestation, and are not the automatic response to economic change.

The Restructuring of London's Docklands

My second example of the intersection of cultural and economic speculation is provided by the recent attempts to regenerate London's Docklands. Like the Don Valley, the Docklands once provided an example of a localised 'structured coherence', in this case one centred on the enclosed dockyards themselves and on a whole complex of port-related industrial activity. As in Sheffield, the labour movement dominated the politics of the area, and the physical segregation of people and place in the bends of the River Thames only added to the sense of community and cohesiveness. As late as 1976 a survey revealed that out of the 6,210 economically active adults living on one of the Dock peninsulas, the Isle of Dogs, 87.5% were in manual employment, and in 1971 83% of all the housing on the Island was still provided by the local authority (Goodwin, 1986b, pp. 14–15). But this community, again like Sheffield's, was to experience rapid change as the Docks closed and as employment declined. Dock employment in the area dropped from 23,000 in 1967, to 7,000 in 1979, to a few hundred in the mid-1980s (Goodwin, 1986b, p. 20), and the related processing and manufacturing industries also suffered a dramatic decline (Brownill, 1990, p. 87). The Docks themselves were closed over a fifteen-year period from 1967 to 1981, a process which left huge swathes of derelict land a mile or two from the centre of London. This prompted planning concern, and in 1974 the Docklands Joint Committee (DJC) was established to draw up and implement a Docklands-wide regeneration strategy (Brownill, 1990; Goodwin, 1986b, 1991).

The DJC, as its name implies, was a joint committee composed of eight representatives from the Greater London Council (GLC), eight from the five Dockland Boroughs, and eight co-optees covering local interest groups. The Committee was delegated powers of strategic planning and development control, but it was never an executive agency and only had advisory and coordinating functions. It had no budget or staff of its own, and its main task was to draw up a framework for redevelopment within which the five Boroughs could situate their own local plans. The DJC hired a team of planners, many seconded from the Boroughs, to use

'the opportunity provided by large areas of London's Docklands becoming available for development to redress the housing, social, environmental, employment, economic and communications deficiencies of the Docklands area and the parent Boroughs and thereby to provide the freedom for similar improvements throughout East and Inner London' (Greater London Council, 1982, p. 31).

The aim was thus to improve rather than to replace existing employment, housing and recreational activities in the area, and to aim these improvements at local people rather than creating a saleable image for outsiders to consume. Work began on the London Docklands Strategic Plan (LDSP) in 1974, and a draft was produced in April 1976. Following extensive local consultation, the final version of the plan was published in June 1976.

Although not a statutory plan, the LDSP sought to guide development within its broad framework for the next twenty-one years. The major components of this strategy were to be a large public sector housing programme, additional manufacturing industry and improved transport infrastructure so that links with the rest of London and the South East could be developed. Both employment and housing were to be matched to the needs and skills of the local population, and office and service employment was seen as secondary to continued industrial growth. But the plan was unable to reverse, or to compete with, broader political and economic trends. The DJC did have some success in coordinating Dockland development (Dockland Boroughs, 1979), but in retrospect its actions were always slightly too little and too late. Just as implementation of the plan began, the DJC ran foul of central and local public expenditure cuts, and with each successive budget in the late-1970s the aims of the plan retreated a little further into the distance. Political as well as economic factors also began to conspire against success, and gradually the Conservative Party began to win the levers of power. In 1977 they gained control of the GLC, and in 1979 returned to Westminster, leaving the five Labour-controlled local authorities isolated in a committee with hostile partners. The hostility was soon felt, and the DJC came under increasing criticism for not achieving the goals of its own strategic plan. Although three years' operation out of a proposed twenty-one does not allow much time for implementation, especially when access to land and finance is limited, the new Government promptly stepped in to impose a new development agency on the Dockland Boroughs.

In September 1979, less than six months after their General Election victory, the Conservatives announced their intention to set up Urban Development Corporations in Merseyside and London Docklands. The intention was made real in the 1980 Local Government, Planning and Land Act (see Duncan and Goodwin, 1987; Goodwin 1986b, for details of the legislative process). In contrast to the DJC, the new body was given executive powers to buy, sell and vest land, and it also gained development functions from the Boroughs. According to the 1980 Act, a UDC could be set up if the Secretary of State considered it 'expedient in the national interest' to secure the 'regeneration' of its area (Goodwin, 1986b, p. 7). It could achieve this by aquiring, managing, reclaiming and disposing of land; carrying out building; providing infrastructure and 'generally do[ing] anything necessary or expedient for the purpose of the object, or for purposes incidental to those purposes' (Goodwin, 1986b, p. 7). It could therefore do anything it felt would lead to regeneration, yet neither 'regeneration' nor the 'national interest' were anywhere defined. In operating these wide powers in Docklands, the UDC emerges as accountable only to Parliament, and is run by a non-elected board appointed by the Secretary of State. In evidence to a House of Lords' enquiry on the setting up of the London Docklands Development Corporation (LDDC) a government representative elaborated on the reasons for these powers, when he asserted that:

'what the government is concerned with here . . . is their assessment which is, if you like a political, economic, social assessment . . . that what is needed in Docklands is private sector confidence which . . . would come with the UDC, and a very large measure of private housing which would come with the UDC and not otherwise, and that they want, as it were, direct government control through a government nominated body – the UDC – for the spending of those large sums in that area' (Goodwin, 1986b, p. 9).

Regeneration is thus based on restoring private sector confidence, by encouraging investment through the preparation and marketing of land and infrastructure. Public subsidy is used as a prime element in this property value 'hype' (Klausner, 1987), but also crucial is the creation and subsequent marketing of a new image for the area. From the outset, the LDDC had identified an 'image problem', and in its initial Corporate Plan (1982, p. 14) it noted that it must change 'the traditional perception of Docklands as a second-rate market'. As Reg Ward, the Chief Executive of the LDDC put it, '[w]e needed to get people to see Docklands with different eyes' (*Financial Times*, 17 July 1986). Docklands thus ceased to be a place and became 'a state of mind' (Davis, 1987, quoted in Brownill, 1990, p. 48). Cultural change, and the building of a mythical urban landscape, became crucial in promoting the 'image', the 'perception' and hence the value of the area.

This process operated at several levels and in a variety of ways (see Brownill, 1990; Goodwin, 1991). At a general level, the area has been promoted as an 'imaginative symbol of the vitality and enterprise of our age' (LDDC, quoted in Goodwin, 1991, p. 271). Indeed, the hyperbole of the LDDC is heavy with images of 'new beginnings', 'new opportunities' and 'renaissance', as the following example shows:

'In the heart of London new life is being pumped into the arteries of the nation's capital. A new living, working and leisure environment born of opportunity is being created. Just as the renaissance heralded the modern world, culminating in an outpouring of creative talent and new thought, so now, in its own way, a new freedom of expression in London's Docklands is fashioning a wholly new environment which heralds the 21st century' (LDDC broadsheet, *The Vision*).

Extensive marketing exercises have been carried out along similar lines, and the image of a run-down derelict industrial landscape has been replaced by futuristic visions of 'the most innovative and important urban regeneration project in the world' (LDDC, 1989, p. 3). £21m were spent on promotion by the LDDC from 1981 to 1989

(Brownill, 1990, p. 40), and through this expenditure 'the richness and diversity of the specific localities of Docklands have been reduced to a commodity to be packaged and sold' (Burgess and Wood, 1988, p. 115).

Specific policy initiatives by the LDDC were also designed to promote the image of the area, in particular its support for private sector housing. In spite of assuring the House of Lords' enquiry that the LDDC's housing intentions were for a tenure mix of 50% owner-occupation, 25% rented and 25% shared ownership, by 1989 housing starts in the area were 85%, 12% and 3% respectively. This is because the LDDC sees owner-occupation as 'the spearhead of the redevelopment of Docklands. It is the most important route to changing the perspective of the area' (LDDC Chief Executive, *Roof*, July/August, 1982, p. 17). Housing is thus seen as 'part of the regeneration task whereas the Docklands Joint Committee looked at housing solely in terms of need' (LDDC Chief Executive, quoted in Ambrose, 1986, p. 225). Again, changing the perspective and image is seen as an important route to regeneration, and housing is seen as one key route to achieving this. This is because housing is not only part of a physical and economic regeneration, but it also helps to change the society and community of an area, another task seen as vital by the LDDC. The initial Corporate Plan (LDDC, 1982, pp. 4–5) emphasised that two of the main problems to be overcome in 'The Regeneration Task' are first, 'too many unskilled workers and old people and second the limitations of existing labour skills, geared to traditional industries (which) are unlikely in the short term to adapt to the needs of potential new industries'. The introduction to this Corporate Plan noted that '[h]ousing was identified as our first priority and our achievements have been highly significant in taking the first steps in repopulating Docklands'. Presumably the need for 'repopulation' was somewhat lost on the 40,000 people who already lived in Docklands.

It was important to the LDDC, however, for changes in the housing and labour markets of the area are the twin spearheads of a cumulative process of speculatively-led development (Goodwin, 1991). In the Corporation's own words:

'We have no land use plan or grand design; our plans are essential marketing images' (*The Times*, 18 November 1986).

And these new images are dependent on changes in both housing and employment. Thus, instead of major developments fitting into a planning framework, such as the LDSP, 'the market place has the opportunity of influencing how and where a development takes place and what form it takes' (LDDC Chief Executive, quoted in Brownill,

1990, p. 53). Planning is replaced by a marketing exercise, and land is simply sold to the highest bidder. To facilitate this the LDDC produced 'area frameworks' detailing infrastructure work, design guidelines and possible uses for major sites. The LDDC saw these as 'marketing tools to secure maximum private investment' (LDDC, 1982, p. 34). But for such a sale to be realised, speculative development must be promoted and underpinned through the building of an extensive 'city myth'; one which re-envisions Docklands, not as a living community of working-class residents surrounded by the derelict properties in which they used to work, but as a symbolic new environment of opportunity and vitality. The area has effectively been emptied of its past meaning by the LDDC, and then recomposed as an aid to financial and property speculation.

Such images, though, are not uncontested. From the setting up of the Joint Docklands Action Group in 1973, the area has seen a proliferation of locally-based community groups at a variety of scales mobilising over a wide range of issues covering, for instance, housing, employment, leisure and recreation, social facilities, transport and childcare (Brownill, 1990; Docklands Consultative Committee, 1988; Klausner, 1985). Protest groups have been formed to fight single issues in specific places, and others act as umbrella organisations for the whole area. The Docklands Forum, for instance, has over sixty affiliated organisations from the local area: and these local groups have explicitly contested the LDDC's vision of Docklands. In the Royal Docks, to give an example, the LDDC saw its provision of a Short Take Off and Landing Airport (STOLport) as 'the type of catalyst needed to change the normal rules of the market place, a project capable of taking advantage of the apparent disabilities of the area in development terms and converting them into major assets' (quoted in Brownill, 1990, p. 53). Its plans for the area thus amounted to the usual marketing images, in this case centred around the STOLport. In opposition to this the local community working through the Newham Docklands Forum and with the support of the GLC produced a 'People's Plan' setting out alternative development plans based on the principle of 'meeting needs and making jobs', and using ideas drawn from the experiences of local people (Brownill, 1990; Klausner, 1985). The alternative plan was presented to the STOLport inquiry, causing the LDDC lawyer to exclaim '[a] people's plan, ridiculous. I've never heard of such a thing' (quoted in Brownill, 1990, p. 127). The plan's employment proposals centred on re-opening the Docks for cargo-handling, ship repair and maritime workshops, and warehouse and storage facilities, thereby utilising local skills; and its housing component was based on large amounts of rented housing, thereby meeting local needs. Brownill concludes that:

'examples of community action and planning alternatives . . . show clearly that the LDDC's was not the only planning philosophy in operation. Yet the LDDC had power, money and land to ensure that its ideology became the dominant one.' (1990, p. 131).

As in Sheffield, new political structures and new agencies have been vital in winning the contest over alternative urban images, and in promoting new cultural landscapes. In each case a planned, community-led version of development has been overturned in favour of one which relies on image and perception to build confidence in speculative developments.

Concluding Comments

The LDDC views the transformations that it has helped to shape in the following manner:

'The original construction of these docks in the 19th century provides the precedent for such an audacious expansion. Now, as then, they are an imaginative response to technological progress, a symbol of the vitality and enterprise of our age' (quoted in Carr and Weir, 1986, p. 7).

For us to understand them, however, these developments must be seen as more than a symbolic and imaginative response. The nineteenth-century development of the Docks, and indeed the steel mills of Sheffield's Don Valley, should be viewed as component parts of the economic and social restructuring of British capitalism. Both environments were specifically created to take advantage of an expanding trading empire at home and abroad. When international changes in economy and society led to a decline in this empire, both environments had outlived their usefulness. The specific localised coherences that Docklands and the Don Valley represented had in fact become as redundant as the thousands of people who once worked there. What is now being witnessed in each city is the rise of another distinctive landscape, not yet coherent in either place, but one based on the service sector rather than on shipping and manufacturing. Such a change has taken place in many other cities, as this volume shows, but the general processes at work have been mediated in London and Sheffield by particular political agencies and relations operating in each place. In each case the local authorities have had planning and development powers removed and handed to an unelected institution. Effectively, an appointed agency is, in each case, replacing the powers of local government in order to carry out a market-led regeneration of each inner city.

But such regeneration is bolstered by the type of cultural speculation which promotes mythical landscapes and new images of each place in order to build speculative confidence. It is necessary to stress the link between economy, image and culture in underpinning an area's localised coherence. Such coherence is partly moulded, and can be broken, by a changing culture and by a transformed 'city myth'. Successive rounds of investment and accumulation are spatially and temporally mediated by distinct cultural relations, but these in turn are the subject of contestation and conflict, and – as the two examples show – often have to be secured through particular political agencies and practices. The transfer from one localised coherence to another is thus fraught with difficulty, and attempts will be made to ease such a transition through the introduction of new political structures and the production of new cultural landscapes. This is what is currently taking place in Sheffield and London, and the outcome of the process is still uncertain and open to further political, social and cultural struggle.

7

Place Marketing:
A Local Authority Perspective

ANDREW DAVID FRETTER[1]

Introduction

Successive United Kingdom governments have been actively involved in selling places for many years through the Invest in Britain Bureau and the British Tourist Authority. National governments have also helped to establish a regional network of dedicated organisations including the Welsh Development Agency, the Scottish Development Agency, regional inward investment agencies in the Midlands and North of England, and Regional Tourist Boards throughout the United Kingdom. Similarly, nearly all local authorities outside of the prosperous South-East England (and some within it) have become involved to a greater or lesser extent in the selling of their areas. This article draws particularly on my experience and knowledge of Birmingham in the English Midlands and Gwent in South Wales, to present a pragmatic local authority perspective.

The Evolution of Local Authority Place Marketing

The great growth in local authority promotional activities can be traced back to the mid-1970s when economic recession started to hit the traditionally successful manufacturing and mining areas of the Midlands, the North and South Wales. Before this, the city 'fathers' (and they were by and large 'fathers') and the planning profession saw their role primarily as guiding the development of their areas through the provision of basic public infrastructure (roads, schools, leisure facilities, council housing) and the use of planning controls to direct and to limit private sector development.

The 1960s and early-1970s brought relative wealth, prosperity and low employment to most parts of Britain, and this reactive approach

163

was therefore generally sufficient. However, from the mid-1970s onwards, increasing competition from imports and shrinking home markets, together with inability to compete in overseas markets, caused many traditional manufacturing sectors to decline at a scale unprecedented in post-war years. The effects were often concentrated because of traditional grouping of certain industrial sectors. For example, rapid decline in clothing and footwear industries in Leicester, motor vehicle industries in Birmingham and coal and steel in South Wales brought devastating effects in the late-1970s and early-1980s. Many other traditional manufacturing sectors were similarly affected, bringing widespread deprivation.

In many respects local authorities were unprepared to deal with the situation. They did not have the experience, the expertise nor the resources to respond effectively or quickly enough to prevent great devastation. Many areas surrounding the prosperous town centres and central business districts became run-down or completely derelict. Unemployment tended to become concentrated in particular wards of inner cities or in the mining and steel towns of North England, Scotland and the South Wales valleys. As a result, these areas became generally unable to attract new investment and the spiral of decline continued.

For example, Birmingham lost 200,000 jobs in the thirteen years prior to 1984, more than the whole of Scotland and Wales put together! Hundreds of acres became derelict in areas that only a decade before had been the manufacturing heartland of the country. Investment in industrial and commercial property, in plant and machinery, in research and development and, perhaps most significantly of all, in training, had declined to appallingly low levels. Unemployment was over 20% in 1985, with over 100,000 people officially unemployed. This figure was over 40% in a number of inner-city wards. In South Wales, meanwhile, 90,000 jobs were lost between 1978 and 1987. Manufacturing industry was particularly badly affected, losing nearly 70,000 jobs, a drop of 33%. The number of jobs in coal mining fell by 70%, with a loss of 18,000 jobs between 1980 and 1988 as twenty-three collieries in the region closed. The number employed in the major South Wales steelworks (Llanwern and Port Talbot) fell by 13,500 between 1978 and 1988, a drop of 61%.

Many of the early responses of local authorities and central government were unsophisticated and piecemeal in their nature. This is perhaps not surprising, as there was no 'conventional wisdom'. Ideas, in the main, had never been tried. There was therefore great scope for novel ideas and new approaches, which from a practising professional's point of view made it a very exciting and rewarding time.

Immediate responses majored on business support, often grant aid and basic business advice, particularly aimed at encouraging start-ups, along with property development, including advance factories. But from the mid-1970s onwards the selling of places was starting to become big business. For many years this focused on the simple promotion of the attributes of a particular place to attract either tourists or inward investors. It used the basic tools of promotion – literature, advertisements, videos and exhibitions – but rarely in a targeted fashion. More recently, though, these activities have 'come of age'. The 1989 Local Government and Housing Act has forced local authorities to prepare a programme of their economic development activities and to consult the local community on its content. Whilst many of the more enlightened authorities were already taking a strategic and corporate approach to economic development work, this Act has had the effect of encouraging all to do so. The requirement to consult is focusing minds more directly on the needs of their customers or clients, or, to put it another way, is encouraging local authorities to take a market-led approach.

At the same time 'marketing' is starting to replace the concept of merely 'selling' (see also the chapter by Holcomb). Selling is trying to get the customer to buy what you have, whereas marketing meets the needs of the customer profitably (or, in local government terms, 'efficiently' at best value for money). Inevitably, this requires a much more sophisticated and more comprehensive approach affecting many local authority functions. Place marketing has thus become much more than merely selling the area to attract mobile companies or tourists. It can now be viewed as a fundamental part of planning, a fundamental part of guiding the development of places in a desired fashion.

It should aim to ensure that urban activities and facilities are related as closely as possible to the demands and desires of targeted customers and clients. It involves the establishment of a new relationship between the local authority (members and officers) and its customers. It calls for a demand-orientated approach rather than the supply-led approach of traditional urban planning. As such, it requires a more *flexible* approach to development plans. Above all, it requires the pro-active pursuit of the desirable rather than the reactive prevention of the undesirable.

The Essential Elements of Place Marketing

Vision

The first prerequisite of successful place marketing is a clear understanding of what is desirable, of what you want to achieve.

165

There is a need to look beyond the short-term to develop a long-term vision of what you want your city, town, county, region or whatever to be. But more than this, the vision must be shared by all those who have a stake in the outcome. This involves everyone within the 'lead' organisation, every other organisation with related objectives, functions, or responsibilities, and all elements of the community. The best way to secure a shared vision is undoubtedly a broad participation in its development, but in practice the local authority responsible for a particular area will probably have to take the lead. This could be the County or District Council, although ideally the vision should still be established jointly or at least with some 'local' agreement.

In Birmingham, the vision of the city 'fathers' was embedded in tradition. Birmingham has always been a proud city: proud of its heritage, proud of its products, proud of the spirit of enterprise of its people, and proud of its tradition of success (see the chapter by Lowe). The city has been a world leader for over two hundred years. Many of the leading lights of the Industrial Revolution lived and worked in Birmingham. It was the first United Kingdom city to have public gas lighting, produced the first copper coinage, had the first letter copying press, the first secondary school, the first municipal bank and the first municipal orchestra of any industrial city. By the late-nineteenth century, under Joseph Chamberlain, the city had become widely known as the best governed city in the world. It was therefore unsurprising that this brash city grabbed new ideas, intent on being the first to develop them. Hence, Birmingham wholeheartedly took to Buchanan's ideas expounded in 'Traffic in Towns' and built a massive motorway-width inner-ring road, reflecting too its other tradition of being 'motor city'. Thereafter, the city 'fathers' enthusiastically implemented the other false panacea of the 1960s of high rise flats. Similarly, in 1964 the Bull Ring – now due to be demolished – was the first indoor shopping centre to be opened in the United Kingdom. But this bold 'traditional' vision permeated all of the organisations in the city, and was supported by many of its people. Birmingham still sets out to be 'the first, the biggest and the best'.

Gwent, on the other hand, has had a more complicated history. Prior to 1974 what is now Gwent was the old County of Monmouthshire plus the County Borough of Newport. Monmouthshire itself could never make up its mind as to whether it was English or Welsh. Its motto *Utrique Fidelis*, 'loyal to both', reflects this. The geographical diversity of the county adds to its divisions. The beautiful rolling hills of the Wye and Usk valleys, together with the rugged beauty of the Brecon Beacons, sit alongside the old mining and steel communities of Gwent valleys and the port and industrial town of Newport. Although the last coal mine closed in 1989, the fabric of the buildings and the attitudes of the people still reflect this heritage.

Gwent is also more complicated because of the two-tier local government system, with one County Council and five Borough Councils. Add to this the strong influence of both the Welsh Development Agency and the Welsh Office and it is clear that one organisation's vision will not be sufficient to guide marketing activities. A shared vision thus becomes essential and the processes by which this is achieved of paramount importance.

Recognising this, Gwent County Council initiated a joint market research exercise entitled 'Gwent Image and Target Markets'. All Borough Councils, the Welsh Development Agency and a number of private sector companies are joint partners. Only once we have a shared understanding of our strengths and weaknesses can one hope to develop a shared vision and work together to achieve it. Alongside this exercise has been a major review of the 'Gwent Structure Plan', setting the framework for future development to the year 2006. Again, consultation, working where possible towards a consensus view, has been the aim.

Whilst there will probably never be total agreement on the details of development and priorities, I believe that we have now developed broad agreement on a shared vision: a vigorous but selective quality growth strategy for the county. Hence, within Gwent County Council's new 'Economic Strategy', the mission is to 'work with all other organisations to make Gwent one of the major growth centres in the UK over the next five years, whilst protecting and enhancing the environment'.

Know Yourself

One of the foundations of success for marketing any commodity, be it a product or a service, is to know exactly what you have to offer. In trying to attract inward investment or tourists, local authorities are offering both a 'product' in terms of all of the attributes of the area (land, communications, skilled labour, training facilities, quality of life) and the service that they (and their partners) can offer to potential investors and tourists.

An honest and comprehensive analysis of strengths, weaknesses, opportunities and threats is essential. In practice, it is very difficult for someone working so closely with the 'product' to analyse it objectively. Independent consultants will often be helpful. Only when you know your strengths and weaknesses will you begin to know whether you have what the customer wants. But it is not only actual strengths and weaknesses that have to be analysed, it is also perceived strengths and weaknesses. How do your potential customers see you?

This is undoubtedly one of the main hurdles for both Birmingham and Gwent. Both have major image problems which are, to some extent at least, no longer justified. Image has not caught up with reality. Birmingham is making great efforts to rid itself of its concrete jungle image, of its heavy and dirty industrial past, the stranglehold of its inner ring road and its dependence on vehicle manufacturing. It is difficult, however, when no less an architectural expert than Prince Charles (!) describes the new Hyatt Hotel, next to the International Convention Centre, as a 'missile silo' and the 1970s Central Library as an 'incinerator'. Similarly, supposedly unbiased quality of life surveys continue to place Birmingham near the bottom of the league table, and hence do little to inspire economic confidence and inward investment.

Gwent and the South Wales valleys as a whole are no longer dominated by mining, steel industries and slag heaps. Indeed, as stated earlier, the last mine in Gwent closed in 1989 and the major greening operation of all of the valleys is almost complete. Employment opportunities in these areas have been transformed by inward investment, much of it from overseas companies making high-technology products. For example, Gwent now has fifty-one overseas companies including seven Japanese, fifteen EC and twenty-five from the United States. Yet the area still tends to be depicted in the 'How Green Was My Valley' idiom in television dramas, with all Welsh people so often portrayed as beer-swilling, singing male rugby players. Old habits die hard!

Define Your Customers

Knowing who is your real target audience, your real customer, is essential to formulating a successful marketing strategy. A local authority's customers will be extremely diverse, and it is not sufficient to define your customers merely as mobile companies and tourists. Indigenous businesses and population are also relevant customers. It is no use attracting new industry if you cannot keep what you already have. Helping them and inward investors to attract key workers must form an essential element of your marketing strategy. Furthermore, if local people and businesses are proud to live, work and hopefully prosper in your area, then they will be your greatest ambassadors in helping to increase awareness, improve image and understanding of opportunities. You need to know not only which industries are most mobile today, but which will be tomorrow. What type of industry is most likely to be attracted by your product? And, most important of all, do you want it?

Once you have decided which sectors are most likely to be attracted, you need to decide whether you stand any chance of attracting complete company moves, or just branch plants. Will it include their higher value headquarter, research and development and marketing operations? Where are these companies currently located and who are they? Who within the organisation should you be aiming to influence and how? This targeted direct-marketing approach is likely to achieve much greater success than the scattergun approach of advertising and exhibitions, but only if the original market research is of a high quality.

A further essential set of customers are the influencers of company moves, investments and developments: namely, the estate and re-location agents, the banks, financial institutions, accountancy and consultancy firms. They probably have a different level of current awareness and have different needs. Who are they? What are their needs? How can you influence them? These questions need precise answers. Only once you have defined your customers can you understand their needs, prejudices and motivations. Only then can you present your product in the best possible way, or start to improve your product so that you stand a better chance of attracting the right kind of investment in the future.

Adapt and Improve Your Product to Customer Requirements

Once you have clearly defined your desired customers and their needs, you may have recognised deficiencies in your current product. Many aspects of the product will take a long time to improve, but others can be changed more quickly. Significant product improvements include infrastructure, investment, training and skills development, and property availability.

Other product improvements can be more fundamental and reposition a place into an entirely new market. For example, once Birmingham's 'managers' had defined one of their prime customers as business tourists, they set about creating the product: firstly the National Exhibition Centre in 1976, and more recently the £148 million International Convention Centre which opened to great acclaim on 2 April 1991. Supported by massive investment and growth at Birmingham's International Airport, these developments have totally changed the market within which Birmingham competes. Originally understocked with hotel bed spaces, there have been twenty-four major new hotel developments over the past few years, including Hyatt and Sheraton.

Not to be outdone, Gwent is also repositioning its place in the inward investment market with both the 'Second Severn Crossing', which will be open in 1995, and the 'Usk Barrage' and associated development in Newport, which will totally transform the 'Usk Riverfront' over the next five years. Garden Festival Wales, held in Ebbw Vale in 1992, attracted almost two million visitors, most of whom went home with a dynamic and positive image of Gwent, but more importantly it marked the reclamation and redevelopment of one of the last sites of dereliction in the county. Private sector investment, most notably in the proposed Severnside International Airport, is adding impetus to a significant repositioning of the Gwent product, which will enable Gwent to attract higher value inward investment, with better quality development and jobs. Headquarter operations with research and development and marketing functions can now be realistically targeted.

Know Your Competitors

Nearly all local authorities and a number of 'big gun' regional bodies are extremely active in trying to attract investment. Yet too little time is often allocated to understanding the competition. How can you hope to win business if you do not understand how to beat the very people who will be your greatest threat, not only in competing for new business but also in potentially wooing away what you already have?

An area's competitors will depend not only on the similarity of products but also on the choice of customers. Who else is targeting the same industrial sectors? What can they offer that you cannot? Better location, cheaper labour, better grants, lower manufacturing costs, more skilled labour, better quality of life? How are they improving their product? What can you offer that they cannot? Answering such questions will usually require detailed market research, which is probably best done by experts in the field. For example, in South Wales, the local authorities, including Gwent County Council, together with the Welsh Development Agency have recently launched the 'Information Technology (IT) Wales' initiative. But this was only done after extensive research undertaken by consultants into the IT relocation market to identify our own strengths and weaknesses against those of our competitors. What factors influence the relocation decisions of IT companies? In what order of priority? How do Gwent and South-East Wales compare with more traditional IT locations and competing areas? What are the chances of success? What sub-sectors of IT are likely to bear the best fruit? Who are the key targets? This type of detailed market analysis will make a significant difference, not only to how we approach the market but also to how, over time, we adapt our product to fit the needs of our desired customers.

Find a Real Point of Difference

Having defined yourself, your customers and your competitors you must find or create a real point of difference. Why should your targeted customer move to your area rather than to that of your competitors? What are your unique selling points?

This is again standard marketing practice, but is an area where most local authorities fall down completely. Just look at an assortment of adverts, brochures or videos. If you took the name off the front, would you know the place that was being described? Almost everywhere is described as the centre of England, the United States or Europe; everywhere has a track record of success and has excellent communications, as well as a skilled and adaptable workforce with a high quality of life (see the chapter by Sadler). You must offer a real point of difference which will make your product stand out in the customer's mind. It should give a clear simple reason why you should be the area chosen rather than anyone else's. You must make yourself distinguishable from the crowd.

This will often be reflected in a logo. For instance, Birmingham is 'the Business City'; Leicestershire is 'the successful location'; Gwent is 'the first County in South Wales' and 'the Gateway to South Wales'. Perhaps the most famous and most successful example is the 'Peterborough Effect'. Unique selling points can also be reflected through campaign themes to get over more precise messages: for instance, in Birmingham, 'the Investment is Working'. But it is very difficult to summarise an area's unique selling points in such a short, sharp way. Selling places is somewhat more complex than selling washing powder!

Here in Gwent we have now adopted a novel campaign theme of 'the Perfect World'. This is not meant to be a completely 'honest' description, as clearly nowhere can be perfect: rather, it is a tongue-in-cheek way of trying to grab the potential customer's attention, to make our promotional material stand out from the crowd. Having hopefully grabbed the customer's attention, we can hammer home our defined unique selling points.

Whilst unique selling points will usually initially be described as *attributes*, they should be presented as *benefits* to the customer. For example, Revlon do not make cosmetics, they sell hope; Black and Decker do not describe themselves as selling drills, but rather marketing holes; domestic coal is not sold as an efficient heat producer, but rather as creating a romantic and cosy ambience. Similarly, Gwent is not being sold as a low-cost location with high government grants which could be viewed as negatives, but rather as a successful, highly profitable location which provides a very high quality of life. These are the benefits that a company receives if it relocates to Gwent.

Having defined the benefits derived through your unique selling points, you must then describe them fairly and honestly. There is nothing more damaging than giving promises that you cannot deliver. Disappointed customers are often lost for ever, and your credibility in the market place will take a lot of winning back. Clearly, there is always a temptation to promise too much. All areas have a tendency to ignore the bad points and emphasise the good, but if this goes too far – and is widely recognised as mere hype – more harm than good will be done. You must always ensure that reality and created expectations meet.

One Voice

Within any area there is usually more than one organisation responsible for promoting awareness, image or opportunities of that area. It is tempting to believe that the more voices there are, the louder and more potent the message: but unless they are all saying the same thing in a co-ordinated manner, more harm than good can be done. Confused messages weaken the argument. On the other hand, if the recipients receive one clear message from a number of sources, not only are they more likely to believe it but they are also likely to be impressed by the fact that a particular 'place' has 'got its act together'.

Again, this is where so many areas have been let down. One would think that where you have such a strong unitary authority such as in Birmingham, it would be easier to ensure that everyone speaks with one voice. The City Council has certainly been very active in the 'image stakes'. Each of their major development projects have their own marketing campaigns – the National Exhibition Centre, the International Convention Centre, National Indoor Arena, International Airport and Heartlands. The Council spent £1.5 million on the Olympics bid in the mid-1980s; it runs the Super Prix each August, the only international motor race in Britain on the public highway, and they also run three major local festivals – a media and TV Festival, a Readers and Writers Festival and a Jazz Festival. The Birmingham Convention and Visitor Bureau, one of the biggest in the country, actively markets the area for leisure and business tourism. And the Economic Development Department runs a targeted campaign to attract inward investment, which also involves the promotion of lifestyle to attract key workers, whilst other departments promote their own services.

Birmingham, then, is tackling its image problem on a broad front; from tourism to lifestyle and culture, sport to economic development. On the face of it, this must be a good thing. Yet it is also, I believe, one of the barriers to further improvement in Birmingham's image. There are simply too many images and messages being put across. Whilst this may not cause too great a problem if the target markets are different for each campaign, it can lead to confusion where audiences overlap. At the moment there is no one clear, consistent message being put across like 'Glasgow's Miles Better', providing an umbrella under which the individually targeted campaigns can fit. The City Council has started to tackle this issue through a major piece of market research which will examine all of the markets within which Birmingham is trying to position itself, and will examine all customers, users of the facilities (residents and visitors), and all potential investors (mobile companies, existing companies, developers, financial institutions, and influencers).

The problem is even more complicated in Gwent. Not only is the County Council actively involved in promoting the county, but there are five Borough Councils, the Welsh Development Agency, the Welsh Office, the Wales Tourist Board and now the Training and Enterprise Council, all of whom have roles to play. Taken together, these organisations have spent several million pounds promoting awareness, image and opportunities in Gwent and Wales generally. Yet market research has shown that awareness of Gwent is very low, and that the image, if it has one at all, is indeed based on out-of-date associations of South-East Wales as a mining and steel area. Great efforts are now being made to ensure that we present a clear co-ordinated message of what Gwent has to offer, and a series of joint marketing initiatives are being developed. It is too early to say how successful these will be, but at last, I believe, we are starting to ensure that we are all fighting on the same side.

Conclusion

Local authority place marketing in the United Kingdom is now a big and sophisticated business, with many similarities to the marketing of any product or service in the private sector.

It involves presenting positive images to boost private-sector confidence and the direct selling of relocation and development opportunities. However, it is still only a part of a successful planning and economic development strategy for an area. Inward investment will not solve all the problems. Indeed, in many respects, it can only ever be a short-term strategy. The high levels of inward investment, especially from the United States, Japan and Germany, cannot continue for ever. Increasing competition for this investment and the globalisation of markets may well turn against the United Kingdom. Indigenous industry and commerce will always form the backbone of a local economy, and their needs and potentials must not be ignored. It is their growth which will provide a more stable self-sustaining economy.

We must also guard against *privatism*, whereby the private sector is seen as the principal or even the only agent of urban change. Private sector investment will always be concentrated in the more attractive urban areas where profits can be maximised. Poorer areas and the people within them (unless they have the right skills) will be ignored. The public sector and the local community themselves will always be the main agents of urban change in these areas. The selling of cities must be seen as merely one part of a balanced approach to ensure that our urban areas are developed in a desired fashion.

Note

1 Andrew David Fretter is Head of Economic Development at Gwent County Council. He was previously Head of Marketing in the Economic Development Unit at Birmingham City Council. However, the views that he expresses in this article do not necessarily reflect the views of his current or previous employers.

8

Place-marketing, Competitive Places and the Construction of Hegemony in Britain in the 1980s

DAVID SADLER

Introduction

In the course of the 1970s and 1980s social science came increasingly to focus upon the significance of *place*, in part as one (largely belated) reflection of the real social and political changes associated with accelerated internationalisation of economic activity. At a time of intensified global restructuring, the roles which particular cities and regions might play in a rapidly shifting world order were more and more evidently open to question. This was especially apparent to politicians as they struggled to develop local and/or national economic strategies in the face of some powerful international currents. The objectives and limitations of such policies (glossing over, for the moment, their diversity) were arguably expressed most clearly in the competitive marketing of the benefits to potential investors of one locality as against any other. A process of 'place-marketing' became increasingly significant, through which the competitive ethos of the market place became translated into a burgeoning 'place-market' (Robinson and Sadler, 1985). Many localities effectively competed with each other within the constraints of a capitalist economy for a share, however meagre, of the investment and jobs apparently on offer, whether from internationally mobile capital, the creation of a new generation of small businesses, or any other source. This marketing operation involved the construction or selective tailoring of particular images of place, which enmeshed with the dynamics of the global economy and legitimised particular conceptions of what were 'appropriate' state policy responses.

In the United States, for instance, much attention was focused on the notion of the city as a 'growth machine' capable of influencing distributional processes to secure differential advantage. For Molotch (1976), the prime imperative of any locality was growth. This objective provided the basis for consensus amongst a politically-motivated local elite, which typically consisted of local businesses and local financial services associated with land. These groupings were orchestrated via 'boosterism', the aggressive marketing of a city or region through activities such as advertising and sports sponsorship, in an attempt to gain a greater share of finite national prosperity. In this fashion, he argued, 'organised effort to affect the outcome of growth distribution is the essence of local government as a dynamic political force' (Molotch, 1976, p. 313). In a similar vein, Logan (1978) considered the competition of places to be a significant cause in the differentiation of society, so that the stratification of place ranked with class and status as bases for collective action.

For Cox and Mair (1989: reviewing a further statement by Logan and Molotch, 1987), however, this 'growth machine' concept required substantial revision. They criticised the privileging of distribution relative to production (implicit in the argument that growth coalitions sought to gain a share of fixed national growth), and argued instead for an approach which started from the notion of 'local dependence', a relation to locality resulting from relative spatial immobility. This spatial constraint, they argued, was a necessary precondition for the formation of local business coalitions. In many ways their argument matched Harvey's concern (see, for instance, 1982) for the analysis of capital as relation not object, and in particular his emphasis upon the place-specific character of devaluation. More specifically, Harvey (1989a) examined the transition from a 'managerial' approach to urban governance in the 1960s to an 'entrepreneurial' one in the 1970s and 1980s. Basing his analysis on a case-study of Baltimore, he identified the speculative character of public–private partnership and the way in which urban entrepreneurialism adds to, or facilitates, the geographical freedom of capitalist investment. Similarly, for Swyngedouw (1989, p. 31), the new rhetoric of 'the local' had to be seen as deeply embedded in processes of global accumulation, in a 'fragmented mosaic of uneven development in which competitive places try to secure a lucrative development niche'. This altered terrain of spatial policies was not just potentially deeply unstable, but also associated with highly selective cultural representations. Entrepreneurialism, for Harvey (1989a, p. 14), opened up and depended upon a range of mechanisms for social control: 'the ideology of locality, place and community becomes central to the political rhetoric of urban governance which concentrates on the idea of togetherness in defence against a hostile and threatening world of international competition'. The emphasis in such accounts was hence upon place-based forms of political organisation which are vulnerable (and themselves contribute) to capital's global ebb and flow.

Precisely *how* such place-based ruling-class alliances are constructed, and how they maintain ideological dominance or hegemony over labour, is another question. For, as Cox (1989, p. 81) argued (and to stress the point again), labour is particularly vulnerable to growth coalition ideologies which are 'potent tools in the hands of capital'. Yet it is by no means obvious how this condition has arisen, and why labour should be so amenable to the mythology of the local, capitalist, solution. Answering such questions is one of the main tasks of this chapter. But first, lest there be any misunderstanding, it should be stressed what this chapter is *not* about. It is not meant to be taken as an argument that people living in such places became 'cultural dupes', part of some grand capitalist scheme. Nor is it intended as a statement against local bases of political organisation as a foundation for more progressive forms of policy formulation, nor as a suggestion that such locally-constituted political practices are necessarily reactionary: far from it. It remains the case, though, that place-based coalitions of interest targeted at economic renewal via a particular kind of programme, one emphasising capital's success in the contest for meanings of 'place', have been particularly common in recent times. If capital is not always to determine such issues, then it is instructive to examine precisely how it structures (indeed, arguably, rigs) the agenda, just as it is important to examine the lessons of other kinds of locally-based political forms which have attempted to re-write the script. The analysis in this chapter therefore takes production as its central starting point, and investigates in greater depth some of the processes which underpin class alliance formation and hegemony through the terrain of place. It seeks to identify some of the mechanics of place-marketing which have underpinned capital's hegemonic thrust.

Such issues are addressed with particular reference to the goals of successive post-1979 Conservative governments in the United Kingdom. These objectives were based on the power of the market, backed up if necessary by a highly coercive state. A national political environment was created wherein the role of business assumed a heightened ideological significance. Put simply, capital gained a clear ascendancy over organised labour (and it is in this sense that the chapter describes a move towards, rather than the definitive establishment of, a new form of hegemony; one that is perhaps partial and incomplete but certainly was a distinctive feature of 1980s-style politics). Yet the dominance of the Conservative Party in the 1980s was associated too with a massive social, economic and political split within the country. The clearest North–South divide of all (accepting the fuzzy geographical expression of this concept for the moment) was in political terms. The evidence below is drawn from one side of this divide, the old industrial regions such as northern England, South Wales and parts of the Midlands. There, the issues of restructuring and transition were most compelling and intractable, and the selling of places was most actively pursued as part of an attempt to impose an alternative form of hegemony to traditional, working-class values. There, the new ideology faced some of its most deeply rooted challenges. The marketing of place and its relation to broader questions of political strategy and social change were therefore particularly apparent.

Three components to the tentative construction of a newly-emergent hegemony via the selling of place and places are identified (and note again that place-marketing was only one, albeit important, element in this process). The first of these components involved packaging places almost as a commodity to be bought and sold – not just their physical existence as land, but also their historical and cultural significance. This is demonstrated through an examination of the advertising publicity generated by Urban Development Corporations (which were created in 1981, 1987 and 1988 as a key prop to the Conservative government's philosophy of self-help regeneration), focusing in particular on its manipulative version of the past. The character of place-marketing activity, then, in terms of promotion and publicity, both concealed and revealed some of its political origins and purposes. This close investigation of some of the surface forms of place-marketing is then followed by a consideration of the widespread disregard or disdain for alternative policy directions and for other conceptions of place. 'There is no alternative' became a catchphrase of the 1980s, but one with some very real and disturbing implications. Place-marketing, it will be shown, rested very heavily upon a deliberate and morally questionable exclusion of alternative analysis and critique. Finally the day-to-day 'art of politics' is analysed, to exemplify and evaluate the maintenance of coalitions within which place-marketing strategies evolved. This is ultimately a central task in this chapter: to re-assert, through the re-integration of place with political strategy, the significance of the *political* to the social construction of place.

Place-marketing:
The Packaging and the Selling of Places

'The marketing programme is being driven by in-depth research and analysis into key industrial sectors with a view to clearly understanding the perception of the "product" that is Sheffield' (Sheffield Development Corporation publicity, 1989).

One route through which dominant class images can be constructed, reinforced and replicated is the medium of advertising. Yet there has been surprisingly little academic analysis devoted to the advertising industry in general, despite its significance in the contemporary capitalist world. Total worldwide expenditure on advertising was estimated at $150 billion in 1987, with the largest agencies (and even larger holding groups) operating truly global networks (Perry 1990). The advertising industry has concentrated on the effectiveness of its campaigns in terms of marketing techniques, but much less academic research has been undertaken into the growth of the industry and its impact in spatial terms, although there are some indications that these may be significant. For example, Clarke and Bradford (1989) analysed the use of space by advertising agencies in the United Kingdom. Whilst the agencies investigated reported that they operated largely in aspatial terms, Clarke and Bradford argued that the concentration of advertising in higher spending areas (through, for instance, the buying of media time), and (less frequently) the 'tailoring' of advertisements to particular regional situations, did have spatially uneven implications. More generally, though, the whole question of spatial variation in advertising impact and delivery has been relatively neglected, despite recent emphases on the growing geographical significance of the service sector.

On the other hand, there have been several analyses of the 'messages' encoded within the advertising of particular places or regions. Such promotional issues are in many senses the classic component of local economic development strategies (see especially Ward, 1990). For example, economic downturn in the later years of the nineteenth century encouraged Luton Council to initiate a new industries programme as far back as 1899. By 1939 85% of the county boroughs and 35% of the municipal boroughs and urban districts in the United Kingdom were engaged in some kind of promotional or development work. Place-advertising in the United Kingdom was also investigated in the late-1970s by Burgess, who found two common themes in such literature:

'first, the economic benefits that would be available to companies locating in the area and second, the substantial improvement in the quality of life that would be experienced by executives and key personnel who decide to move with the company' (Burgess, 1982, p. 7).

Creating a particular image therefore posed quite a few problems: one local government officer was reported as saying that 'many of the things that I have said about this town, I have said about elsewhere' (quoted in Burgess, 1982, p. 6).

A more recent trend, though, has been towards the distinctive

179

packaging of particular localities around a series of real or imagined cultural traditions. In the 1980s South Tyneside became 'Catherine Cookson country' in honour of the fiction writer born locally, whilst Middlesbrough similarly sold itself as 'Captain Cook country'. The London Docklands Development Corporation set out to create a major publicity campaign based upon images of the East-End drawn not from 'real-life', but from the manipulation of popular television. From their analysis of this, Burgess and Wood concluded that:

'places have become products offering emotional and economic benefits to their "consumers". Thus, the richness and diversity of the specific localities within East London have been reduced to a commodity to be packaged and sold' (Burgess and Wood, 1989, p. 115).

Such packaging was important, not least because it could point to the objectives of those doing the parcelling. In this context the example of the London Docklands Development Corporation (created in 1981) was doubly significant: as the focus of considerable national political commitment by the Conservative Party, and as a forerunner of a further round of Urban Development Corporations which were designated in 1987 and 1988. The advertising of these later organisations repays close attention.[1]

Apart from obvious promotional devices such as the availability of government grants and good transport links, three common themes were evident. The first rested upon an appropriation of industrial heritage as something to be fostered, emulated or re-kindled. This was enshrined in many slogans. The Black Country looked forward to 'an industrious future from an industrial past' whilst 'the talents of Teesside stem[med] from a century-long industrial and commercial operation'. Bristol was 'rich in industrial history', Trafford Park had seen the 'unstoppable surge of industry', Tyne and Wear 'produced some of the great entrepreneurs of the last century', Manchester was once 'the most important industrial city in the world', Sheffield 'used to be the greatest steel-producing centre in the world'.

A second focus was upon the opportunity for profitable private investment – as in earlier campaigns – coupled with sound, largely private-sector led, planning for the future; and it was this latter aspect which represented the novel element. The Black Country offered 'the best investment in land since the gold rush', Teesside aimed to create a 'compelling climate for investment', Bristol 'a new spirit of enterprise', whilst Tyne and Wear's task was 'to respond to the needs of the marketplace'.

Thirdly, another common theme – echoing an earlier round of activity – was the promotion of a greater 'quality of life' compared to other parts of the United Kingdom. Often, quite specifically, this meant 'better than' the South East. It was also frequently coupled with green or environmentally-sensitive proposals for the regeneration of the area. Thus, the Black Country rejoiced in the motto 'The Green Country', where 'the industry of today and the future will blend in with a healthier, greener environment' which was 'pleasant for families to grow up in, and for people to travel to work in'. Trafford Park offered 'the best of both worlds', combining lower costs of living and higher disposable incomes, 'even if salaries are lower than London's'. Teesside too enjoyed 'excellent access to its scenic wealth', and Sheffield proposed to establish an 'Urban Ecology Park' alongside the River Don.

Whilst these three themes were common to practically all of the UDCs, there were also subtly differing emphases, even among the four major ones (Black Country, Teesside, Trafford Park, Tyne and Wear). The two North East agencies focused particularly strongly upon the promotion of both self-help and entrepreneurship. Teesside's main slogan of 'initiative, talent, ability' was borrowed from a speech by Margaret Thatcher given in the area in September 1987: '[w]here you have initiative, talent and ability, the money follows'. Tyne and Wear also had a clear emphasis in favour of letting the market decide:

'The reality of the late-1980s is that we live in a market economy, and it is the resources of that economy that will create the future. We are not a plan-led organisation. Our role is to guide and focus market forces.'

Trafford Park too concentrated upon these themes, emphasising in this case (perhaps paradoxically) the strengths of its existing industrial base. 'Join us in the Big League' was its invitation, with the implication that what other companies had chosen must be worth following. Finally, the Black Country, of all of the four agencies, most emphatically made a connection between the two ideas of heritage and opportunity, extolling a social obligation to restore pride to the area (see also the chapter by Lowe):

'The Black Country earned wealth for the nation. It's time the wealth came back. Tomorrow's Black Country will be an asset to the nation, and will return to the region its self-reliance and self-respect.'

These varying emphases and common themes were significant. They suggested some of the ways in which place-marketing strategies selectively appropriated particular aspects of a locality's culture to further a series of pre-determined and in large measure *national* political goals. Just as in the adoption of countryside 'traditions' and in the re-enactment of history through rose-tinted glasses in industrial museums, both as a characteristic of middle- or service-class values (see Thrift 1989), the created heritage of these old industrial working-class districts was also highly partial and selective. Many of the issues of social conflict and squalid living conditions were carefully erased from memory, creating instead a misleading sense of conflict-free growth which somehow benefited all sections of the population. The Industrial Revolution was effectively de-politicised; the Victorian industrialist heralded as the paternal capitalist which he (and the word is used advisedly) all too rarely was.

The promise of a better, greener future was also subservient to this task. It was not the process of industrialisation *per se* which was held to be responsible for the squalid environmental conditions of these areas, but rather the *decline* of manufacturing industry. What was presented as a solution to this problem was a new generation of industry and industrialists, often openly couched with reference to some carefully selected Victorian values. In this respect the frequent evocation of 'market forces' and the related aversion to planning indicated clearly the national political goals to which the diverse cultural heritages of these areas had been appropriated (or, perhaps more accurately, hijacked). From a partially re-created past came a politically-guided prognosis for the future.

'There is no Alternative':
The Privilege Given to Markets and the Denial of Debate

A further mechanism of hegemony in 1980s United Kingdom society rested in the assertion that there was only one solution, and that was through the market-place. The role of the state was to be determined by the market (to use the language of the New Right), which in practice meant a radical re-definition of the character of, rather than of the scope for, state policies. Cushioned by an over-whelming parliamentary majority (which was partly sustained by carefully-crafted income redistribution policies), alternative economic analysis which might entail a different conception of the role of place in everyday life was, quite simply, ridiculed without substantive engagement or dialogue. To raise questions over the direction of strategy, locally as well as nationally, was held to be counter-productive to the broader goal of 'freeing the hand' of market forces (to use a phrase favoured by government ministers).

In this scheme of things, appearances were all important. Business confidence was held to be a vital precursor to fresh, 'rejuvenating' investment, even if it was apparent during the later years of the decade that the economy was far from dynamic, and that the (state-facilitated) wave of consumer demand (which had in any case sucked in an increased volume of imports, leading to heightened trade deficits) was fading fast. In continually stressing success, despite rising evidence to the contrary, much of the potential terrain of debate was effectively closed off before it even reached the agenda. This was recognised, for instance, by the Town and Country Planning Association, which focused on the much-vaunted 'revival' of the North of England during the late-1980s:

'Terrified of being seen to "talk the North down", industry in the region has been talking it up with a vengeance. Anybody who dares to cast doubt on all this euphoria is immediately branded as a "whinger" and suspected of anti-Government leanings. . . . [Yet] it is our view that the recent signs of an upturn are not a significantly reliable base on which to found talk of a real revival in the North' (Town and Country Planning Association, 1989, p. 3).

Such reactions as that described here – cast in terms where to question was to upset the balance – were far from confined to northern England. In Wales the impact of Japanese investment was also the subject of such counter-debate. The Japanese companies in Wales were taken to task by Morris (1988) for their low skill requirements and under-representation of research and development activities. This viewpoint met with a fierce reply from Jones (1988), a senior executive with Sony, one of the companies in question. Regarding Morris's analysis as 'essentially a depressing and discouraging one', he went on in an attempt to claim the moral high ground:

'Business success requires confidence, and even intentionally academic articles have the potential to damage the job prospects of real people in the real world. What we are talking about here are not jobs and employment in an abstract sense, but the lives and economic prospects of real people living and working in Wales' (Jones, 1988, p. 60).

In other words, even to question the basis of proclaimed success was somehow damaging to it (or at least to some 'real' people). This was an increasingly familiar theme.

An additional twist to this discourse included attacks upon the motives or personal integrity of alternative commentators. In the course of the 1980s there were numerous instances of 'academic' contributions being savaged in this way. The activity of British Coal (formerly the National Coal Board) in shedding doubt upon the credentials of expert witnesses and alternative testimony, for instance, has been recorded elsewhere (Beynon, 1988). Within North East England, espousal of alternative economic scenarios drew a similarly fierce response from politicians, focused not infrequently upon personalities as much as on policies. For example, a report commissioned by BBC (North East) on the experiences of unemployment within the region (Robinson, 1987) led to wholesale condemnation of both the message and, especially, the messenger. Piers Merchant, Conservative MP for Newcastle Central, described the report as 'a diatribe in despair and an exercise in uncontrollable pessimism', and also as 'a depressing and soul-destroying document' (*Hansard*, 9 and 24 March 1987). In his words, he was 'left wondering about the mental health of the compiler' (*Hansard*, 24 March 1987). A parliamentary colleague, Michael Fallon (one of the region's few other Conservative MPs), posted an early-day motion in the House of Commons attacking Robinson's institution (Newcastle University) for a 'highly partial, intellectually shoddy report', and calling on the Vice-Chancellor to 'uphold and enhance tarnished academic standards' (*Hansard*, 24 March 1987). Finally, reflecting Jones's (1988) concern for business confidence, Merchant commented on what he saw as the negative impact:

'[S]uch a catalogue of gloom is exactly what we do not need if we are to attract investment, industry and jobs to the north and the other deprived regions. Instead we should sell all the good points of those regions' (*Hansard*, 9 March, 1987).

Perhaps more worrying is that much of this and other similar assaults rested upon an assumption of superior competence or upon a denial of access to sources of information. Further illustration of this can be drawn from the area of Derwentside in North East England. The social and economic prospects of this part of County Durham, suffering from the closure of the Consett steel works and two major branch plants, were investigated in Robinson and Sadler (1984). Derwentside was the target of a high profile publicity campaign seeking to attract new employment. We argued that 'whatever the merits of the re-industrialisation strategy, the starting point for thinking about the future of the area must be that high unemployment will continue into the future' (Robinson and Sadler, 1984, p. 80). Such a view was in fact shared elsewhere. A joint working party of the Local Authority Associations, for instance, said of Consett that 'the task of industrial regeneration is now looking to be very long-term indeed' (Local Authority Associations, 1986, p. 5).

To the organisation which held itself to be responsible for economic renewal, the Derwentside Industrial Development Agency (DIDA, a local enterprise agency, supported by local business along with local and national government), these comments were anathema. Its chief executive, Laurie Haveron – responding in part to a local television programme based on the findings of our report – described Robinson and Sadler (1984) as 'ill-informed pessimism, largely based on opinions rather than fact'. He went on: 'the authors have little realism to offer, and the unending gloom depicted by them will hardly help us in our continuing drive to attract inward investment' (*Newcastle Journal*, 25 May 1984). He added later that the report 'hasn't understood the real progress, in business terms, which is being made' (*Consett Guardian*, 31 May 1984). Further attempts at eliciting additional simple information were similarly rebuffed by claims of superior competence together with the need to protect business confidence. The merits of an independent jobs audit into enterprise agencies such as DIDA – given that no real impartial basis existed for a serious debate on the effectiveness of policy – were alluded to by Boulding *et al.* (1988).[2] DIDA chairman Mr. Crangle commented simply that it was 'deeply distressing to see work like this' (*Newcastle Journal*, 29 May 1986). The real issues at stake – to do with the effectiveness of policy – became first blurred, then removed from the public agenda (for further details see Hudson and Sadler 1989, pp. 109–116). Alternative commentary and the opportunity for open, mature debate were cast aside.

The privilege accorded to a particular position, then, became a key underpinning of coalitions of interest built around the appropriate step to regeneration. These power bases took strength from assertions that business confidence should at all costs not be shaken, that the motivations of critics were ill-founded or suspect, and that in any case sceptics were incapable of interpreting the right information on which to base alternative judgements. This is certainly not to argue that so-called 'academic' contributions should be immune from the conflict of politics. However, once so thrown into that arena (academic naïveties notwithstanding), it is perhaps reasonable to expect acceptance as stimulus to open debate, rather than condemnation without consideration. It was a mark of both the relative absence of challenge to 'established' values and the strength of this hegemonic thrust that the kinds of issues considered above took on such sharply antagonistic lines in the United Kingdom during the 1980s.

This alone, however, and even when coupled with the image-making machinery of advertising and place-marketing, is insufficient to explain how, in practical terms, such coalitions achieved and maintained dominance. To do this we need to turn to the day-to-day art and craft of politics, and also to some much longer-standing political processes.

The Politics of Re-industrialisation

One of the dominant themes in urban and regional regeneration in the United Kingdom and elsewhere in the 1980s was that of 'partnership', symbolising the heightened involvement of the private sector alongside elected public authorities. In part, at least, such developments were interpreted as a British adoption of the United States version of urban growth coalitions (see, for instance, Lloyd and Newlands, 1988, on Aberdeen). Much of the enhanced role for the private sector nonetheless rested upon national government policies which facilitated the creation of an environment in which corporate interests emerged paramount over those of a broader constituency. Organisations such as Business in the Community (founded in 1979) promoted the involvement of companies in a wholly new network of over three hundred local enterprise agencies which sprang up in the 1980s. A White Paper *Employment for the 1990s* (HMSO, 1988) subsequently formally initiated the creation of a national system of some one hundred 'Training and Enterprise Councils', led by employers and intended to become the local means of delivering training and promoting enterprise. In other ways too corporate involvement in economic planning blossomed: via the Confederation of British Industry's 'Task Forces' in Newcastle and Birmingham, for instance, as well as via re-development consortia such as 'Partnership Renewal of the Built Environment' (established by financial and construction interests).

Such arrangements, undertaken either in partnership with or alongside local government, represented a clear challenge to existing institutional structures. Like the Enterprise Zone and Urban Development Corporation 'experiments', they represented a re-direction of planning, particularly in the interests of property developers (see, for instance, Anderson, 1990; Brindley *et al.*, 1989). They were intimately connected to the apparent resurgence of relatively small groups of notable individuals, drawn largely but not exclusively from the private sector, who dominated local politics and often had privileged access to central government (see, for instance, Shaw, 1990 on Tyne and Wear in North East England). In this way the political agenda was reordered during the 1980s as power became increasingly concentrated both centrally and in the hands of a few (self-appointed) regional 'representatives'.

However, such arrangements were by no means wholly novel (see Sadler, 1990). Regions such as North East England had a long historical record in which key individuals controlled many of the levers of power. The evolution of regional promotion agencies in the North East from the 1930s onwards, for instance, was characterised by consistency among the ranks of leading figures (see Cousins *et al.*, 1974). Lord Ridley was instrumental in the creation of the North East Development Board in 1935, the Northern Industrial Group in 1943 and the North East Development Association in 1944. When the latter two merged to form the North East Industrial Development Association (NEIDA) in 1953, the four key officials of the new organisation were drawn from the leading positions of the old bodies. To cite another and perhaps better-known example, T. Dan Smith, leader of Newcastle City Council in the early-1960s, was a member of NEIDA from 1959 to 1961, of its successor body (the North East Development Council) from 1961 to 1964, and then first chairman of the Northern Economic Planning Council from 1965 to 1970. A 'unified elite', argued Cousins *et al.*, followed a common purpose from the 1930s onwards: 'to persuade the people of the North East to accept a certain range of policies and assumptions about policy' (Cousins *et al.*, 1974, p. 143). The significance of this, they argued, lay 'in alternative approaches which have been repressed'.

Whilst the partnership arrangements of the 1980s were a new twist to urban and regional policy, then, their novelty too should not be over-stressed. Indeed, there are some signs of continuing representation of the 'old' power bases in the new institutions. Similarly, the repression of alternative commentary (as considered above) during the 1980s was far from original, although the terms in which it was cast had undoubtedly shifted. And these elements of continuity point to another significant feature of pro-business coalitions: namely, the limits to their internal coherence. The area of mutual agreement and interest was relatively restricted, and commonality of interest could not be inferred from shared membership of a range of organisations. What did unite these groups, though, was their concern to extol the merits of (and prepare the ground for) capitalist interests in the North East.

Such policies and priorities dominated re-industrialisation efforts during the 1980s, and were epitomised in Derwentside. We have already seen how wary this coalition was of alternative analysis or criticism (and note also that in this case it embraced rather than excluded local government). This was so despite (or, perhaps more accurately, because of) the area's national significance as a kind of role model following traumatic economic collapse from 1979 to 1981 (and note again that the district had a long tradition of industrial paternalism and domination by a few key individuals through the influence of the Consett Iron Company: it is more than just ironic that the area should be the location for one of the United Kingdom's major industrial museums, at Beamish). Yet Derwentside also provides a revealing insight in other ways, for towards the end of the 1980s the coalition of interests began to break down as failure – even on its own terms – became increasingly apparent. In such circumstances the basis for adopting, as well as the very wisdom of, earlier policies were increasingly questioned.

Much of the strategy's rationale had rested on being seen to be successful, in proclaiming effectiveness and in grouping around it both local businesses and local government, orchestrated via a local enterprise agency. In 1989, for instance, Derwentside District Council indicated that on its own estimation a total of 4,500 new jobs had been created since 1980. Of these, it was claimed that 96% were full-time and that the majority were in manufacturing. It was anticipated that a further 1,600 jobs would be created before 1993, the latest in a long line of assertive forecasts. Neil Gregory, chair of the Council's Economic Development Committee, commented optimistically that 'Derwentside now has a stronger and more diverse economy than it has had for many years' (quoted in *Newcastle Journal*, 29 August 1989).

Increasingly, though, it was difficult to relate these and other such figures to the reality of life within the district (see Hudson and Sadler, 1990). One resident, Patrick McNulty, wrote as follows in terms which sharply captured emergent disparities:

'I sometimes wonder if there are two Derwentsides. One which I and others like me live in, and one which the people from the Derwentside Development Scheme [*sic*] inhabit. These people are constantly making statements about the "boom" now taking place in Derwentside. I do look for signs of this "boom" but I'm afraid it continues to elude me. All I see are small business units springing up employing between five and twenty employees, most of whom are on one-year adult schemes, ET [Employment Training], or two-year youth schemes, YTS [Youth Training Scheme]. Anyone fortunate enough to have a permanent position is usually low-waged and non-unionised. . . . While I agree that the people in the job creation agencies must not become pessimistic, I do, however, believe that they ought to be frank and objective and not mislead the people. If the Government is led to believe that Derwentside is now "booming" and "prosperous", then it will see no need for the massive assistance which is urgently needed to save this area from total and complete demise' (in the *Durham Advertiser*, 14 September 1989).

One of the fiercest critics of the re-industrialisation programme in the early- to mid-1980s was John Kearney of the Derwentside Unemployed Group. In 1986 at a meeting in Durham he remarked that:

'Worklessness is a fundamental crisis for the Left and for the Right, and any kind of politics which doesn't accept this is a sham. What we're facing is a crisis of political organisation, where the structures which we have are no longer capable of solving our problems for us. Labourism is currently about bigger bribes for industrialists to relocate. The price of maintaining Labourism is to further victimise the victims.'

Within Derwentside this 'political organisation' hinged upon a traditional Labourist local authority and an aggressively oriented development agency, but as the gap between claims and reality widened, so the District Council began to re-evaluate its conception of the re-development process – partly spurred on by the publication of government figures shedding doubt on the nature of change. The 1987 employment census (published in 1989) indicated that employment in the District had climbed from 18,800 in 1981 to 22,700 in

1987 (compared with 28,000 in 1978). Of the 4,000 net new jobs, over 3,000 were in the service sector and only 500 in manufacturing (see **Table 1**); and about 50% were part-time. Much of this was clearly starkly contrary to earlier expectations raised within the district.

TABLE 1
Employment Change in Derwentside, 1981–1987

Division		1981	1987	Change 1981–1987
0,1	Agriculture	700	500	−200
2	Extraction	700	800	+100
3	Metal goods	2,200	2,300	+100
4	Other manufacturing	1,900	2,400	+500
5	Construction	1,000	1,200	+200
6	Distribution, tourism	3,600	4,500	+900
7	Transport	1,300	800	−500
8	Financial services	600	1,100	+500
9	Other services	6,700	9,000	+2,300
Total		18,800	22,700	+3,900

(Source: NOMIS)

In the process of re-evaluation, a number of elements of the re-industrialisation strategy were seriously questioned for the first time – in particular, the appropriate relationship between appointed enterprise agency and elected local authority.[3] One local government officer with long experience reflected on this as follows:

'DIDA [the enterprise agency] began as a complete song-and-dance routine, as a comprehensive economic development unit in its own right. It became very much an animal of its own. But in the last couple of years, the District Council has resumed more of the mantle of economic development, and pulled back some of the power. DIDA has been left with the role of industrial policy for the time being, but that's also under appraisal. Industrial policy is still the Council's emphasis, but there's an increasing recognition that it's not all the answer. The approach to economic development is no longer solely centred on industrial policy, but also on things like tourism, and community businesses.'

Such statements were echoed by a senior colleague:

'This Council is under new management: at the political level and at the officer level. Two years ago a new Council was elected which sought to change what had gone before. Now we're going out to sell ourselves more to the public and to see what they want of us.'

There was also a consideration of how and why the earlier strategy evolved as it did:

'Ten years ago it was simple crisis management. We had to get new jobs in the District, and they had to be in the manufacturing base. Now it's slightly different. We can look around and question what kind of future we want, what kind of development the District wants.'

But at the same time the composition of the Council, along with the relations between officers and members, came under strain:

'The problem we have in the organisation is this. Some of the members and officers have very quickly grasped what we're trying to achieve. But there's also a lot of people who are suspicious, or downright hostile, to change. So it's very hard to move the whole organisation forward together.'

Even reinterpretation of the past, then, reflected the severe constraints imposed upon local government by the very nature of its power. These were, of course, crucial to the kind of policy adopted (or, perhaps more accurately if also more harshly, to the abdication from responsibility) in the early-1980s. Then, crisis management had been the order of the day: piece-meal local solutions being the preferred answer, because the electorate expected local politicians to do something for their area, despite the legal and financial constraints imposed by central government which severely limited what could be done. The kind of policy which had been adopted – and which was proving increasingly ineffective even in terms of its own stated objectives – therefore rested not so much on a basis of rational choice, but rather was a simple reflection of the narrow political and intellectual scope for alternatives. This restricted area did not come about purely or simply by chance, but had been deliberately encouraged and fostered.

Conclusions

This analysis has considered three elements to the construction of place-marketing strategies in Britain as these emerged during the 1980s: a representation of place to serve specific political goals; an emphasis on market forces and an associated disavowal of the possibility of alternatives; and a process of local coalition formation around a heavily-constrained and ideologically-laden agenda. What emerges is that the construction of apparently competitive places was one central feature of class domination; place-marketing was one embodiment of a broader phenomenon. In the United Kingdom in the 1980s the idea that places as well as people could be competitive became central to an increasingly powerful ideology, despite some notable attempts at opposition (such as the 1984–1985 miners' strike, or the work of more progressive local authorities such as the Greater

London Council). It is significant that many of these alternatives, including more than a few campaigns against the collapse of traditional sources of employment and bases of social existence, foundered on the rocks of territorial fragmentation. Place became one further basis of political division within society, intersecting and reinforcing those of class, race and gender, and a basis of division that acted (and in part was actively used) to foster the construction of hegemony.

Situating place-marketing in this context enables the political dimension to come back into the social construction of place. The developing hegemony incorporated a particular conception of place – its value to capital – and alternative policies which might have considered the meaning and value of place both to labour and as community were notable (with a few exceptions such as those outlined above) for their absence. Place, in other words, should be seen as both a social and a political construction. Localities, cities and regions are not necessarily objects in their own right (except in the purely administrative sense) but are rather part of spatially-grounded social processes of production and consumption, with meanings which are contested rather than inherent or given. Places are therefore not necessarily competitive: it is only a specific political packaging of the concept of place which makes them seem to be so. This recognition is essential both for understanding the contemporary geography of the United Kingdom and for attempts to transform it.

Such transformative efforts also need to be aware of the deep and complex significance of labour and community history, especially in old industrial regions. There is no denying the significance of working-class culture, for instance, but a great deal of difficulty remains in defining exactly what it is, and even more in considering how some parts of it might relate to progressive political ideals. It is especially important, therefore, to appreciate not just the conditions under which history was created and the ways in which it has recently been claimed, but also the limitations imposed by that history. In his influential survey Hewison (1987, p. 10) criticised the 'heritage industry' for producing 'fantasies of a past that never was', for its reassertion of undemocratic social values, and for blocking the capacity for creative change. Such thoughts apply not just to the realms of museums, the countryside tradition and the arts, but also more broadly and much more immediately to the present day. Growth or re-industrialisation coalitions in the United Kingdom in the 1980s wandered perilously close to selective re-interpretation. There is no denying the solidity of sense of place, of 'belonging' somewhere, on which much working-class (and nationalist) culture rests, but the transformation or appropriation of this feeling into a particular political project of whatever kind is a different matter altogether. It is one which needs to be exposed and evaluated very carefully indeed as an essential prelude to alternative policies.

Notes

1 The following section draws upon an interpretation of publicity materials produced by the four major Urban Development Corporations created in 1987 (Black Country, Teesside, Trafford Park, Tyne and Wear) and by three of the later round of 'mini-UDCs' in smaller, central urban areas (Bristol, Manchester and Sheffield).
2 Although not published until 1988 the paper was widely circulated in the region before then in response to a series of requests.
3 The following section draws upon research involving interviews with officers and members of the local authority that were conducted by the author during 1989.

9

John Wayne Meets Donald Trump: The Lower East Side as Wild Wild West[1]

LAURA REID AND NEIL SMITH

Mythic Frontiers, Economic Frontiers

Over the past two decades, North American cities have been transformed fundamentally by the processes of urban restructuring and gentrification. In the media these changes have been represented through the highly suggestive language and imagery of the Western frontier; frontier language and history have been appropriated to construct the meaning of contemporary processes. Realtors, developers and gentrifiers portrayed as 'urban cowboys' – rugged individualists, driven in pursuit of civic betterment – tame and reclaim the dilapidated communities of the downtown urban frontier. At their hands, city neighbourhoods are transformed as residences are rehabilitated and new luxury apartment complexes are constructed for incoming middle- and upper-class residents. New boutique landscapes of consumption emerge catering to their gastronomic, fashion and entertainment demands, and new landscapes of production are created with the construction of new office buildings: the workspace of the residents of the 'new' city.

Frontier mythology is being reapplied in the context of urban restructuring and gentrification to shape the ways that we think about and interpret an entirely new and dramatic set of processes transforming the city. It presents the city as a challenge, but an elite challenge for white America with all the overtones of manifest destiny, connecting sufficiently with a different experience of the city – that of the working class, poor, immigrants and 'non-traditional' households – just sufficiently to explain it away. For there is a clear underbelly to the restructuring heralded by gentrification, a different

193

urban frontier. In an insightful analysis of the broader contours of restructuring, Koptiuch (1989) refers to the rapid immigration of Koreans, Filipinos, Dominicans, El Salvadorands, Indians, Chinese and others as the 'third worlding of America'. She observes that a 'little-explored interior frontier has been created right at the West's core, abutting the communities of the exotic others-come-home and the communities of those internal others who have long struggled for inclusive parity in the hallowed pronomial phrase "we the people"' (Koptiuch, 1989, p. 12). This has induced anxiety in 'the discourses of hegemonic purveyors of culture and politics', and has generated a 'panic provoked by the defamiliarisation-effects of what seems to many an increasingly alien nation' (Koptiuch, 1989, p. 8). In response, much as in the nineteenth-century frontier of the American West, this 'third worlding' has unleashed a neo-imperialist onslaught on the inner-city by a host of institutions, notably the Church, the media, educational and governmental bodies.

Signifying a profound reversal in geographic, cultural and economic terms of postwar urban abandonment, the social meaning of the gentrification process is increasingly constructed through the adaptation of the frontier mythology. And yet the mythology is neither monolithic nor impregnable. Unable to mask completely those aspects of the urban transformation that are conflictual, the borrowed frontier imagery serves both to legitimate and to eradicate conflict arising from both the constructed meanings of gentrification and its material consequences. The violence and conflict that accompanies the process is both rationalised and at the same time denied, hidden, and in the name of progress, civilisation and the general good, left unreported.

The significance of the frontier goes beyond the invocation of a playful set of images and language for their own sake. It works to facilitate and to legitimate the gentrification process in which working-class communities are destroyed and their residents displaced as new landscapes of production and consumption are created for the needs of a new class of residents and the needs of the real estate industry. The imagery operates to rationalise the violations incurred in the transformation of neighbourhoods by establishing the superiority of reinvestment over the previous experience of abandonment and decline. It also functions to project an image of consensus regarding these changes, insinuating ruling ideas about race, class, gender, community and urban development as universally accepted.

The pervasive acceptance of the frontier myth as metaphor for urban restructuring reveals the ideologies of racism and class privilege that structure popular and political language, thought and action. Implicit in the frontier imagery is the highly insulting designation of

the working-class, poor, female-headed households and Latino/
Latina and African-American 'natives' of the downtown neighbourhoods as part of the savage, urban wilderness. It is implied that not
only are these groups responsible for the decline of the city, but that
as part of this wilderness of their own creation they are culturally,
socially and morally inferior to the 'civilised' world outside. The
frontier myth then functions to facilitate and to legitimate the
gentrification and restructuring of their neighbourhoods by projecting
an image of consensus that these changes are superior and progressive, whilst excluding the voices and destroying the communities of
the residents of neighbourhoods being gentrified. But, the defence
of terrain and the many bitter struggles over housing needs, development issues and the use of public space in neighbourhoods under
siege challenge the cacophony of gentrification myths for those who
care or strive to listen. These acts of resistance explicitly reveal the
class- and race-based conflicts that arise in neighbourhoods targeted
for reinvestment as residents struggle to maintain control of housing
and their communities.

Yet, whilst the social and cultural representation of the city as new
frontier is highly ideological, this myth may be undergirded by an
economic truth. This relationship between representation, ideology
and the economics of material reality is cynically captured in Limerick's
rendition of the conquest of the West in which she posits that

'[if] . . . Hollywood wanted to capture the emotional centre of Western History, its
movies would be about real estate. John Wayne would have been neither a gunfighter
nor a sheriff, but a surveyor, speculator or claims lawyer. The showdowns would
appear in the land office or courtroom; weapons would be deeds and lawsuits, not
six-guns' (Limerick, 1987, p. 264).

Just as the outward establishment of the Western frontier was at its
core an economically motivated process of geographic expansion, the
exploitation of a 'frontier of profitability' (Smith, 1986) may be driving
the rapid restructuring of inner-city neighbourhoods. Unacknow-
ledged in the popular press's celebration of the 'new frontier' as
cultural triumph and social progress are the real gains in the conquest
of downtown: the massive profits reaped by those speculators
and developers who buy, sell and rehabilitate the abandoned and
dilapidated property and land.

This paper attempts to explore the construction and meaning of
the new urban frontier myth, and to understand its part in the
gentrification process. In particular, we want to expose its economic
underpinnings and begin to understand the origins and operation
of the *economic frontier* that is a cornerstone of the gentrification
process. This economic frontier, or the frontier of profitability,

represents the dividing line between disinvestment and reinvestment and can be located geographically and historically in gentrifying neighbourhoods. Disinvestment from a neighbourhood, the relative withdrawal of capital in all its forms from the built environment, is a prerequisite for urban restructuring and gentrification. Ahead of the frontier, properties are still undergoing disinvestment and devalorisation. Behind the line, gentrification and capital reinvestment in previously disinvested landscapes and structures have begun.

Specifically, we look at the capitalisation of the frontier of profitability in New York City's Lower East Side (see **Figure 1**) through an analysis of the spatial and temporal patterns of capital flows in the shift from disinvestment to reinvestment. We begin our project by elucidating the connections between this economic aspect of gentrification and the reconstruction of the social and cultural meaning of neighbourhoods as place. We argue that the gentrification of the neighbourhood was facilitated, if not quite caused, by the arts and entertainment industry through their adaptation of the frontier imagery to create an '*avant garde* ambience' attractive to investors in the arts and in real estate industries alike. This abstraction and manipulation of the social reality of the Lower East Side was abetted by the City's government which contributed to the neighbourhood's recreation via a variety of programmes and policies influencing the use of housing and public spaces. We conclude by connecting these components of the restructuring process with a preliminary research agenda for tracing the onset of actual physical transformation in the neighbourhood's housing stock. As part of ongoing research, we look at the temporal patterns of rehabilitation projects for a sample area and study their relationship to the frontier of profitability.

By establishing the connections between the shifting frontier of profitability, the physical transformation of the neighbourhood and the construction of the new frontier mythology as an ideology, we want to expand our knowledge of the mechanisms of the gentrification process. This investigation is intended to contribute to the empowering of the residents of gentrifying communities against speculators and developers who have turned whole neighbourhoods into 'new frontiers'.

The Lower East Side as Wild Wild West

The construction of the new frontier, its attendant imagery and its impact on restructuring is both complex and geographically varied: it adapts to place as it makes place. In New York's Lower East Side two industries were primarily responsible for recreating the neighbourhood as a culturally and economically 'attractive' place. The real

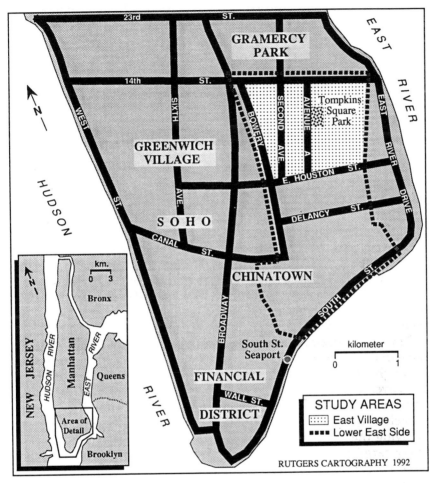

FIG. 1. New York City and the Lower East Side.

estate industry, traditionally central in the marketing of place, promoted the northern part of the neighbourhood as the 'East Village' to capitalise on the image of the neighbouring stable, and already gentrified, Greenwich Village. But of central importance in establishing the new cultural and social meanings of place was the arts industry, which – through its language and images and also its physical presence – succeeded in converting the dilapidation and squalor of the neighbourhood into *ultra chic*.

The City was also a player. In the construction and execution of its housing policy, attack on the drug industry and control of parks and public spaces, the City aimed less toward providing basic services

197

for existing residents, and more toward evicting locals and homeless people and subsidising development for the real estate industry. New York's housing policy, under the jurisdiction of the Department of Housing, Preservation and Development (HPD), was largely structured in the 1980s to promote the gentrification process. This was to be achieved by auctioning off the massive city-owned housing stock, composed of 'in-rem' properties, acquired mostly as a result of foreclosures from private landlords for non-payment of property taxes. By the early 1980s HPD held over two hundred such properties in the Lower East Side and as many vacant lots. Trying to stimulate the neighbourhood's real estate market, and exploiting the area's emerging reputation as a centre of *avant garde* art, HPD proposed the renovation of sixteen of these properties for artists. For the remaining buildings, a cross-subsidy programme was proposed in which developers were to be offered properties at below market rates and given tax breaks in return for setting aside a vaguely specified 20% of newly-built or rehabilitated units for low-income tenants (Bennets, 1982; Carroll, 1983). These policies were fought and successfully blocked by community activists, who argued that their neighbourhood's housing stock was being given away for development projects that would not serve the needs of the community.

The City's housing policy, however thwarted, signalled its endorsement of the social and economic restructuring of the neighbourhood. It sought to aid the real estate industry in the fulfilment of these objectives by mobilising policy and police support to clear the streets and public spaces of the neighbourhood of 'undesirables': the homeless, the drug-users and sellers, and the kids (local and visiting) 'hanging out' in neighbourhood parks. The Lower East Side was a major target for all of these programmes. In 1984 the City launched 'Operation Pressure Point', a drug crackdown targeting the southern core of the Lower East Side in which 14,000 arrests were made in eighteen months (Greer, 1985). Ostensibly intended to stifle the drug industry, Operation Pressure Point was widely understood in the neighbourhood as part of a gentrification strategy. Rounding out its strategy to reclaim the neighbourhood for 'higher and better uses', the City has sought to evict the homeless and other 'undesirables' from local parks. Starting with Union Square, in the outer north-western corner of the area, $3.6 million of public money was dedicated to its renovation in 1984. A nightly curfew was introduced in other City parks to prohibit the homeless from sleeping overnight and establishing encampments. Whilst the City has sought to 'take back the parks', the strategy was opposed most noticeably in Tompkins Square park in the Lower East Side, where on 6 August 1988 some four hundred police 'rioted' in response to an anti-gentrification demonstration. Armed with riot gear, mounted on horses and accompanied by two helicopters, the police were met by about three hundred demonstrators protesting the gentrification of the park. Seeking to quash the protest swiftly, the police forced people out of the park at midnight, but were met with hurled stones and insults. After an infantry charge by mounted police, a vicious battle erupted over a four hour period in which the police chased and brutally assaulted people, seemingly at random (Carr, 1988; Ferguson, 1988; Gerritz, 1988).

Since 1988 there have been periodic clashes between the police and protestors, with the struggle broadening out to include squatters' rights in abandoned housing, homeless organisations and the wider housing movement. In December 1989 the police evicted some three hundred people living in an estimated one hundred shanties in the park. In spite of repressive police tactics, the park once again became home to nearly two hundred homeless people occupying the park in the spring of 1990. A second eviction came in the summer of 1990, and since the third in June 1991 the park has been closed, fenced off, and heavily guarded.

It is not that the whole community endorses the use of the park as a shelter or an all-night playground, or supports squatting. Clearly there are conflicts over the use of the park, as there are over possible solutions to homelessness. But there is a widespread if not universal tolerance of these uses derived from an understanding of the connections between homelessness, poverty and gentrification, and from an awareness that the City has made no real attempts to address these problems (Carr, 1988). Equally, as the numbers of homeless people in the Park have increased since 1989, the unsympathetic opposition has became more vocal.

The struggle to control who uses the park has developed as a forum to express perceptions about the gentrification process: gentrification has been likened to both 'class war' and 'genocide' by protesters, an attack on the working class and poor motivated by the real estate industry's search for profits. Such an alternative scripting of the gentrification process aims at shattering the frontier myths constructed and purveyed by the media, the City and the real estate industry. Gentrification, they argue, is *not* for the good of all and is *not* a progressive development from the perspective of the community and its residents. For them, it means homelessness, displacement, expensive and inaccessible housing, and a challenge to the cultural diversity, practices and tolerance that have been the mark of their neighbourhood.

The Frontier of Profitability

The connection between gentrification and profitability elucidated by local resistance in the Lower East Side provides the key to understanding the relevance of the frontier imagery. At the core of the gentrification process are the profits derived by the real estate industry from capitalising on neighbourhood decline and disinvestment. Property owners disinvest from their buildings when alternative investments promise higher returns. Whilst capital is systematically being withdrawn for investment elsewhere, repairs and maintenance are minimal if at all, and over the years the buildings become dilapidated and ultimately uninhabitable, abandoned by property owners and tenants alike. The consequent economic devaluation of the building brings about a reciprocal devaluation of the land that it occupies, and eventually produces a 'rent gap' (Smith 1979, 1987; Clark, 1987; Badcock, 1989). Developers thereby not only glean the money that might have gone to repairs and upkeep, but also create the opportunity to profit from a whole new round of investment. It is highly ironic, and also symptomatic of the perverse logic of the capitalist economy, that in reinvesting in the neighbourhoods which they had moments before pushed into decline the real estate industry should be praised by the media for their civic selflessness.

The geography and history of this reinvestment are crucial to understanding the way in which gentrification takes hold of a neighbourhood. Developers do not randomly invest or necessarily seek out those properties where the rent gap is greatest (Marcuse, 1986). Uncertainty as to the momentum, speed and success of their speculations tempers action and shapes strategy, but the gentrification frontier is a very real presence in the minds of developers. And it is to discerning the location and spread of this 'frontier' that we now turn.

Reconstructing the Frontier

In order to construct the geographical and temporal patterns of the frontier, it is necessary to select an indicator that allows us accurately to locate the 'point' where disinvestment is succeeded by reinvestment in a given neighbourhood. Specifically, we want an indicator that highlights the earliest signs of reversal or 'turning points' in the experience of disinvestment, signifying the onset of the broader processes of gentrification and urban restructuring. We use *tax arrears data* for residential properties, believing these to be a very sensitive indicator of the earliest stages of reinvestment (Salins, 1981). In declining neighbourhoods, landlords and building owners often withhold property taxes as part of their overall strategy of neighbourhood disinvestment. Withholding these payments provides property owners with guaranteed access to capital for investment elsewhere. The City forecloses on buildings with serious 'tax delinquency' levels, and we might expect that reversals of disinvestment trends (reinvestment) would be discernable in the pattern of tax arrears. Perceiving the opportunity for substantial reinvestment, landlords and building owners will aim to retain possession of those buildings whose sale prices are expected to increase. This may involve the repayment of a sufficient level of back taxes to avoid the initiation of foreclosure procedures by the City. Repayment of tax arrears can therefore function as an initial form of reinvestment.

There is considerable empirical research on disinvestment (Sternleib and Lake, 1976; Lake, 1979; Salins, 1981), but Marcuse (1984) may have been the first to connect this research with a more theoretical perspective on disinvestment. In New York City foreclosures or 'in-rem' proceedings effectively begin against buildings that are twelve or more quarters (three years) in arrears, and this provides a responsive measure of investment and disinvestment (Salins,1981; Williams, 1987). City-wide tax delinquency levels have declined steadily from a high in 1976, when over 7% of residential properties were in arrears (Williams, 1981). The subsequent decline reflects the easing of the recession and fiscal crisis of the mid-1970s, but more specifically it reflects the rapid inflation of real estate prices in New York from the late 1970s throughout the 1980s alongside the burgeoning urban restructuring and gentrification. Nonetheless, delinquency and disinvestment still corrode a large section of the City's housing stock, concentrated geographically and economically in the oldest neighbourhoods where the housing stock is predominantly rental: large tenements and multi-unit buildings.

The Lower East Side is one such neighbourhood. Through the analysis of trends in residential tax arrears for the Lower East Side, we can trace the geographical and historical patterns of the reversal of disinvestment and thereby identify the gentrification frontier. Tax arrears data for the City are collected by the Department of Finance and organised at the census tract level for which summaries of various categories of arrearage levels are provided. We examine delinquency levels of greater than twelve quarters, given that this is the official threshold for initiating 'in-rem' proceedings.

Disaggregating the tax arrears data geographically, a series of turning points indicates the geographical and historical location where disinvestment is superseded by reinvestment. This we identify as the year of peak arrears (twelve plus quarters) for each building (tax lot) in the study area. Arrears data for tax lots is provided at two geographic scales: census tracts and tax blocks. Beginning at the census tract level, we identify turning points for each of the twenty-four usable tracts in the Lower East Side.

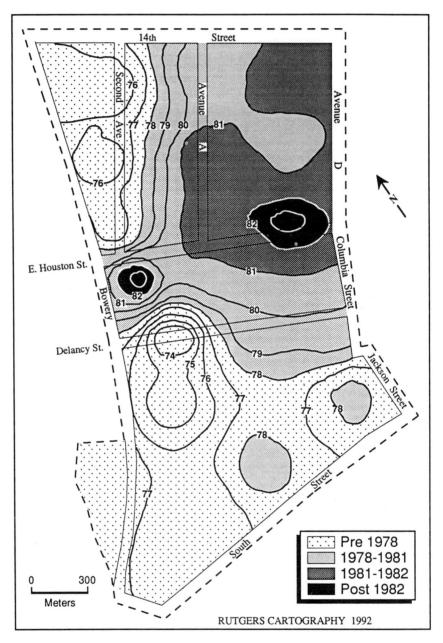

FIG. 2. The shifting Gentrification Frontier in the Lower East Side (by tax
arrears per census tract).

Findings

Every tract had 'turned' by 1985, with the earliest reinvestment occurring in 1975. These turning points were mapped using the 'Surfer' mapping package (Golden Software) which generalised the data via an 'inverse distance squared' method to produce a chorographic map illustrating the origin and spread of gentrification. In **Figure 2**, annual 'contour' lines join points with the same chronological turning point. By way of interpretation, significant space between contour lines indicates that gentrification is diffusing rapidly whereas barriers to gentrification are indicated by steep contour lines. The frontier is most evident where there are no enclosed contour lines (no peaks or sink holes). Peaks, with later years at the centre of enclosed contours, represent localities of greatest resistance to gentrification, whilst sinkholes, with early years at the centre, represent localities opened up to reinvestment in advance of surrounding areas. The major pattern is a reasonably well-defined west to east frontier line, with the earliest encroachments in the north-west and the south-west of the study area.

In geographical context, the gentrification frontier has advanced eastward into the neighbourhood from Greenwich Village, Soho, Chinatown and the Financial District. Each of these areas experienced gentrification prior to the Lower East Side, as early as the 1950s and 1960s in Greenwich Village and more recently elsewhere. The early reinvestment from Chinatown in the south-west of the Lower East Side can in part be explained by the influx of Taiwanese capital to the community in the mid-1970s, and later from Hong Kong, both of which have led to Chinatown's outward expansion to the north and east. Nevertheless, the apparent extent of early reinvestment indicated for that section is somewhat surprising, and may also be exaggerated by boundary effects in which data has been overgeneralised.

The gentrification frontier met with resistance in several localised areas. Delancy Street, the site of one such peak, is a wide thoroughfare leading to the Williamsburg Bridge which connects Brooklyn and Manhattan. A commercial strip, its noise, congestion and impassibility at its eastern end may have deterred reinvestment. More generally, the peaks can be thought of as indications of the limits of the gentrification process in the neighbourhood. The eastern and southern edges of the area are flanked by large public housing projects that could be expected to dampen the real estate market. In addition, these areas coincide with the poorest communities, the Latina/Latino neighbourhood, and the focus of Operation Pressure Point (the police organised drug crackdown of 1985).

This same procedure can be repeated at a higher geographical scale, using the number of actual housing units rather than lots. This

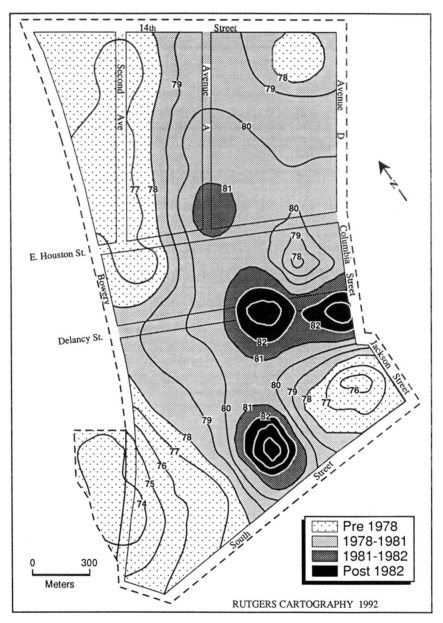

FIG. 3. The Shifting Gentrification Frontier in the Lower East Side (by housing unit arrears per census tract).

procedure offers more refined maps of the frontier, since it responds to variations in the neighbourhood's housing stock that we might expect to influence investment decisions. A multi-storey building on the western edge of a census tract, for example, anchored by more stable and already gentrified adjacent buildings, is likely to be perceived as a less risky investment than a similar multi-storey building several blocks east, where deterioration and disinvestment may not have been reversed. Conversely, a large tenement of sixty apartments and an adjacent brownstone with twelve may be treated very differently by speculators in the early stages of reinvestment. The frontier map generalised from tax lot data should more accurately illustrate how reinvestment potential varies between geographically disparate localities in the neighbourhood and according to the density of units rather than to the number of buildings.

Figure 3 shows the gentrification frontier, calculated in terms of housing units, at the census tract level. There is a high degree of correspondence with **Figure 2**. As before, the earliest encroachment of the frontier of profitability is in the north-west and the south, and it migrates eastward. But there are also significant differences with the first reconstructed frontier. The south-west does not emerge here as an early locus of reinvestment, and the gentrification frontier in the north-west moves quite slowly eastward until 1981. The 'resistance' to gentrification indicated by the map can be explained by reference to internal variations in this northern part of the East Village. There is a striking demarcation of levels of disinvestment, deterioration and abandonment east and west of Avenue A, which coincides geographically with the slow-down of the shifting frontier between 1977 and 1980. To the west, the neighbourhood always retained a level of commercial viability, catering to the needs of the local Ukrainian and Polish residents, and as such may have been perceived as a less risky property market. To the east, in what is known as Alphabet City, the majority of residents were and still are Hispanic, vacancy rates and abandonment are higher and the drug market flourishes. The risks of reinvestment here were far greater.

Somewhat puzzling is the early reinvestment coincident with the western end of Delancy Street. From here, reinvestment radiated south, west and east, but was blocked to the north. This can perhaps be explained by the proximity of this district to the southern section of the Bowery: a commercial centre specialising in the sale of industrial kitchen equipment, but also a resilient 'skid-row' where flop-houses, homeless shelters and cheap hotels are concentrated. Other so-called undesirable land-uses are also found here, including gas stations and vacant lots. It would seem that this contextual understanding of both the housing stock and local land-use patterns

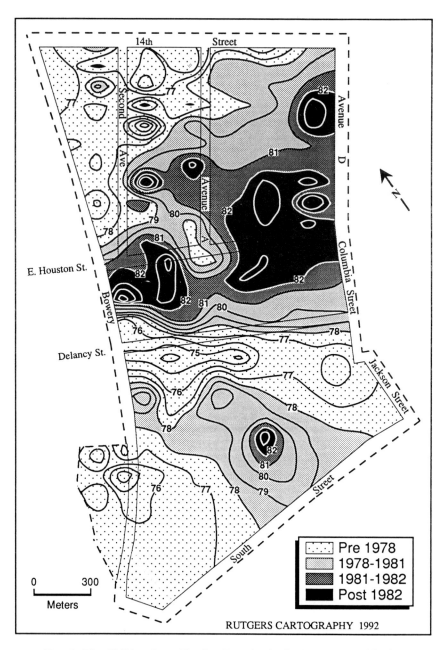

FIG. 4. The Shifting Gentrification Frontier in the Lower East Side (by housing unit arrears per tax block).

provides a more concrete explanation for this particular barrier to the frontier.

Figure 4 shows the frontier calculated in terms of housing units at the tax block level, the highest possible scale of resolution. In comparison with the convincing patterns and gentle curves of the less spatially disaggregated census tract maps, we see a much more complex, fragmented and localised pattern of reinvestment. As before, the earliest turning points are in the north-west and in the south, with the eastern section turning later, but now there is a more widespread distribution of areas that 'turned' later. By 1978–79 the frontier had begun to consolidate somewhat in the northwest, after which it moved east with greater strength. In the south the frontier advanced east fairly steadily between 1976 and 1979. The very dense barrier of resistance between East Houston, Delancy and the Bowery is evident again, although this time we are picking up internal variations in the investment potential at this localised scale, with some areas 'turning' before others. Another major area of resistance is found at the eastern end of East Houston Street, a locality containing many public housing units and flanked by the East River Drive, a heavily used and noisy highway.

From these maps it is apparent that the economic frontier is not continuous, spatially or temporally, but proceeds in a somewhat disjointed, stop–start fashion, influenced and shaped by local variations within the neighbourhood. In turn, its form and advance are also influenced by events external to the neighbourhood, particularly those impacting capital flows and investment decisions such as housing and land-use policy, drug response, interest rates and the like. The fragmented nature of the gentrification frontier does not, however, detract from its meaningfulness. The advance of the Western frontier from which the imagery derives was also a highly uneven, unequal and spatially differentiated process which left behind in its wake many pockets of resistance. In part the greater complexity of **Figure 4**, as compared with **Figure 2**, results from the more detailed data employed. Both maps represent generalisations of the gentrification frontier, but at different scales. The decision concerning an appropriate level of generalisation is not a technical one but a conceptual one.

So far we have looked only at reinvestment manifested in the redemption of tax arrears. We have assumed that financial transformation is geographically and historically correlated to the onset of actual physical transformation of the built environment, but that remains to be shown. The rehabilitation of structures may be the most immediately apparent and most trenchant indication of gentrification, but – as we have argued – the onset of reinvestment predates actual physical improvements. As part of ongoing research we intend to look at the spatial and temporal correspondence between the financial frontier and the physical frontier marked by building rehabilitation.

Only a pilot analysis for a single tract is yet available, but it does suggest the correlation of financial and physical frontiers and the feasibility of their reconstruction. Information on rehabilitation and construction activity is obtained from the City's Buildings Department, where building permits are granted for projects that meet the City's approval, and it can be shown for individual census tracts that the trend in physical rehabilitation follows closely the pattern of tax arrears (as measured by housing units). This evidence suggests that a 'lag' of two years or less exists between the onset of financial reinvestment and that of physical rehabilitation. A similar relationship is apparent if tax arrears data measured in terms of tax lots is used. Although the pilot study is limited in its scope it does suggest a strong spatial and temporal relationship between earlier and later stages of reinvestment.

Conclusion

Our project in this paper has been to elucidate the connections between the social and cultural representations of gentrification and the economic geography of the process. By exploring and giving a context to the imagery, language and symbolism that surrounds contemporary urban restructuring, we have attempted to illuminate some of the ways in which ideologies of gentrification are both constructed and propagated, as well as indicating something of how these ideologies then contribute to the reshaping of the city.

The representation of the Lower East Side as a frontier – as exotic, chic, dangerous, savage – has proven a potent invitation to gentrifiers and consumers in general. Yet this carefully constructed appearance is deeply laden with some very fundamental contradictions at the heart of capitalist society, and it serves to disguise hegemonic realities of class, race, community, reproduction and profit which constantly reshape lives and landscapes. The cultures and communities of the Lower East Side lie outside white, middle-class America and its definitions of civility, which in turn attempts to appropriate, supplant and dominate the area as an internal frontier.

As part of the struggle over class interests and privilege, property and land values are transformed as capital flows back into previously disinvested neighbourhoods. With the stakes thus raised, the momentum of gentrification is powerful. Yet it is important that we do not view this reworking of the city as a deterministic, unidirectional process. Neither the reinvestment nor the reimaging of a place guarantees that gentrification will occur. Besides the unpredictability and the influence of external events such as recession, changes in policy and the investment climate, internal resistance waged by organised residents presents a considerable deterrent and powerful influence on the pace and direction of reinvestment. However, the resultant of the forces of attack (operating by means of representation and reinvestment) encountering the forces of resistance (operating by community organisation) cannot be predicted. There is no precise correlation between the frontier of profitability and the geography of resistance. But by furthering our understanding of how ideology, culture and economics combine and work in tandem to construct gentrification, it may be possible to bolster popular resistance and to enhance the abilities of local communities to gain some control over their own lives and futures.

Note

1 This research was partially sponsored by the National Science Foundation, grant number SE 87-13043. The present paper draws on research reported in Smith, Duncan and Reid (1989).

10

Local Hero!
An Examination of the Role of
the Regional Entrepreneur in
the Regeneration of Britain's
Regions[1]

MICHELLE LOWE

Introduction

On Thursday 26 October 1989 Radio One's *Newsbeat* Programme ran an item on 'the tallest tower in the world' shortly to be erected by the Richardson brothers at the Merry Hill Shopping Centre at Brierley Hill in the West Midlands. The tower (the purpose of which was unspecified) would not, it was claimed, require planning permission as it was to be erected in an enterprise zone. A subsequent article in the journal *Planning* did, however, suggest that the Richardsons would need the approval of air traffic controllers (*Planning*, 1989).

This *Newsbeat* item, my chapter suggests, illustrates an extreme version of the potential power of the 'regional entrepreneur' in moulding the contemporary urban landscape. The chapter focuses on the activities of the Richardson brothers at Brierley Hill in the West Midlands. Their proposal to build a two thousand foot tower (see **Figure 1**) is the latest in a long line of plans that they have pursued which, they claim, aim broadly to regenerate the area of the West Midlands known as the Black Country. The Richardsons are an example of a new genre of 'regional entrepreneurs' who are currently having a fundamental impact on the human geography of the United Kingdom. This chapter examines the role of 'regional entrepreneurs' like the Richardsons in the regeneration of Britain's regions. It begins by positioning these individuals within a framework

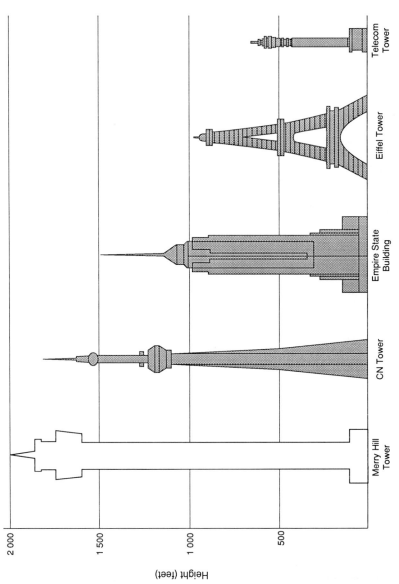

FIG. 1. The Proposed Merry Hill Tower in Relation to Other Tall Buildings.

of an 'economic localism' that emerged in Britain in the 1980s. It then outlines the activities of the Richardsons as 'regional entrepreneurs' and demonstrates how these individuals have been encouraged by the ideology and policies of the current Conservative administration. The chapter shows how people like the Richardson brothers draw specifically on local image and local culture to underpin their activities. Finally, it concludes by questioning the contribution of this type of 'regional entrepreneurship' to local economic regeneration.

'Economic Localism' and the Rise of 'Regional Entrepreneurialism'

The 1980s witnessed a significant shift in the nature and amount of regional aid available in Britain's declining regions. In line with their belief in the free market and the enterprise culture, Mrs. Thatcher's successive governments systematically dismantled the system of regional aid which had been operational in various forms since the 1930s. From the 1979 review of regional policy onwards the map of regional aid was 'rolled back' (Martin, 1986), and in particular the 1988 amendments – which involved redrawing the financial map, whilst leaving the geography of the assisted areas the same – signalled an explicit shift to a new ideological rationale: the economic and political promotion of an enterprise culture for Britain's regions.

This changing face of regional policy has coincided with an emergent 'economic localism' in Britain's regions. There appears to be a general agreement that broad-brush packages and policy measures are inappropriate for dealing with contemporary regional problems. In contrast, there has been the growth of a plethora of new policy initiatives and innovations designed to alleviate the plight of particular areas and particular labour market groups. Central government, local government, the private and the voluntary sectors have all pursued this type of strategy.

Within the framework of economic localism can be placed the activities of Councils like Birmingham City, discussed elsewhere in this collection (see Fretter, this volume), who have increasingly sought to lure investors and conference-goers through their new emphasis on culture and conventions. It is also under this head, however, that a new form of 'regional entrepreneurialism' has emerged. Encouraged by the incentives offered in certain key areas by central government and by the model of entrepreneurialism currently being peddled by many local governments, as well as being financed by monies of the type provided by a flourishing corporate sector, a number of individuals are currently making their mark by directing the nature of regional regeneration within their particular localities. In short, the way has been left clear for:

'... a person of vision, tenacity and skill (such as a charismatic mayor, a clever city administrator, or wealthy business leader) to put a particular stamp upon the nature and direction of urban entrepreneurialism, perhaps to shape it even, to particular political ends' (Harvey, 1989a, p. 7).

It is the place of such individuals in the regeneration of Britain's regions that this chapter will begin to examine.

Set against the context of economic localism outlined above a number of private sector key individuals have found a niche in a contribution to local economic development. Indeed, such 'regional entrepreneurs' became flag bearers for Mrs. Thatcher's 1980s vision of regional regeneration. Not surprisingly, they have become the targets for a great deal of media attention. John Hall, for example, the most famous of these characters, the man behind the massive Metrocentre at Gateshead, played a pivotal role in Mrs. Thatcher's cleverly managed 1987 election campaign. His centre, situated on a site on the banks of the River Tyne, provided an image of a Britain – and perhaps more significantly from the point of view of vote-winning, a view of a North East – pulling itself up by its bootstraps. Hall is currently developing a major leisure/living complex around his home, Wynyard Hall, near Billingham in Cleveland.

John Hall, albeit the best known, is only one of an increasing number of individuals who have emerged in some of Britain's regions as leaders of a new form of regional regeneration. Within the same retailing/leisure framework Eddie Healey has instigated the development of Meadowhall, a new shopping and leisure experience close to the M1 on Sheffield's outskirts, whilst Ernest Hall in Halifax has bought and redeveloped Dean Clough Mills, the former home of Crossley carpets, into an integrated complex for work and living reminiscent of those established by nineteenth-century industrial paternalists (Campbell-Bradley, 1987). At Merry Hill in the West Midlands the Richardson brothers, on whom this paper will focus, are stamping their identity clearly on the regeneration of the Black Country via their (now infamous) tower, shopping and leisure complex and new office spaces.

What is fascinating with regard to the theme of this volume is the way in which such people appeal to a sense of place, an image of regional culture, in the promotion of their activities for 'selling their place'. This chapter will return to this question shortly. In the meantime it will outline the activities of the Richardsons as a contemporary case study, as well as providing some more detail on the way in which central government currently provides a healthy climate for such people to pursue their activities.

The Richardsons at Merry Hill

Don and Roy Richardson, twin brothers (see **Figure 2**), have developed an out-of-town regional shopping centre, at Brierley Hill near Dudley in the West Midlands (see **Figure 3**). At the heart of the Black Country, the area which most recently gained publicity through the dramatisation of David Lodge's novel *Nice Work*, the Merry Hill development rose like a phoenix from the ashes of industrial decline and is currently into Phase Five of its operations.

FIG. 2. Don and Roy Richardson.

The Richardson Family background is in heavy truck distribution in the West Midlands region. Their business bought and sold trucks and lorries nationwide, and was very successful. In addition, the family had also been involved in the acquisition and redevelopment of steelworks sites in the Black Country area. In many ways, then, these brothers were and still are archetypal 'local heroes'.

As well known local characters and businessmen, the brothers were approached by Dudley Metropolitan Borough Council who wanted them to become involved in the Dudley enterprise zone in order

215

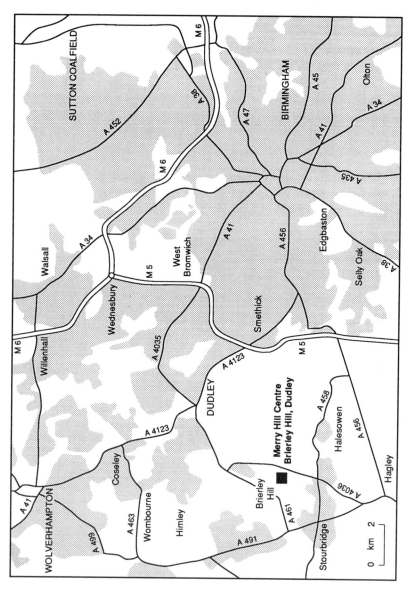

FIG. 3. The Location of the Merry Hill Centre.

to 'create activity' (Richardson, personal communication, 1989). Following this, Don and Roy made efforts to purchase the initial 108 acres of the zone from the British Steel Corporation at Round Oak and exchanged contracts on that land in November 1982. At this point, development progress by the twins was overtaken by events. The closure of the main Round Oak Steelworks (comprising 110 acres) was announced, enterprise zone status for that land followed in the shape of Dudley (Round Oak) Enterprise Zone, and Don and Roy eventually purchased that site, also from British Steel. This purchase, together with that of another 84 acres in the vicinity, meant that the Richardsons' land now totalled approximately 300 acres (and the brothers have added to that in little bits and pieces since). The first site to be developed by the Richardsons in the enterprise zone was 'The Wallows' industrial estate, situated on the former marshalling yards for the Round Oak Steelworks. During the life of the steelworks this was at one time the biggest private railway in the country, owned by the Earl of Dudley. Ignoring advice to leave well alone, the twins developed the site and there are now sixty to seventy industrial units on that estate. Private and public companies, international companies and individuals are all represented.

The centrepiece of the Richardsons' involvement in the Dudley enterprise zone, however, is the Merry Hill Centre (see **Figure 4**). By 14 November 1989 Merry Hill retail had reached Phase Five. Phases One, Two and Three respectively comprised retail warehousing of a pseudo-postmodern variety. MFI were the first tenants, and they were quickly followed by Toys R Us, Comet, a 'drive thru' Macdonalds, a free-standing Pizza Hut and the first Children's World in the United Kingdom. Aptly, perhaps, these first developments were nicknamed 'shed city' by critics of the scheme in the local press. In later stages, though, there was a shift in the style of the development. Thus, Phase Four was the first phase of upmarket retail, a double-storey connecting mall, and this has subsequently been followed by Phase Five, a retail fashion-based mall of over one million square feet incorporating several shops and major stores. Designed on two levels with decked and surface car parking serving both levels, Phase Five includes a finance court in which all of the major banks are represented, several building societies, a stockbroker and the latest up-to-the second information from the stock exchange. Sainsburys, Debenhams, Next and BHS, amongst others, are all currently trading. Phase Five, then, is:

'Probably the most important and certainly at this stage the most expensive part of the development' (Richardson, personal communication, 1989).

Thus, in the words of Don Richardson:

'If you look from Phase One to Phase Five, you can see how it has moved upmarket. It has moved from the retail park concept, which is extremely successful, to Phase Five, which compares favourably with similar developments all over the world. . . . On the exterior we have got marble, Italian marble in large areas completely covering parts of the development. That is mixed with mirror glass . . . and it all gives a pleasing effect. . . . Inside the malls you have got marble as well and intricate plasterwork, up-market concept all the way through . . .' (Richardson, personal communication, 1989).

The Merry Hill Centre was given its biggest boost yet in August of 1989 when Marks and Spencer announced its decision to open a huge new out-of-town store at the complex. Despite local controversy surrounding the Richardsons' decision to relocate several of their existing tenants in order to accommodate the flagship store, the move seems set to establish Merry Hill firmly in the retail shopping hierarchy of the West Midlands.

Not content with their success in the retail field, however, the Richardson brothers have topped the whole Merry Hill scheme with a monorail system threading through the current complex and scheduled to link Merry Hill retail with further Richardson developments currently under construction in the vicinity. The monorail is aimed to be an attraction in its own right, and this combined with other plans for the site is intended to make the Merry Hill Centre a substantial regional attraction:

FIG. 4. Advertisement for the Merry Hill Centre.

'The monorail is Swiss construction costing approximately 20 million pounds. . . . The first in the world was Sydney Harbour in Australia, and this is the first one in the northern hemisphere . . .' (Richardson, personal communication, 1989).

In addition to the development already on site, there is a planned waterfront leisure and business scheme on an adjoining site of 80 acres which will complement the Merry Hill scheme. 'Merry Hill: The Waterfront' will be a major leisure complex. The latest proposals include at 'Waterfront East' office development, and at 'Waterfront West' two hotels, conference and exhibition centre, a fun park, waterworld, ice rink and other sporting facilities.

Moreover, the Richardsons will not rest with the transformation of the West Midlands region which they have orchestrated by their Merry Hill scheme. Their latest project involves one of Britain's most famous industrial landmarks. The former tyre factory, Fort Dunlop, alongside the M6 in Birmingham, is to become the centrepiece of a multi-million pound scheme creating thousands of new jobs. The eight-storey Fort Dunlop building is reputed to be the largest brick industrial building in the Midlands, and the Richardsons currently have it ear marked for a combined business, industry and shopping complex. In addition, the brothers are intent on building a twenty-first century capital city for the Black Country, and are currently searching for a local site on which to do so.

All of the above activities are presented and marketed as part and parcel of the private sector's response to economic decline in the West Midlands region. However, whilst it is clearly the case that the Richardson brothers as 'regional entrepreneurs' have played a substantial facilitating role in the development of the Merry Hill centre to date, as well as in plans afoot for the former Fort Dunlop factory, it is possible to argue that schemes such as these are only flourishing in the United Kingdom at the present time because of the support (overt and otherwise) which was offered to this specific type of entrepreneur as a central part of Mrs. Thatcher's political project (and which continues to be offered under Mr. Major). It is to this controversial aspect of 'regional entrepreneurialism' that this chapter will now turn.

Central Government and 'Regional Entrepreneurs'

Central government support which has fed the Merry Hill development in the Black Country is unmistakable. Dudley Metropolitan Borough, as this chapter has already suggested, has two enterprise zones, both of which have been developed by the Richardson twins. The Dudley Enterprise Zone was designated during the first round of enterprise zones by the Department of the Environment in July

1981, and the Dudley (Round Oak) Enterprise Zone followed, itself being designated in October 1984, after the closure of the Round Oak Steelworks. The fact that the Richardson's Merry Hill developments are located on these enterprise zones has given them leeway not only in terms of their monstrous tower, but has also afforded the centre exemption from rent, rates and taxes over a ten year period. Furthermore, the Richardsons have also been aided by the fact that they were given an 'urban regeneration grant' of £3.25 million towards the cost of transforming the 36 acre former site of the Round Oak Steelworks. Thus, 'the Richardson brothers are the recipients under this Whitehall handout designed specifically to encourage inner-city sites without the need to go through the local authority' (*Planning*, 1987). Current developments at Fort Dunlop also qualify for substantial financial assistance as the proposal forms a key part of the new 'Birmingham Heartlands' Project.

It is interesting, then, to examine developments like these in the context of current central government regional and urban policy and to bear in mind how schemes such as Merry Hill have been substantially encouraged by the present political and policy situation. What is clearly of central significance with regard to Enterprise Zones and Urban Development Corporations is their focus on partnerships between the public and private sector. Indeed, Harvey has suggested that:

'the new entrepreneurialism has, as its centrepiece, the notion of a "public-private partnership" in which a traditional local boosterism is integrated with the use of local government powers to try [to] attract external sources of funding, new direct investments, or new employment sources' (Harvey, 1989a, p. 7).

In the case of the schemes under discussion here, the rhetoric used to promote them is almost identical to that utilised to sell/promote concepts like the free-market enterprise zones. More specifically, their most important aspect is considered to be the impact in terms of growth and regeneration that will result for their surrounding regions. Merry Hill, for example, is reputed to 'act as a catalyst for the regeneration of the Black Country as a whole' (Harris, 1989). In this way, schemes like these find themselves inextricably bound up with the rhetoric of current government ideology and promise surrounding the enterprise zone experiment. In fact, it is this imagery that is one of the more appealing aspects of a study of 'regional entrepreneurial' schemes, and it is to the appeal to image that the Richardsons make in their attempt to 'sell their place' that this chapter will now turn.

Regional Culture and 'Regional Entrepreneurs': The Richardson Brothers and the Black Country

Don and Roy Richardson are 'local heroes' in a very real sense. Born and bred in the Black Country, the area which they are now promoting, the brothers emphasise regional identity in the promotion of their projects. Indeed, this chapter will now suggest that these individuals are exemplary of the curious intermingling of regional culture, regional identity and regional regeneration which is occurring during an era largely dominated by increased globalisation (Harvey 1989; Swyngedouw 1989). Perhaps the strong regional accents of these individuals alone are signal to this effect? It is to the relationship between regional entrepreneurialism and local culture that this chapter will now turn.

In his examination of economic restructuring and local economic development strategies in Bristol, Bassett highlights the importance of 'images and ideologies' in the construction of local economic policy. Thus he suggests that:

'It is likely that different coalitions with different objectives will embrace different perceptions of the local area, and will tend to draw upon different local cultural traditions and different ideological elements in order to construct a distinct spatial identity for the area. Pro-growth and urban renewal coalitions, for example, may draw upon elements from the area's past to present an image of an historic or cultural centre adapting to new growth conditions . . .' (Bassett, 1986, p. 167).

In the case of Bristol, Bassett argues, 'a set of distinctive ideological elements and a particular image of the city' (Bassett, 1986, p. 170) were drawn upon. Hence, for example, 'local Conservative and Labour leaders linked the port scheme to Bristol's glorious maritime history, . . . [whilst] many of the media promotions (symphony concerts in Brunel's railway terminus and steam excursions in period costume) presented a selective view of the past in a way designed to appeal to potential investors and managerial elites' (Bassett, 1986, pp. 170–174). Within this framework there has recently been a great deal of attention devoted to the linkages between the specific features of local areas and local government economic policy (Duncan and Goodwin, 1988; Halford, 1991). However, linkages between other locally-focused responses to facilitate economic regeneration and the specific characteristics of local areas have hitherto been largely ignored. Most recently, though, Sophie Watson has drawn upon what she terms 'the new symbolic representations of de-industrialised regions' in her discussion of economic restructuring and local economic regeneration in Australia. Watson argues that the creation of new forms of identity for old industrial regions is central to their response to deindustrialisation (Watson, 1991). Within this vein of developing work, then, it is possible to situate this study of regional entrepreneurship. The entrepreneurs discussed throughout this chapter have consistently emphasised images and features of their local areas in their own attempts at 'selling their place'. So, as is Watson's contention in her work on Sydney and Wollongong, the same can be said of the appeal of John Hall, Ernest Hall, Eddie Healey and the Richardsons to the identity of the North East, Halifax, Sheffield and the West Midlands respectively: 'within each there is an attempt to produce and maintain a unique sense of place and to create a myth of identity' (Watson, 1991, p. 61). This chapter will now examine this contention in more detail with reference to the Richardson brothers and the Black Country.

The Black Country and the Richardsons

'The complex has five stations with names such as the Boulevard, Grand Central and Times Square and local connections are catered for by the aptly named Round Oak and Merry Hill' (Rhodes, 1989a).

'Merry Hill is a dream transformed into reality. It has been part of our lives since childhood, first as a gigantic steelworks, then as a depressing redundant industrial site and now, as you see it, an exciting and vibrant shopping centre which offers value-for-money quality goods in a superb environment' (*Express and Star*, Merry Hill Supplement, 1988).

'The Black Country is not so much a place as a state of mind. . . . It is a certain pride in the achievements of the past. It is not wanting to be swept away or to become . . . just an annex of Birmingham. This is a remarkable and very independent place. . . . Pride in the Black Country is on the increase. . . . People are proud of it' (Rhodes, 1989b).

'In terms of image their perception of image is not just a physical image, . . . it's something much more dynamic. The Richardsons are arguing that we are the Black Country, we have always been entrepreneurs, we don't need anybody else to help us, we are doing it for ourselves, we are turning ourselves around. Merry Hill is a great beacon. I would definitely give the Richardsons the benefit of the doubt, that their first loyalties definitely lie with the Black Country. They definitely have got a Black Country identification and they definitely identify with Brierley Hill because of their childhood. They think that what they are doing is just fantastic' (Sparks, personal communication, 1989).

In recent years there has been a significant attempt to 'sell' the Black Country, the region of the West Midlands which reputedly got its name from the blackening of the landscape which occurred as a result of the burning of a thousand furnaces. Thus, for example, local Black Country papers have run articles on local culture from Aynuk and Ayli (local comic characters) to faggots and grey peas (local delicacies), whereas roadsigns on entering any of the five boroughs which collectively comprise the Black Country are emblazoned with the name of the region – until recently an identity to be buried – together with a couple of cogs as a symbol of local industrial heritage.

The famous Black Country Museum and the planned Black Country World complex are all testaments to this effect. Nowhere in the region, however, is such imagery played on so strongly than is the case in the Richardson story. The brothers' vision of Merry Hill is one of the centre as the new symbol of Black Country success in the same way that the steelmills and chainmakers formed the image of the vital Black Country past. More specifically perhaps, the brothers trade on a sense of identity which is felt with the Black Country rather than with Birmingham. To people outside of the region, the Black Country and Birmingham are perhaps one and the same. In stark contrast to this, though, Don Richardson is quick to emphasise the strong and fierce pride felt in the Black Country as distinct from any feeling for Birmingham, and this he does in an extremely 'thick' Black Country accent:

'I have always thought that Birmingham is a big powerful city, and good luck to it. You know I hope that Birmingham succeeds and goes from strength to strength. The Black Country, adjoining Birmingham, consists of four boroughs: and the boroughs don't pull together unfortunately, and they are regarded as Birmingham's back yard and there are a lot of derogatory remarks made about the Black Country. ... It is written off by places like London and the South East. As far as the Black Country is concerned, the people in the Black Country are amongst the most honest, if not the most honest, in the country. They are certainly hard-working. That's been proven over the last hundred and fifty years; you've got the tough rough trades like steelworks. They have been prepared to get their jackets off and work at it and produce for their wages. That can't be said for people in other parts of the country, certainly not so much as it can be for people in the Black Country. ... The danger is that people forget what the strengths of the area are. There is a lot of strength in this area. The West Midlands and Birmingham and the Black Country have all got their good points and none of those are appreciated outside the area' (Richardson, personal communication, 1989).

Don goes on to demonstrate how Merry Hill is part of his vision of a revitalised Black Country:

'The way we see the Black Country is [that] the Black Country has to be brought up in the eyes of the rest of the country, and the world for that matter. One of the ways to do it is to do spectacular developments like this. The designation of an enterprise zone in Dudley has given us the opportunity to do something that nobody else has ever thought of and nobody else had ever attempted, and it was difficult to do because the area had been busy over the last hundred years, the ownership of the land had been fragmented and it has been difficult to build up a site of the order of two to three hundred acres, so you couldn't do a big development until this opportunity came along. Our object, all the time we have been interested in property, for around twenty five years, has been to build a large site and do our own thing. We have been able to do that here because of the relaxation in the planning restrictions' (Richardson, personal communication, 1989).

The appeal to the Black Country region's past is a central part of the Merry Hill scheme: its promotional literature is full of references to the Round Oak Steelworks, and the local press, as the quotations at the beginning of this section demonstrate, have been quick to capitalise on this aspect of the Richardson's activities. This appeal – to local history and to an associated local identity – has most recently been manifest in the Richardson's determination to emphasise heritage as an aspect of the 'Merry Hill: The Waterfront' development which is currently under construction. As is the case with heritage emphases more generally, however, such projects are clearly selective in their interpretation of local history (see for example Hewison, 1987). Moreover, and as Watson found in her work on Australia, 'it is the myth of unique identity which is what counts' (Watson, 1991, p. 68). This mythology of unique identity can also be recognised in the activities of the Richardson brothers. Thus the Richardsons identify with the Black Country as opposed to Birmingham, but this has not been at the expense of their recently moving their activities into Fort Dunlop, a factory on the other side of the West Midlands conurbation.

In summary, then, we can suggest the Richardsons are 'sentimental about the past (and local culture) but that they do not let that sentiment stand in the way of profit' (Hattersley, 1989). This contradiction is perhaps best exemplified by the way in which details given in the local press about the new Phase Five operations constitute a curious interplay of local, regional, national and international culture:

'It could be Dallas but it's Dudley and you wonder what Aynuk and Ayli would reckon to the place. You wouldn't come here for a tin bath, sheep's head, packet of firelighters or a bluebag for your Monday wash, that's for sure. Mike Johnson from the London-based consultants McColl who have designed this feat of fantasy say [that] they took the local character into account when they drew up their plans. There is a water theme running through Phase 5 of the complex, all sizes of fountains and pools, that was inspired by the nearby Dudley canal. Even Handel's water music was serenading shoppers as they passed through. . . . The six collonaded malls are all colour-coded so that people can identify where they are. Marble from Sicily covers the floors' (Rhodes, 1989a).

It is some of these more dubious aspects of the new wave of regional entrepreneurship that this chapter will now begin to elaborate on.

'Regional Entrepreneurship' and Local Economic Regeneration

This chapter has thus far situated the Richardson brothers as regional entrepreneurs within the context of an emergent economic localism in 1980s Britain. It has also suggested that these individuals draw specifically on local histories, cultures and images to underpin their activities. Here the chapter will begin to raise a series of questions concerning the contribution of this type of regional entrepreneurship to local economic regeneration. Focusing on the activities of the

Richardson brothers to date (leaving aside, that is, the up-market extravaganzas which they have planned for the future), it is possible to offer a few pointers as to the role of such 'local heroes' in the regeneration of Britain's regions. Here, then, analysis will begin by discussing the issue of job creation. It will then examine how far the rhetoric which such schemes embody in terms of their apparent ability to regenerate not only their immediate vicinity but also their surrounding area is matched by the reality of what has actually been achieved. Finally, and somewhat more ambitiously perhaps, the chapter will tie this study of regional entrepreneurs to a developing body of work which analyses the expanding tourism and leisure industries through critical eyes (see for example Harvey, 1989b; Urry, 1990).

The Number and Type of Jobs Created

'There are already about 3,000 people working in stores open on the existing part of Merry Hill. There will be something like an equivalent of about 2,000 full time jobs going in the next phase' (Benjamin, 1989a).

'More than 5,000 new jobs when the new phase of Merry Hill opens in two years time. The waterfront is expected to create about 5,000 jobs for office, shop and restaurant workers when it is opened in about two years time. This does not, of course, count the work for building workers in the meantime' (Benjamin, 1989b).

As demonstrated by persistent revelations from the press concerning the Richardson's contribution to job creation in the West Midlands region, there is a great deal of controversy surrounding the amount and type of employment that is created in shopping and leisure schemes such as that at Merry Hill. Traditionally, the debate has focused on the comparison between the number of full-time jobs for men (provided in manufacturing industry) set against the number of part-time female jobs (available in the new retail, catering and leisure trades). Brought to the fore in the starkest sense at Merry Hill where, prior to the enterprise zone experiment and in its heyday the Round Oak Steelworks employed nearly 5,000 men, the argument goes that the replacement of these jobs by the type provided by the Richardson brothers is not an equal exchange, the latter being largely part-time jobs for women.

Of course, in reality the trade-off between these two types of job is not as simple as it may at first seem. More significant is the *quality* of employment provided in such new and expanding types of work. Indeed a great deal of work remains to be done on exactly what are the *qualitative* as well as the *quantitative* employment impacts of schemes such as Merry Hill. Even if the scheme is assessed at base level – in terms of the sheer numbers of jobs created – there appears to be a good deal of contradictory evidence. Don Richardson himself is unsurprisingly generous in his 'jobs created' estimates, and he speaks of the strong commitment that he maintains to providing local people with work, once again echoing the emphasis on local connections brought out in the previous section of this chapter:

'Our main contractor, which is Tarmac, have indicated that there are something of the order of seven or eight hundred jobs in Phase Five. Some of these jobs are highly paid. . . . Everything we have tried to do we have tried to do locally, we have tried to give them the first opportunity to enjoy the prosperity in the area and to create jobs for people in the locality. If you look at the main contractor, Tarmac, Tarmac are a local outfit, based in the Black Country. We have awarded them three large contracts on the site so far. . . . There are several thousand jobs already on site in the retailing sector' (Richardson, personal communication, 1989).

Local councillors, in contrast, are not quite as supportive. Furthermore, the argument is put that, once again, it is the *conditions* of the jobs together with their *long-term economic impacts* which should be considered:

'Firstly, the absolute numbers of jobs created have not reached their target. No way have they created 10,000 full-time or full-time equivalent jobs in that enterprise zone. Number two, a lot of the jobs are in fact part-time jobs. Now I think that part-time jobs have an increasingly important role to play in the economy, and to oppose part-time jobs *per se* is not to be just sexist but to be stupid. The quality of the jobs, however, is not good; in terms of careers, in terms of monetary rewards, in terms of conditions and ultimately, from an economic point of view . . . pushing supermarket trolleys around is no substitute for making steel. Now I am not being sexist in this, I am saying that you are not getting value-added production in that location whereas you did have it before with the steelworks. That cannot be totally satisfactory economics. . . . You are looking at the economic substance of jobs as well as the social and employment substance. . . . I am not convinced on the jobs scene, it has undoubtedly made a contribution, but not by the targets I would set if I was in central government' (Sparks, personal communication, 1989).

Ultimately, of course, the bottom of the numbers of jobs created question can only be reached with reference to an in-depth analysis of the schemes themselves, and even then (and as is normally the case with such evaluations) it would be difficult to take spin-offs elsewhere into consideration. The questions raised here, however, do suggest that there is a need to consider the overall amount of regional regeneration provided by such schemes. Such a question must necessarily focus on the appropriateness of the type of 'regeneration' which is forthcoming to all rather than only some sectors of the population, as well as on some of the 'spread' effects of regeneration in the locality as whole.

Regional Regeneration?

'Already many Wolverhampton shoppers have experienced the safe shopping environment provided at Merry Hill Centre with over 10,000 free car parking spaces and a major high street under a climatically-controlled roof. I'm sure many more will' (Alasdair Fulton, Marketing Manager, Merry Hill Centre, *Express and Star*, 4 November 1989).

'Does Mr. Alasdair Fulton, the Marketing Manager of Merry Hill not realise that his wonderful Merry Hill is killing our town centres and our community spirit? It is fine for people who have transport and can afford it, but what about the old, infirm and ill who need our local shops? It would appear that no-one cares' (J. Towe, Compton Road, Cradley Heath, *Express and Star*, 13 November 1989).

An inherent contradiction of schemes like that at Merry Hill is encapsulated in the above correspondence which took place in the letters page of the local Dudley newspaper, the *Express and Star*. The exchange clearly indicates that the 'success' of the Merry Hill scheme must be measured in terms of its impact on regenerating the Black Country region as a whole, and not just in terms of its role relative to a select few areas or to improving the lives of only a few people. This does perhaps seem a tall order for any such scheme to achieve, but it must be remembered that the regional entrepreneurs under discussion here specifically market their activities as forms of 'regional regeneration'. In contrast, critics of such schemes have consistently suggested that their ethos is not akin to regeneration for all citizens. Rather, it has been argued that such schemes are likely further to enhance social polarisation which has emerged in many places during the 1980s. Moreover, it is argued that such schemes will have a detrimental rather than a positive effect on surrounding areas.

Firstly, it is significant that the few studies of shopper profiles carried out at centres like Merry Hill (notably a Metrocentre survey by the Oxford Institute of Retail Management) have suggested that it is only certain types of individual who visit such schemes. It could reasonably be expected then that a similar profile will develop at Merry Hill. At the Metrocentre, for example, there are very few shopping groups headed by older unemployed people. In contrast, the centre attracts large numbers of younger age groups (especially those from 25–44) and people with young children (Davies and Howard, 1987). The vast majority of the centre's shoppers travel there by car. Indeed, the design of the types of scheme under discussion here is specifically focused on the 'mobile shopper', and it is revealing that Merry Hill's promotional literature consistently emphasises '10,000 free car-parking spaces'. As one consequence, a large number of complaints concerning the difficulty which staff without cars experience in getting home from the centre have adorned the letters pages of the local press. To take this line of argument further, the employment of the centre's own security guards (a 'selling point' in Merry Hill's promotional literature) can only mean the maintenance of 'private spaces' controlled and monitored by the centre's own 'police force' from which it will be possible to exclude 'undesirables' and (more significantly perhaps) the possibility of industrial disputes. It was notable, for example, that in the run-up to Christmas 1989, striking Ambulancemen were seen in every other local shopping centre in the West Midlands conurbation but Merry Hill.

Secondly, and maybe more relevantly, is the impact of such schemes on both surrounding centres and (less tangibly perhaps) surrounding communities. In the same way that the Metrocentre has affected the centre of Newcastle City, it appears that Merry Hill will have a detrimental impact on surrounding shopping centres. Indeed, there is ample evidence suggesting that this is already the case, particularly with closures in Dudley town centre closely mirroring openings at the new Merry Hill Phase Five development. In turn, of course, such factors have implications for the overall regional employment impacts of centres like Merry Hill (see above). Such questions, concerning the impact and effects of the activities of regional entrepreneurs, like the issues of job creation discussed above, clearly require much further investigation. The evaluation of such things is, of course, far from easy.

It is certainly possible to damn developments such as Merry Hill through an emphasis on the inequalities which such schemes seem to engender. However, it is equally possible to suggest that at the present time a Merry Hill is 'better than nothing'. Any evaluation of regional entrepreneurship and its effect on regional regeneration must be realistic. Thus, here we are not discussing a trade-off between manufacturing and service jobs, but more realistically we should be assessing whether a Merry Hill is better than nothing at all. Furthermore, and less tangibly perhaps, it is important to remember that a mainstay of such schemes is not so much their immediate impact on the regeneration of their area as their longer term impacts and their effect on the confidence of the people of the area which will arguably (and also in the longer term) lead to regional regeneration. This issue is a notoriously difficult one to get to grips with. Having said this, though, there are in my personal experience a number of individuals who live in the West Midlands and who think that in terms of job opportunities and 'areal image' the scheme at Merry Hill is marvellous. Critics would doubtless argue that the development is merely providing these people with 'a mental refuge in a world which capital treats as more and more place-less' (Harvey, 1989a, p. 14), and a great deal of work evidently needs to be done to untangle these tricky questions.

The Tourist Gaze?

'Tourists from all over Britain and the rest of the world already flock to Merry Hill and their numbers will rise dramatically thanks to a new initiative. Dudley Council is to invest £5 million in tourism over the next five years. . . . Travel firms and coach operators are being encouraged to visit both Dudley and Merry Hill by the newly-formed Dudley tourism association' (*The Merry Hill Experience*, March 1987).

Centres such as that at Merry Hill raise a number of questions concerning the recent development of tourism in Britain. As Urry (1990) suggests, such activity is no longer confined to a fortnight's holiday by the seaside: rather, it is increasingly allied to everyday experiences in urban as well as rural locations. The 'local heroes' studied here are united by a common thread, that is the appeal to tourist revenue in some way, shape or form. At Merry Hill, a 'water-way' theme, monorail and planned waterfront attractions (which are to include a decommissioned submarine) combine to conspire to attract visitors from far afield.

The *raison d'être* behind such 'spectacles' is considered to be the way in which they will attract tourists and finance from outside of the region in which they are located. Indeed, the schemes both current and planned by John Hall on Tyneside, the Richardsons in the Black Country, Eddie Healey in Sheffield and Ernest Hall at Dean Clough Mills in Halifax are all directed at luring in such activity. There are clear potential macro-economic problems which could result from such a strategy. What is significant in the case of Merry Hill, however, is the way in which such notions underlie current local authority support for the Richardson's activities, whereas in the past local councillors were rather more sceptical. What must be ascertained is how far do such tourist emphases divert attention from real social and economic problems which still exist despite the actions and involvement of the 'local heroes' existing in the locality. There is clearly much work which needs to be done on this. Most significantly here perhaps, and also from the standpoint taken earlier in this chapter concerning the selective recreation of images, it is possible to argue that in the same way that theme parks such as Disney World in Florida have 'shrunk and distorted the world' (John, 1989), so developments like Merry Hill have 'drawn a veil over real geography through their construction of images and reconstructions' (Harvey, 1989b, p. 87). What is of course pivotal here is the way in which many local authorities have been taken in by this new version of the classic 'zero-sum' game. Indeed, it is in this way and in emphasising tourist benefits that Dudley Metropolitan Borough Council justifies its change in strategy from opposition to support of the Merry Hill scheme:

'For some considerable years now we have had the attractions of the zoo, and the castle has been there many years further back than that. . . . The Merry Hill Centre fits in as the latest tourist attraction that Dudley has. Certainly, when the leisure side is developed as well, to increase the visitor experience. It *does* complement what is going on in the Borough anyway. I think when people come to visit an area shops are normally integrated in some way into their experience. . . . People are attracted to areas where, hopefully, there is a future and where things are going on . . . where there are a lot of initiatives happening' (Smith, personal communication, 1989).

'The other reason why there has been a change of policy is that it is also the case that Merry Hill will bring in a lot of people to the Borough, and that does support the tourism drive. . . . In my experience, the more attractions you have in a place the better . . .' (Sparks, personal communication, 1989).

Some evidence, perhaps, to support Harvey's view that even the most resolute and avant garde municipal socialists will find themselves in the end playing the capitalist game (Harvey, 1989a).

There is clearly space for much future research on the impact of schemes and initiatives developed by regional entrepreneurs on regional regeneration. There is also a pressing need to incorporate the study of the activities of such individuals into an analysis of the changing face of United Kingdom urban and regional policy. Such questions clearly set an agenda for future research on the area which I have set out to highlight here.

Postscript

In 1990 the Richardson brothers sold their Merry Hill retail developments to Mountleigh Group, based in London, for an undisclosed sum (Wells, 1990). The Richardsons still retain ownership of the Merry Hill Monorail and the Merry Hill 'Waterfront East' and 'Waterfront West' developments. In May 1991 Michael Heseltine hinted that he is prepared to amend the planning regime in the Dudley Enterprise Zone in order to reassert planning control over the 2,000 foot Merry Hill Tower (*Planning*, 1991).

Note

1 I would like to acknowledge the help of the following people in the preparation of this chapter: Sean Cross, Merry Hill Centre Management; Penny Hadland, Department of Planning and Architecture, Dudley Metropolitan Borough Council; Steven Masters, Brierley Hill Library; Don Richardson, Richardson Developments; Jim Smith, Department of Economic Development, Dudley Metropolitan Borough Council; and David Sparks, Chair of the Economic Development Committee, Dudley Metropolitan Borough Council.

11

Architecture as Advertising: Constructing the Image of Redevelopment

DARREL CRILLEY

Introduction

While critical urban discourse in the 1980s began to pay greater attention to images and representations *of* redevelopment in the city, exposing the semiotics of image construction (Burgess and Wood, 1988), analysing attempts to revision the image of the industrial city (Watson, 1991; see too the chapter by Holcomb), and demystifying the pervasive tendency for redevelopment to be scripted as the taming of a hostile urban frontier (Smith, 1988; see too the chapter by Reid and Smith), the images projected *by* the city, particularly its architecture, attracted less sustained attention. This was despite the upsurge of interdisciplinary debate over the aesthetics of the postmodern city and a tendency to read landscapes through the metaphor of 'text'. My chief contention here is that the postmodern architecture of redevelopment – with its facadal displays, penchant for recycling imagery and theoretical rationale in semiotic theory – is fully incorporated into the ideological apparatus of place marketing, playing a major role in mediating perceptions of urban change and persuading 'us' of the virtues and cultural beneficence of speculative investments. The focus is upon two closely linked enclaves composing the redevelopment landscapes of New York and London: London's Canary Wharf and New York's Battery Park City. These projects are connected by several similarities: both proceeded almost in tandem in the 1980s, impelled by a speculative imperative to accommodate the restructuring of financial services; both attempt to condense the functional diversity of the city within their precincts; both are private initiatives featherbedded by legal and financial public subsidy (ranging from tax abatement packages for developers, through discount land sales to provision of supportive infrastructures); both are under-pinned by a globalised property developer (Olympia and York, hereafter O&Y) working in unequal partnership with quasi-public redevelopment authorities (the London Docklands Development Corporation (LDDC) in London, and Battery Park City Authority (BPCA) in New York); and both are key components of a socially and spatially uneven redevelopment process that has recaptured the central city for the privileged social strata of a so-called post-industrial society.

In this chapter I propose to demonstrate that they are being sold and legitimated *by* and *through* their architecture. In both cases there are carefully prescribed visual assemblages, each proclaiming a restoration of tradition. It is the past, both recent and distant, popular and patrician, which is used to authenticate current undertakings. For, in a society disillusioned with the urban prophecies of modernism and its functionalist logic, and where the nostalgic impulse grows stronger, pervading cinematic genre (Jameson, 1984) and nurturing the heritage industry (Hewison, 1987), designing urban futures to look like the past is an expedient route to popular consent.

Canary Wharf, a commercial megastructure scheduled to provide a 12 million square foot 'city within the city' when completed in the mid-1990s, represents the apotheosis of the union between aesthetics and grandiose speculative property development. Its developers, global conglomerate O&Y, have engineered an elaborate cultural physiognomy for the project, in which a diverse, historicist architectural aesthetic, supposedly derived from the patrimony of London's town planning traditions, plays a key role. It serves a double publicity purpose: it is a key weapon in the bid to pre-let unfinished office space in a location (the Isle of Dogs) considered 'peripheral' to its intended market, firms in the higher echelons of financial services of the City of London; and it helps to engineer public consent for a project effectively withdrawn from public control, since – by presenting us with a decorous, expertly managed mini-city, cloaked in the respectable trappings of architectural culture – O&Y hope that, rather than being accused of oppressing us with the brutalism of urban renewal, they will be complimented for edifying us through their promise of an authentic urbanity.[1] Battery Park City, an exclusive and affluent mixed-use project located on the lower west side of Manhattan, combines the New York precedent for Canary Wharf (the World Financial Centre, also developed and managed by O&Y) with a collection of up-market residences. It too relies upon its design for publicity effects. Here, beginning in 1980, the designers engineered a marketable aesthetic by abandoning earlier modernist visions for the site (a 92 acre landfill) and redesigning the project as a facsimile of older New York neighbourhoods. Like Canary Wharf, this was intended to imbue the neighbourhood with the instant aura of tradition and familiarity, and in the process tap into the gentrification market dominating lower Manhattan in the 1980s chiefly by catering to the middle-class tastes there being expressed. As a prelude to these case studies, however, it is helpful to analyse the nature of architecture's current identification with advertising.

Architecture Becomes a Form of Advertising

Architecture has once again become a form of advertising as city governments, powerful corporations and innumerable redevelopment agencies show a renewed interest in manipulating the built environment pro-actively. There is more to this than cultivating the skyline or pursuing assiduous conservation policies. Rather, it is an integral part of the incorporation of cultural investment and policy into urban growth strategies, as cities struggle to attract inwards investment by amassing the correct mix of cultural or 'soft' infrastructure, whether it revolves around a 'livable' environment adorned with public art works, a major new institution (such as Liverpool's 'Tate of the North') or the creation of 'cultural districts' (Whitt, 1987; Bianchini, 1989).

Architecture, as much as expensive city marketing campaigns, is mobilised to transmit a catching, idiosyncratic image of urban vitality. This is evident at all levels in the urban hierarchy. Thus, whilst French leaders in the 1980s have commissioned an innovative coterie of modern architects to design Paris's spectacular *Grands Projects*, conveying its cultural dynamism (see the chapter by Kearns), Birmingham City Council has been acquiring its own architectural show pieces (including Britain's largest international convention centre) and collection of public art trophies in a bid to establish its status as Britain's second city (Moore, 1991; see too the chapter by Fretter). Just as significantly, it has adopted a City Centre Design Strategy recommending a comprehensive package of aesthetic improvements calculated to overturn the image of Birmingham as a dehumanising and unfathomable concrete metropolis. It proposes all manner of changes from recladding the facades of the city's innumerable slabblocks, through guidelines on how to convert tall buildings into memorable landmark buildings, to a conservation strategy to preserve Birmingham's so called 'Quarters'. As the urban designers stated, the goal was to remake civic identity:

'The intention is to put Birmingham on the map as a European City – *the* European Second City. Development must be unique to, and appropriate to Birmingham – not just a collection of inappropriate trans-Atlantic copies or anonymous tired solutions that can be seen anywhere' (Tibbalds *et al.*, 1990, p. 2).

Other British cities have more modest goals. Thus, lacking the basis for an image rooted in conservation and heritage, Swindon's boosters projected the alternative image of a 'high-tech' city by legislating in favour of late-modern architects such as Norman Foster (Bassett *et al*, 1989). Meanwhile, further along the M4 corridor Cheltenham has meticulously restored the regency ambience of British spa towns in its bid to attract relocating companies and state agencies (Cowen *et al*, 1989). Similarly, in gearing up for competition to attract visitors and investment, Glasgow complemented its notoriety as 1990 'European City of Culture' by rehabilitating its tenement buildings to add a 'Victorian' motif to the city, only to feel the tension of trying to accommodate an expansion of office space within such conservationist constraints (*Architects Journal*, 30 May 90).

Urban Development Corporations, so often accused of promoting image over substance, have also used their powers of aesthetic and development control to nurture distinctive images. Redevelopment of London's Docklands, for instance, has been conceived and executed within a framework of conservation and contextualist design (except on the Isle of Dogs), premised on the notion that old wharves, dock basins and conservation areas would be the most valuable environmental asset at the disposal of the LDDC in its bid to propagate the image of a 'Water City of the 21st Century' (LDDC, 1987). By 1990 the Corporation presided over approximately twenty conservation areas and literally thousands of listed structures ranging from dock side cranes to Victorian wharfs, all of which were steadily being incorporated into a series of heritage trails. Elsewhere, the LDDC's use of architecture to build a glitzy image has proved more pernicious and controversial. In 1988, for example, it rejected (by way of a belated architectural competition) a modernist scheme for a new Isle of Dogs neighbourhood centre, evolved for and in consultation with the local council. The reasons were overwhelmingly aesthetic, centering on the fact that it did not match the future envisaged for Docklands. A functional and flexible prototype for a new town hall, it was patronisingly dismissed as 'primitive and foreign', more suited to the context of a Third World city and simply not 'up-market' enough for the enterprise world of the Isle of Dogs (Beigel, 1989). In its place stands a glitzy postmodern alternative.

Nor is architecture's publicity value restricted to specific aesthetic effects. Buildings designed by culturally consecrated architects also function as 'symbolic capital', signifying the cultural nobility and taste of the patron. Once architects have gained recognition, status for the owners of their buildings resides as much in association with the name as in the building's appearance. While large corporations have long used architects as 'designer labels' to command a premium for their buildings, redevelopment agencies have also sought to improve their cultural credentials through support for known architects. Canary Wharf includes a full inventory of big name architects, for instance, and at Battery Park City an approved list stipulated that architects involved must be 'world class' and 'internationally recognised'. Again, the LDDC is an instructive case of attempted cultural upgrading. Long criticised for presiding over a cultural 'wasteland' or 'architectural zoo', the LDDC has recently attempted to combat such negative imagery by touting its commitment to 'cultural regeneration' and to design quality. Thus, an arts action programme has been launched (LDDC, 1990a); an 'urban design' department and design review board established; a guide to the architectural merits of the Isle of Dogs published (LDDC, 1990b); a series of design symposia hosted at the LDDC; and an architectural awareness programme launched, advertising Docklands icons. In the latter case regular press releases have emphasised buildings or areas within Docklands that have received critical acclaim from the architectural meritocracy, the LDDC hoping to enhance its public profile by association with architecture carrying official credentials.[2]

The cross-over of architecture into advertising is not a wholly client-centred phenomenon. Rather, it is facilitated by an architectural paradigm permeated by the ethos, techniques and language of marketing. Thus, theoretically, leading proponents of postmodernism have argued that architects learn their imagery from the findings of market research and its segmentation of mass markets into a plurality of 'taste cultures' and 'semiotic groupings' (Scott Brown, 1980; Jencks, 1987, pp. 55–62). Practically, architectural firms have adopted a more formal and ordered approach to advertising their services, deploying marketing consultants, glossy brochures and celebratory monographs to package and to propagate their styles (Kieran, 1987; Gutman, 1988). And, just as place advertisements frequently project optimistic images of the urban future, so too have architects become increasingly adept and technologically sophisticated in simulating the 'virtual reality' of redevelopment projects (Pawley, 1990). Traditional marketing techniques like brochures are now supplemented with a mixture of futuristic models, holograms, audio-visuals and computer mock-ups helping potential clients and the public to visualise completed developments. In the case of London's Canary Wharf, for instance, the computer-aided designs of Carlos Diniz were vital to the marketing of this vast commercial mini-city. As Diniz himself comments:

'When the architects and developers came to us, the Canary Wharf project was only very sketchily designed. We helped dream up the neoclassical and postmodern architecture the architects felt was appropriate to this ancient waterway. We drew our inspiration from such sources as Canaletto's famous seventeenth-century oils of the Thames and from prints of the river in Victorian times to show how the presently run-down section known as the Isle of Dogs could be transformed into an instant urban scenario' (Carlos Diniz, *Los Angeles Times*, 20 November 1988).

More significant than architects appropriating and replicating advertising strategies, though, is that buildings themselves are designed to 'read' as gigantic outdoor advertisements. The most obvious instance of this is the historicist phantasmagoria of contemporary festival retail malls. At sites such as New York's restored South Street Seaport or London's Tobacco Dock, we witness an aestheticised maritime history, with nostalgic architectural compositions establishing the cultural theme and ambience conducive to up-market consumption. In both these instances emblems of mythologised piracy, mercantile conquest and nautical adventure establish a fictive nexus between retailers and consumers. Whilst advertising generally indicates appropriate geographical contexts for consumption (Sack, 1988), here architectural codes actually form the context nurturing consumer desire. The development and standardisation of such motifs is akin to a form of market research. As Gottdeiner suggests, the 'uniformity of motif choice is the architectural analogue of the marketing research search for canonical forms of appeal to be engineered into consumer products and product logos themselves which have been proven to stimulate consumption' (Gottdeiner, 1986, p. 294). Such procedures are in fact grounded in architectural theory.

It is particularly noteworthy that a leading theorist-practitioner of postmodernism, Venturi (1966; *et al.*, 1972), urged architects to concentrate on surface decoration of complexly programmed buildings, rendering the built environment a 'billboard' displaying the images and meanings selected by architects. According to Venturi's programme, architecture was to be quintessentially symbolic manipulation and 'rhetorical applique' rather than producing forms. Although few architects have followed his specific advocacy of learning from commercial vernaculars of the Las Vegas strip or Levittown, his programme to restore the fictional and symbolic qualities to architecture has become *the* leitmotif of postmodern architecture (Klotz, 1987; Ghirado, 1990; Mcleod, 1988). Under the rubric of postmodernism, architecture and its cognate aesthetic practices (such as urban design and public art), has primarily become a matter of the systematic, purposive manufacture and marketing of commercialised meanings. Buildings are viewed pre-eminently as communicative texts, which like advertisements are culturally encoded with popular meanings: and, just as all advertising follows the ancient art of rhetoric in always being an attempt to persuade potential customers that a particular commodity is worthy of purchase, so does the imagery of architecture seek to persuade the public of the virtues and propriety of the property capital commissioning it.

Moreover, just as modern advertising rhetoric supersedes the merely utilitarian and informative to fill the 'hollowed-out' world of consumer goods with meaning, drawing upon the existing normative order of society (Williamson, 1978; Wernick, 1983; Leiss *et al.*, 1986), so too do contemporary architects strive to give urban redevelopment projects an acceptable cultural alibi by using an already resonant architectural symbolism. Against modernism's aesthetics of shock, defamiliarisation and estrangement through an anti-ornamental architecture of new materials and spatial complexity, the influential postmodernism of Venturi *et al.* (1972) seeks to familiarise by systematically learning from the city 'as it was and as it is', using recognisable design codes and pattern languages. Framed within an acceptance of consumerist *status quo*, the intention is to create a mildly educational, entertaining architecture with popular commercial appeal. In fact, the architecture so derived is a powerful and tangible adjunct to place advertising. As I will show, it too bolsters attempts to overturn negative perceptions of 'marginal' redevelopment areas and erase stigmatisation of landscapes of industrial decay or those ravaged by the visual blandishments of modernism.

Further, in the redeveloped city, architecture's mode of reception resembles that of advertising insofar as 'the public' are interpellated as spectators at a series of triumphant architectural displays. The modernist emphasis on the spatial experience of architecture and optimistic faith that architecture could be an effective instrument of social change, meeting the pressing needs of a reductive, physiologically-defined model of humanity, stands renounced and replaced by a deification of the two-dimensional, visual apprehension of the city (cf. Colquhoun, 1985; Holston, 1989). In the practice and discourse of postmodern architecture, surface appearance and visual effect is paramount as buildings are designed from the outside in, from the vantage of an external gaze (cf. Harries *et al*, 1982). Accordingly, within the scenographic enclaves and panoramas of the postmodern city, 'the public' are positioned as consumers of visual imagery who passively receive meanings (double-coded, multivalent, 'high' and 'low') prescribed for them by architects. The 'univalent' built environment of modernism is refashioned with unprecedented stylistic eclecticism, conveying vernacular traditions, historicist motifs and exotic allusions. It is the content of these images and their ideological significance in the context of the redevelopment projects considered here that now requires detailed analysis.

Darrel Crilley

Canary Wharf: an Architectural Cinerama of London

In the bid to transform Canary Wharf into London's 'new business city in the east' and to establish its public acceptability, architecture has been a principal instrument. Not the least significant part of this is the dramatic impact of its showpiece – an 800 ft obelisk shaped skyscraper clad in stainless steel – on the skyline of London. Towering over the vernacular of east London, this monumental landmark is an unequivocal if aesthetically controversial reminder of O&Y's presence in London. Awareness is not acceptance, though, and towards this end O&Y have deployed a more subtle design rhetoric to mollify and to humanise their gargantuan development.

Its form and appearance relies on the art of dissimulation: making the megastructure appear to be what it is not. Thus, its deliberate anti-modernism expresses a desperation to avoid the 'project look' of La Défense in Paris and to deny its own genesis as the product of a single speculative investment. Every aesthetic power is mobilised to make this complex an enchanting and familiar place, simulating the diversity of a historically developed city within a compressed time frame of under ten years. It is composed of a network of artfully designed public spaces, giving coherence to over twenty-four buildings, each packaged with a different architectural style, thus giving the impression of several discrete initiatives rather than a corporate monolith. As an O&Y architect explained, it will be a 'district', not a 'development':

'Canary Wharf I think will not just feel like a large development. One will hopefully go from area to area within the larger district such that it becomes like the West End where you go from one address to the other: from Grosvenor Square to Piccadilly will be mirrored in the transition from Westferry Circus to Cabot Square. The public spaces form a linked network, but each has a different emotional appeal that can be related analogously to London's Squares' (Vinay Kapoor, personal communication, 1990).

Canary Wharf's public image is thus to appear as a seemingly natural, organic extension of London. This is orchestrated by commissioning a controlled architectural diversity and through an analogical design process which conscientiously tried to discover what makes London's public spaces memorable, and then to incarnate them in new oversized forms. As the leader of the design team suggests:

'We were looking at the character of London . . . soaking it up, observing both its formality and informality. Examining what were the big pieces you remember and trying to work out why it was that you remembered them. Because, when it came to Canary Wharf, we knew that was what we would be trying to do. To make places that were both memorable and comfortable' (Tony Coombes, quoted in O&Y, 1990a).

Architecturally, it is an unprecedented display of patronage for stylish design. To attract several tenants from the world of high finance and persuade the public that it is a genuine mini-city, O&Y had to commission a variety of named architects offering the choice necessary to cater for the different 'taste cultures' of business. Accordingly, Canary Wharf has the full inventory of styles and architects from the past decade, all loosely connected by the 'common language' set by Skidmore, Owings and Merrills's (SOM) design guidelines for the entire site. Each practice vends its own 'ism'. Thus, Cesar Pelli's steel-clad obelisk, a consummate example of 'romantic modernism', faces SOM's ornate neo-Edwardian revival; both of these face the sleek surfaces of Troughton Mcaslan's version of neo-modernism; across Cabot Square stands a granite-clad structure from Kohn Pedersen Fox, doyens of skyscraper design in the 1980s; and further west, I. M. Pei's modernism is juxtaposed with the regency classicism of West Ferry Circus. Even more academic architects such as Aldo Rossi, the acclaimed Italian rationalist, have offered their own packages. And one of Britain's 'big three' architects, Norman Foster, has submitted provisional designs for the skyscrapers to follow the existing tower. Not since the Rockefeller Center was built in the 1930s (cf. Balfour, 1978; Tafuri, 1980) has a developer staked so much on producing a cultural cladding of such fashionability and on juxtaposing such divergent architectural factions. For advocates of architectural pluralism such as Jencks (1987), Canary Wharf is accredited as a reconciliation between competing architectural stances; a resolution of contemporary 'style wars'.

No matter that this architectural diversity is entirely superficial, each practice adding a set of optional gift wrappings to a standardised package determined by the exigencies of producing expansive floor plates, it is still portrayed as evidence of aesthetic neutrality, even tolerance, and benign commitment to environmental excellence. Further legitimacy accrues when the scenography of the most ornate facades pays homage to tradition. As one architect explains:

'No. 3 Cabot Square recalls the spirit, human scale and texture of the traditional buildings of London. Rather than referring specifically to any one building, architect, style or historical period, it captures and distils the essence of the entire region' (Adrian Smith, quoted in O&Y, 1990b).

The grandiloquence of this claim is only surpassed by the suggestion that Canary Wharf as a whole is a microcosm of London. Sir Roy Strong, for example, former director of the Victoria and Albert Museum and now cultural apologist for Canary Wharf, has advanced it as a 'major reassertion of the tradition of the London square' (quoted in LDDC, 1990b, p. 53) and as a development 'more truly in the idiom of this metropolis than any for a very long time' (Strong, 1989). In the one third of the 72 acres devoted to 'public space', meticulous attention has been given to engineering the right kind of emotional response by modelling them on the typologies and detailing of London's existing public realm. O&Y's strategy treats the city as a repository of history from which an 'essence' can be discerned and incorporated as the animating principle of an entire new mini-city. According to one planner this ensures that architects do not produce 'rootless postulations of new urban forms' conflicting with the desire to produce a 'normal city' (Coombes, 1990). The result is a montage of shapes (esplanade, circus, square) and finishes (hard, formal landscaping and soft, gardenesque spaces) designed to evoke our collective memories of places such as St James's Park and Trafalgar Square. Other architectural conventions such as street-level arcades connecting buildings, cornice lines and street walls clad in stonework are also intended to recreate the feel of London's historic cityscape. Hence, in the same way that advertisers seek to persuade us of the virtues of their products by imbuing them with a meaning drawn from already understood cultural codes, this strategy tries to habituate us, 'the public', to the corporation's view of the future by cultivating a nostalgia for our most cherished and familiar city environments. It reuses typologies already carrying historically-accrued connotations of democracy, citizenship and 'the public domain', so as immediately to establish the civic credentials of the megastructure.

Attention to finer visual and tactile details strengthens this benevolent self-image of O&Y as conservator of civic conviviality. Every conceivable experience and sensory delight is apparently planned for in this attempt to aestheticise everyday life. This includes an assortment of custom-designed street lights and furnishings commissioned from the world's finest craftspeople; artistically programmed water displays and fountains; 1.5 miles of waterfront promenade; manicured gardens and mature vegetation imported from Europe, including more than 120 species of trees and shrub; and spaces of quiet serenity mixed with spaces of grandeur and theatricality. As the promotional rhetoric avers, Canary Wharf aspires to be the container for a 'complete urban community' mixing the proximity and solidaristic social interaction of 'country town life' with the vitality of 'city life at its best' (O&Y, 1990c). Here is a 'strategy of admiration' (Laswell, 1979), then, bidding for acceptance through exuberant display, cultivation of the picturesque and urban theatricality.

In a double sense, therefore, architecture forms the image of Canary Wharf. On the one hand the very patronage of high-style architects, provision of nominally public space and painstaking attention to detail, are intended to serve – along with the promotion of free-to-the-public arts and events programmes (cf. Crilley, 1992) – as signs of cultural goodwill and public accountability. This general aestheticisation works to distance the project from a previous era of urban renewal, purporting to retrieve the spirit of enchantment fostered at the Rockefeller Center (the first landscaped, multi-block, skyscraper city) as a legitimating guise for speculative investment (Tafuri, 1980). Canary Wharf does not so much cultivate a 'strategy of awe' as a 'strategy of admiration' (Lasswell, 1979). It is not experienced as a space of domination and power, but is intended instead to arouse a sense of comfort and play. On the other hand, the aestheticisation is of a particularly historicist kind. Though a gigantic piece of pastiche in the sense that Jameson (1984) uses the term – claiming to speak through the architectural syntax and lexicon of London, mimicking older forms and styles devoid of ironic or satiric intent – there is nothing 'random' about its use of the past. It does not cannibalise all former architectural styles, but intimates more methodically that it is an advance in the struggle to restore a distinctly London tradition of design. As an early promotional campaign announced, 'by complementing the personality of London, Canary Wharf makes that personality its own. Here the modern meets the traditional, as Canary Wharf respects the touchstones of the city whose great legacy it will perpetuate' (Canary Wharf Development Group, 1986).

Battery Park City: More 'New York, New York'

Unlike Canary Wharf, Battery Park City's design has received virtually unanimous accolades from architectural commentators, bestowing iconic status upon it in the history of postmodern urbanism. Its traditional design schema, drawn up by Cooper-Eckstutt Associates[3] in 1979 as part of a last-ditch rescue operation to save Battery Park City Authority from financial default, is hailed as a multi-faceted 'triumph' of urban design. For design critics (cf. Goldberger, 1986; 1988a; 1988b; Wiseman, 1987; Fisher, 1987; Gill, 1990; Johnson, 1990), it embodies numerous canons of postmodern urbanism: a close-grained mixing of uses replaces the 'decontaminated sortings' of modernist zoning (Jacobs, 1985, p. 36); the street is reinstated as a basic organising element, celebrating the city at pedestrian not automative scale; the endless flowing planes of modernism are tamed with a revitalised public realm composed of figural, room-like spaces, fantastic atria, a one mile waterfront esplanade, and a contemporary version of the Renaissance piazza; public art is not a minimalist abstraction plonked in an open space, but a series of site-specific works providing usable objects (tables, benches) and recognisable iconographic content; and, not least, in contrast to earlier megastructural visions for this landfill, it does not stand apart from the fabric of the existing city but is sensitively woven back into it through contextualist design principles. Paul Goldberger, influential architecture critic for the *New York Times*, epitomises the adulatory mood. Designating it a 'realist's' city in contrast with the idealism of earlier 1960s plans, he is unequivocal that 'Battery Park City . . . is close to a miracle . . . [I]t is not perfect – but it is far and away the finest urban grouping since Rockefeller Center and one of the better pieces of urban design of modern times' (Goldberger, 1986). Wiseman (1987) concurs: BPC is the 'standard by which any future urban design must be measured'. In fact, such rhapsodic assessments have been so hagiographic and so untempered by critical assessments of the project's elitism (the eradication of low-income housing from the project, its insulation from normal procedures of public review, and its public subsidisation via tax abatements are never mentioned) as to obviate any need for a pro-active publicity campaign on the part of the BPCA; they simply distribute the most congratulatory articles.[4]

Moreover, if architectural celebrations of BPC's public spaces function as an effective advertising campaign for the BPCA, then these opulent spaces in their turn serve as emblems of 'New York Ascendant'. In the propaganda of urban boosterism, nothing signifies ascendancy more convincingly than the model public of the late-capitalist city strolling, contemplating, consuming and soaking up the ambience of these picturesque display cases.

The design strategy itself foreshadows that adopted at Canary Wharf. It too bids for commercial and political legitimacy by drawing on memories and metaphors of a New York already embedded within popular consciousness. And it too is exalted as the antipode to modernist urban renewal. In place of historical erasure, the design of BPC begins from the principle that collective memories of old New York should be the animating theme. As the BPCA (1979) argued, the design was premised on 'the judgement that few examples of postwar development can match the city's older, more established neighbourhoods in stability, diversity and architectural character'. The first major step in fabricating an historicist aesthetic was to extend New York's famous grid. Just as Canary Wharf is staged as a theatre of memory, reinterpreting traditional London typologies, so too is BPC to be connected to New York, physically and psychologically, by extension of its neutral, rectilinear grid system. Doubtless the grid also served its time-honoured purpose of expediting a market-led process of land development (cf. Marcuse, 1987), but it also had the special significance at BPC of laying the basis for three-dimensional links with the remainder of New York.

Cooper and Eckstutt systematically surveyed New York's most popular neighbourhoods at both macro and detailed levels, extracting an essential set of qualities capable of endowing BPC with a familiar and traditional feel.[5] 'We wanted it to look as though nothing was done', claims Eckstutt; a sentiment shared by Cooper, who saw the task as being to 'normalise that bit of dirt through shifting the focus of design back to the city as it is' (*Battery News*, 6.11.89). Their intention was not literal reproduction, however, but to evoke nostalgia. As Eckstutt explained:

'I'm not interested in reproduction at all. My work is much more about interpreting the nature of places. While everyone who walks down to BPC believes it is part of New York, they realise they've never been to anything in New York like it before. The esplanade may have all kinds of feelings of the Carl Schulz or Brooklyn Heights esplanade or Riverside park, but it is not like any of those. Yet you feel rooted in [New York] because there are familiar dimensions, familiar materials and furnishings' (*New York Newsday*, 30 June 1986).

This strategy is therefore metonymical in the sense that it does not simply copy New York's architectural past, but selectively borrows elements from it that, by recalling the ambience of popular outdoor spaces, will hopefully stimulate the sense of place and history associated with them.

BPC is indeed less a replication of New York than a synthetic, idealised version of it (see **Figure 1**). Compressed into BPC is the look of inter-war apartment buildings combined with the ambience of Brooklyn Heights; the exclusivity of Gramercy Park juxtaposed with the populism of Olmsted's park design. The design borrowings, pervading the entire complex, are both literal and abstract. At the grandest scale, the four towers of the World Financial Center are designed with sculpted tops, set-backs, and a graduated mix of granite and glass, so as to blend in with the Downtown skyline. Smaller details of the complex are similarly intended to connect with familiar spatial routines and experiences: the opulent marble lobbies are intended as memory devices to connect us with the experience of other, well-known sumptuous interiors such as the Woolworth or Chrysler building. Indeed, the vast circular stairway of the Winter Garden (the World Financial Center's key 'public' space) is intended as a 'hangout' modelled on the public use of the steps of the 42nd Street library or of the Metropolitan Museum of Art; and outside, the retrieval of classic New York City park furniture including old-style trash cans, Olmsted's original lighting fixtures and grey hexagonal paving stones, connects with park design throughout the City.

In the residential neighbourhoods masonry street walls connected to vest pocket parks evoke the elegance of 1920s apartment houses

FIG. 1. Battery Park City: An Amalgam of New York's Architectural Past.

City Tales :
FEEDING PATTERNS

All Night Fruit

by DAVID RIEFF

photograph by Steve Hill

Alone among the great cities, New York is a haven for single people. There is something unnatural about being alone in, say, Los Angeles; there is something almost ideal about being alone in New York. Solitude, like work, becomes a drug, an attitude, an aesthetic stance. In the small hours of the morning, however, even the hardiest solitary may falter a bit. Some iron, if simultaneously ironic, universal law decrees that single people grow ravenous after midnight; food, after all, being the best consolation when the face resting on your pillow is the same one resting on your shoulders. ✍ Outside New York, the hungry solitary is most often reduced to seeking out some all-night diner of somewhat sinister aspect or condemned to the microwaved mercies of the Seven-Eleven. For those who remain fond of things that grow in the earth or drop from the trees, this is not a happy situation. There is, of course, junk food in New York. But there are also Korean fruit stands. ✍ The Korean fruit stands started popping up about fifteen years ago. They arrived just in time. Earlier groups of immigrants were abandoning the trade, and it was an open question as to whether New Yorkers would be condemned to wandering in all-night supermarkets in neighborhoods of dubious virtue. Single people trembled. Then, like the

Seventh Cavalry, like a cool breeze, like a check in the mail, the Koreans arrived, opening their impeccable emporia all over town. It was the greatest event in the lives of single people since the invention of the VCR. ✍ The poet, W.H. Auden, once said that though he was not an American, he was reasonably confident he was a New Yorker. I have a somewhat similar feeling about the Koreans. It is, after all, largely thanks to their stores that New York remains, reasonably safely, the all-night town that it must be to go on being New York. The paradox is that it was the arrival of a new group of foreigners that allowed the city to remain not as it is but as it was. ✍ At two a.m., I sometimes leave my apartment and walk to a Korean fruit stand nearby. There, among the mangos, grapes, fresh-squeezed orange juice, and glutinous, hypertrophied muffins, I watch old grandmothers, still psychically in Korea, shelling peas; hardy middle-aged couples obsessing over their inventories; and their adolescent kids, already two-thirds of the way toward another life, toward America, working the counter, basketballs and violin cases resting on boxes of plantains. Eventually, I buy a nectarine, a bag of designer taco chips, or a box of blueberries, and go home, sated before taking the first bite of whatever is in the grocery bag.

Satisfy your urge for food, your hunger for company, or just
your sense of curiosity at any of the new restaurants
and cafés at The World Financial Center. Picnic on the plaza
or grab a snack on the steps in the Winter Garden.

FIG. 2. 'City Tales' Advertisement for World Financial Center
(reproduced courtesy of Drentell Boyle Partners).

constructed on Riverside Drive and Park Avenue. The design guidelines regulating residential development are carefully researched to cater to middle-class tastes, explicitly trying to rival the aesthetic attractions of gentrification.[6] The lineage of these requirements is unmistakable. They work around a coherent if nostalgic perception of the city; an image of intimate streets lined with polychromatic, textured stone facades, shopping arcades and punctuated by tree-filled open spaces. This is the image of inter-war apartment buildings that became defined as the authentic patrimony of the middle classes once the retro-aesthetic of gentrification took hold. It is based on the marketing rationale that the aesthetic most likely to attract people to live in Downtown was not the glass and steel of modernism, but more 'New York, New York'. Essential to this was to simulate the diversity of a neighbourhood that has developed incrementally, again as with Canary Wharf, to avoid any resemblance to 'the project look'. Towards this end, the design review policy of the BPCA mandates scenographic diversity: the preference was that no architect should design more than one parcel, and where this did occur some degree of difference would be demanded; elements of individuality were to be supplied by sculpting the roof-tops; variations within a predominant 'warm earth' colour tone were to be deployed for decorative purposes in special locations; and throughout, the architects were expected to reconcile the contradictory demands that their buildings fade into the background yet display some form of idiosyncrasy (BPCA, 1985). In contrast to the social diversity championed by Jacobs (1985), therefore, the designers of BPC were preoccupied with diversity in a narrowly pictorial regard. Diversity was a visual effect intended to attract an homogeneous class of housing consumers.[7]

The close resemblance between advertising formats and this architectural scenography is evident from early publicity for the World Financial Center (see **Figure 2**). Titled 'City Tales' and prefaced as 'Its more New York, New York', it consisted of a series of advertisements (initiated in October 1988) aiming to stimulate awareness of the Center's newly-opened public spaces and up-scale retail gallery. Seemingly unconnected to the Center, the adverts present short literary vignettes and black and white photographs focusing on recognisable icons of New York public life such as Grand Central Terminal and Korean Delicatessens. Their subtle rhetoric was intended to tap people's warm sentiments about New York and then to transfer this warmness to the Center itself, a tactic underlined by typographics which literally frame New York within its precincts. They aspire to generate an immediate sense of belonging for the Center in the weekly round of up-market consumers, an effect accentuated by their placement in publications such as *Vanity Fair, Time* and the *New Yorker* as well as by the intellectually teasing format of their message.

Towards a City of Scenographic Enclaves

Though I have focused on only two redevelopment projects, when placed in a metropolitan-wide context they presage an important transformation in the forms of London and New York. In the absence of a holistic and forward-looking vision for the postmodern city, its form is characterised by a proliferation of consciously stylised and historicist nodes of the variety presented here. It is a city of self-enclosed architectural tableaux, each designed as relatively autonomous entities with contextual concern only for what is immediately adjacent or directly contributive to value, and each projecting different entertainments and amusements (cf. Boyer, undated). Future additions to London and New York threaten to multiply these scenographic arrangements to the point where the arcade panoramas of the nineteenth century, with their promise of vicarious pleasure in moving images of distant places and past times, are writ large in the city (Buck-Morss, 1989). This notion of 'scenography' is entirely appropriate, as these images are literally the 'cultural clothing' (Harvey, 1987, 1989b) of serially reproduced redevelopment formulae of late capitalism, a proposal amply demonstrated by the structural similarity of O&Y's activities in London and New York. Diversity is engendered by intense inter- and intra-urban competition to offer the most convincing aestheticisation of homogeneous office complexes. At the intra-urban scale it appears that whichever property coalition can offer the combination paying most homage to architectural heritage, deferral to existing physical context, or fitting the prevailing canons of aesthetic taste, stands to gain in the form of speedy planning consent and quick lettings. Brief mention of a selection of proposed additions serves to contextualise the cases considered here.

Thus, following BPC's moves towards completion, New York's 'raw and bawdry' Times Square is slated for comprehensive redevelopment by a comparable tripartite alliance of City, State Urban Development Corporation and private capital. Again, contextualist and historicist architectural imagery is invoked as a cultural rationalisation for 4 million square feet of speculative development which will bulldoze existing buildings under the spurious auspices of architectural crime prevention (Lopgate, 1990). This vast proposal for what is generally considered over-subsidised and overbuilt office space, even by neo-conservative commentators (Huxtable, 1989; Goldberger, 1989), embodies two forms of scenography. The bulky office blocks, destructive of one of Manhattan's rare preserves of low-rise, low-density development are to be mollified by guidelines mandating abstract wall patterns, kinetic roof lines, and state-of-the-art illuminated displays, all in the hope of retaining popular memories of Times Square in its current commercial form. At the same time as it promises to preserve the aura of the 'Great White Way', the project also embodies guidelines for the restoration of nine theatres in the district, intimating a restoration of their 'legitimate' character in the heyday of the 1930s.

In London proposals in the early 1990s for yet more office complexes within the vicinity of the City of London proffer exotic allusions under the banner of classical revivalism. At London Bridge City on the south bank of the Thames facing Tower Bridge, revivalist John Simpson (an architect carrying royal assent) has dreamed up a speculative office complex in the manner of a compressed mock Venice, complete with renditions of Doges Palace and Piazza St Marco and making explicit reference to Canaletto's vistas. A consummate example of catering to the cosmopolitan and image-laden *musée imaginaire* championed by Jencks (1987, p. 128), it participates in this landscape of allusion and simulation. It evokes one of the most famous waterscapes in the world, and in doing so sets out to transform London's utilitarian river into a fantasy of picturesque romance.[8] Meanwhile, at the heart of the City the unloved concrete modernism of Paternoster Square is the subject of a classicist mini-city promising a vintage restoration (also conceived by John Simpson, in conjunction with fashionable postmodernist Terry Farrell). Similar in scale and concept to London Bridge City – 1.25 million square feet of office space on a 3 ha site – it promises a *pot-pourri* of polychromatic classical styles and period dramas, condensing onto a revived medieval street pattern, a contribution from virtually every neo-classicist of fame. Billed as a scheme deferential to the context of St Paul's Cathedral and reparative of modernism's damage to the 'fabric' of the City, it offers a display case reconstituting the City tradition of classical architecture, incorporating allusions to precedents such as Soanes's Bank of England Building (1808) and Lutyen's Midland bank (1924–39).[9]

Now, whilst these architectural advertisements may well present a pleasurable, entertaining and apparently popular corrective to the 'great blight of dullness' (Jacobs, 1985) inflicted by modernism, they – like product advertisements – are ideologically charged. They are ideological in the double sense of being dissimulatory and diversionary. In advertisements, commodities appear as symbolically-charged fetishes, the conditions of their production having been carefully erased from view. This is the tendency of the architecture and accompanying rhetoric considered here. It works to dissimulate the social relations of their production and the selectivity of their social use. Thus, although BPC appears integrated and diverse, this conceals its homogeneity and systematic segregation from New York City's street environment. Similarly, just when O&Y demonstrate the power to level local differences to the abstractions of a globalising property industry, Canary Wharf is presented as a regionally-derived mini-city adorned with superficial but supposedly reassuring signs of locality. Moreover, in step with its obliteration of the last remaining traces of a former Docklands history, epitomised in the comprehensive renaming of the site, Canary Wharf re-presents an anterior bourgeois past in the form of synthetic architectural borrowings. And, finally, the presentation of both projects as unequivocal triumphs of 'public space' suppresses the fact that these spaces are provided on public land appropriated for private profit, are controlled by the management logic of O&Y, and – far from being accessible to all – are socially selective through both the outright control of their 'undesirables' and the exclusivity of their up-market consumption emporia.

At the same time as it denies its own social production, the architecture of redevelopment diverts. In particular it performs an effective screening role conducive to geographical and social myopia. By representing redevelopment as a patchwork quilt of discrete, random and diverse 'projects', it not only denies its systematic unevenness – severing the connection betweeen the spaces of redevelopment and those of underdevelopment – it also deflects the gaze away from the social problems of the areas which abut these scenic enclaves. As Harvey claims when speaking of the formulaic application of postmodernism to downtown redevelopment, they resemble a

'. . . carnival mask that diverts and entertains, leaving the social problems that lie behind the mask unseen and uncared for. . . . The formula smacks of a constructed fetishism, in which every aesthetic power is mobilised to mask the intensifying class, racial and ethnic polarisation going on underneath' (Harvey, 1989b, p. 21).

There is more significance in this than the temporary escape that urban voyagers experience wandering through these scenic arrangements where the city is effectively held at bay. Indeed, these aesthetic benefits are often literally conflated with social benefit, as a concern for physical context goes hand-in-hand with indifference to problems of social and political context. Thus, BPC's 'triumph' was predicated on a resigned abandonment of the potential of the site for subsidised low-income housing and on the consequent entrenchment of social segregation (Deutsche, 1988). At Times Square remembrance of the area's illustrious past, alongside contextualist gimmicks in the form of jazzy electronic signage, is reciprocally related to flagrant disregard for the current multi-ethnic vibrancy of the area based on high-turnover rates and low capital investment. To ensure that the area is shifted to a 'higher and better' use, the current social history of the site is erased only to be re-presented as a carefully orchestrated corporate architectural spectacle. In this respect ethics are aestheticised – the 'good' city becomes the enchanting and visually stimulating city – and postmodern aesthetics becomes a stake wielded by the forces of redevelopment, reinscribing social power in the landscape. Corporate power is no longer only represented by the 'raw' steel and glass citadels of America's urban landscape (Relph, 1987) or the savage fortifications of downtown Los Angeles (Davis, 1990), but has been 'cooked' through architectural culture into an eclectic *mélange* of romantic and even 'witty' skyscrapers, vernacular business 'villages', classicist 'cities within the city' and visually diverse 'neighbourhoods'.

This is not to invoke an unduly crude notion of a new urban hegemony. After all, in the case of Canary Wharf at least there is enough controversy to indicate that the developers' aesthetic strategy has been a far from smooth success (Tibbalds, 1989; HRH, 1988). There is also little enough evidence on how these scenic arrangements are 'read', 'interpreted' and 'decoded' in the course of their social use. Not until Canary Wharf has been open to the public for some length of time will it become evident whether it accommodates spatial practices and readings counter to those intended. It is therefore premature to assume that the images projected by architecture are *in practice* diversionary or that serious historicist design automatically guarantees popular consent.

Even with these qualifications about 'reception', however, the prioritisation of aesthetic issues in the political debates over redevelopment is itself diversionary because of the issues that it 'crowds out'. Elaborate exchanges over the architectural garb of Canary Wharf or over its disruptive impact on the skyline of London threaten to cocoon debate, deflecting attention away from the more disquieting questions of the distribution of material benefits and the propriety of allocating substantial public land and finance to subsidise private speculation (cf. Crilley, 1992). To restore questions of homelessness, housing crises, unemployment, and strategic planning to assessments of Canary Wharf, it has now become necessary to break through the clouds of aesthetic debate. It is salutary, for instance, that the articulation of these issues has been left to the archipelago of local protest groups with least formally empowered voices (cf. Brownill, 1990; Docklands Consultative Committee, 1990). Likewise, in the public inquiry over the architectural costume for London Bridge City Phase Two, only the local and disempowered planning committee challenged the principle of adding more office space to the Southwark waterfront. For the rest of the proponents, the propriety of competing architectural styles occupied the centre of attention. And, in the battle to save London's Spitalfields wholesale market from a relocation imposed by redevelopment (see the chapter by Woodward), aesthetics were used to repressive effect. Esoteric wrangles about conservation contributed to a feeling of marginality and disempowerment on the part of campaigners seeking to save the market from relocation: in fact, their concerns with the disastrous housing and employment impacts of redevelopment were suppressed in favour of concerns for the Georgian character of the area.

In each of these cases, however much developers are criticised, the centrality of aesthetic issues scarcely challenges the fundamental basis of the developers' own power. Owing to the drastic narrowing of notions such as context in these debates, the arguments still serve as effective publicity spectacles. The more that formal issues are fore-grounded in crude polemics over architectural style, the more developers fade into the background and, along with them, the troubling issues of who builds what, where, for whom, with whose finance and at whose expense. As Ghirado comments on a parallel formalist tendency of architectural commentary in Houston, it is yet another way of architecture contributing to advertising:

'. . . builders and developers could not in their wildest dreams have designed a strategy of such academic and intellectual status that it would successfully direct analysis toward trivial matters of surface and away from much more vexing matters of substance' (Ghirado, 1990, p. 236).

Notes

1 Canary Wharf is removed from the normal procedures of planning control and public review by virtue of its location in an enterprise zone under the jurisdiction of an unaccountable, unelected central state quango, the LDDC. A much lengthier critical analysis of the political economy behind Canary Wharf and of the ideological significance of its design is beyond the scope of the argument advanced here. It can be found in Crilley (1992), where Canary Wharf is considered alongside the World Financial Center in the context of O&Y's development strategies. Since only 4.1 million sq ft of Canary Wharf's scheduled space is certain to be completed by 1992 and finer details of the design are still subject to modification, comments here about later stages are tentative.
2 The LDDC has recently distributed a series of leaflets and posters, London-wide, that featured John Outram's award-winning Storm Water Pumping Station on the Isle of Dogs. They are intended to signpost the LDDC's design achievements, pointing to such prominent acclaimed works of late-modern architecture as Nicholas Grimshaw's Financial Times Printworks and also to awards from the planning profession for the redevelopment of Surrey Quays. Ironically, it is an index of the disappearance of public patronage that a comparatively innocuous structure such as the Pumping Station is lavished with praise as a restoration of ornamental Victorian traditions, used as an icon of Docklands's environmental excellence, and attributed an over-inflated significance as evidence of the LDDC's commitment to public patronage.
3 Cooper-Eckstutt, until the partnership broke up in 1986, was an architectural

practice specialising in production of master-plans and urban design codes for large-scale redevelopment schemes across United States cities. Both partners continue to apply the basic tenets devised at BPC to projects that they deal with in their independent practices (Cooper-Roberston and Ehrenkrantz, Eckstutt and Whitelaw).

4 In contrast to the LDDC, the BPCA has been fortunate enough to rely upon favourable press coverage to meet its publicity needs and has never launched its own advertising campaign. Whilst the LDDC has rarely been able to distribute press reports owing to the overwhelmingly negative reaction to Docklands in the late-1980s, the BPCA information programme has consisted of nothing but.

5 Just as Venturi *et al.* (1972) conducted an architectural survey of the Las Vegas strip to derive design precepts and the designers of Canary Wharf launched a careful study of London, so too did the design team for BPC conduct extensive field studies of New York. They walked its streets, observed its parks, photographed street walls and noted the colour of materials. They then put their findings together, devising an analogical rendition of what they had discovered.

6 Statements of design intent issued by the BPCA openly acknowledge learning from the aesthetic allures of nearby Tribeca and Soho, as well as to adapting the design image of successful New York neighbourhoods including Central Park West, Fifth Avenue, Park Avenue and Tudor City. They declare an intent to find a workable combination of design elements traditionally associated with prestigious New York (see BPCA, 1979; 1985).

7 That BPC has become an homogeneous preserve for the affluent is amply demonstrated by the BPCA's own survey of 3,200 residents in 1988. The mean household income was $102,632, with approximately 40% of households having incomes of greater than $100,000 (BPCA, 1988).

8 Details of London Bridge City Phase Two are taken from submissions to a public inquiry into its design held in 1988.

9 Details of Paternoster Square restoration are taken from information and advertisements provided by the Paternoster Square Development Information Group, 142 Warldour Street, London, W1V 3AU.

12

One Place, Two Stories: Two Interpretations of Spitalfields in the Debate Over its Redevelopment

RACHEL WOODWARD

Introduction

In the late 1980s Spitalfields, a district in east London, came under increasing focus from the media. A three-hundred year old fruit and vegetable wholesale market was to be relocated and the fourteen acre site redeveloped. In this chapter I shall examine the ways in which the area and redevelopment plans were presented by two different groups – the developers, the Spitalfields Development Group (SDG) and a group constituted from local residents opposed to the redevelopment plans, the Campaign to Save Spitalfields from the Developers (hereafter the Campaign or CSSD). My purpose is to show the types of representations or portrayals of the scheme and area which were constructed by both parties, and to illustrate how two very different constructions of a place and a scheme emerged from a scenario where confrontational tactics and languages were thought to be crucial in gaining public support by both parties.

An Emerging Conflict

A fruit and vegetable market had been situated in Spitalfields, an east London district nestling close to the eastern border of the City of London, for over three hundred years. As had happened with London's other wholesale markets such as Covent Garden in the 1970s, plans for its removal from the London Borough of Tower Hamlets had been mooted since the end of the Second World War on the grounds that its site was unsuitable for its purpose. In the early 1980s the market became the focus for attention from a development company, later formally to name itself the Spitalfields Development Group (SDG), who saw a potential use in the site as both a retail centre for an area under-served by large supermarket units and as an office complex to serve the then booming City of London and its demands for purpose-built office space within and around the Square Mile. The site's owners, the City of London Corporation, were faced in 1986 with a bid from the SDG for the redevelopment of the site, which they duly rejected on the grounds that such a large and important development should be bid for in open competition. A 'competition' for the site (if we can call it that) was held, with three development companies participating. Each company had to secure a new site for a relocated fruit and vegetable market and had to secure planning permission for any scheme from the site's planning authority, the London Borough of Tower Hamlets. In addition, the development companies – and chiefly the SDG – conducted a public consultation exercise in the area with the aim of ascertaining from the residents of the area a consensus on the form that the redevelopment should take. In 1987 the SDG won the Corporation's tender, and were granted planning permission in the October of 1987.

The SDG held planning permission for the redevelopment of the fourteen acre site into an office complex with retail units and some housing for sale on the open market. An important part of the scheme was the granting to Tower Hamlets of planning gains worth over £20 million for the construction of various community facilities and housing and retail units for use by the local population. The construction of the new development could have started as soon as a new home for the market, on a site four miles to the north east, had been completed. However, because this ancient market had been founded by Royal Charter granted by Charles II, an Act of Parliament was required by law before the market could be moved. This was to be obtained by Private Bill, the City of London (Spitalfields Market) Bill to both Houses of Parliament, and such a Bill was deposited in October 1987. In theory, all that the SDG had to do now was sit back and wait until the Bill had been passed: at that point it was thought to have been a mere formality. They were wrong.

In October 1987 a group calling themselves The Campaign to Save Spitalfields from the Developers (CSSD) appeared on the scene in Spitalfields, the first indication of their presence being leaflets distributed door to door in the area calling for mass opposition to the redevelopment scheme. The Campaign was not a spontaneous formation of residents previously silent or inactive in the area, but was comprised of people who had been very active in the local Labour

Party over the preceding ten years, and of people working within the non-statutory voluntary sector in the area. It had a core membership of around fifteen people, all of whom lived in the area, and most of whom were white and self-described as middle class, in contrast to the majority of the population of the area who were Bengali in origin and self-defined as 'working class' (see Woodward, 1991, for a full examination of the Campaign). The Campaign's aim was the prevention of the relocation of Spitalfields Market, and one of their first actions as a group was to submit to Parliament a petition against the City of London (Spitalfields Market) Bill, thus forcing a debate, both in the parliamentary chamber and in Private Bill Committees, on the future of the area and on the plans for the market. The terms in which the Campaign were to phrase their opposition are apparent in the quotation below, taken from an information leaflet distributed around Spitalfields in early 1988:

'. . . now the Developer's plans have come *unstuck*. They had been hoping that their plans woud go ahead unopposed, but now they've got a *fight* on their hands – with *local* people who have had ENOUGH of massive office blocks and expensive shops; local people who want to REMAIN living and working in Spitalfields' (CSSD, 1988a).

This group were unequivocally opposed to the relocation of the market, establishing in their portrayal of the redevelopment scheme the idea of a group of developers whose sole aim was the conversion by wholescale redevelopment of a poor inner urban area in opposition to the wishes of local residents.

Spitalfields's future, in term of function and appearance, became the subject of heated argument between the SDG and the Campaign. At this time the debate was not conducted face to face as it would be in later months and years at meetings between the two groups. Rather, two ideas on the form and function of the area were developed, in relation to each other, via publicity material distributed to the people of the area. If we examine the two discourses, the two constructions of what redevelopment and Spitalfields meant, we do not perhaps arrive at any startlingly new or original theorisations about the construction and then the selling of places. What we do have, and what I shall illustrate here, are two 'little stories' which illustrate the ways in which that sale can be conducted and ideas constructed.

A War of Words

The SDG were first to step into the ring over the fight to convince the populace of Spitalfields of the most suitable form that a redeveloped Spitalfields should take. It is only right that I should start with their ideas, for these set the tone for the debate and provided many of the concepts therein. The first SDG brochure to explain to the people of the area their plans for a redeveloped site announced that:

'Anyone who lives or works in Spitalfields will know that for a long time there has been talk of moving the market to better premises, further out from the City centre. But how is it to be done? What will happen to the site? And how will a change of use affect the surrounding area: an area with a very special architectural and social heritage as well as very special social needs. Answering these questions isn't easy. It means finding a scheme which will respect what already exists in Spitalfields – a community of people as much as it is bricks and mortar – but at the same time generate enough money to help finance itself and cover the immense costs of providing the market traders with a new home' (SDG, 1986).

This quotation, from the SDG's first publicity brochure distributed in the area, established certain parameters for the debate that was to follow. We are presented with the idea that there is a common understanding amongst the local residents about the redevelopment. The quotation leads us to assume that sufficient publicity had been produced over the preceding months to inform all residents about the plans. This was something that the Campaign hotly disputed at every opportunity. Above all, two ideas are contained within this quotation. The area is presented as having 'special needs', a euphemism for social problems. The idea presented is that this development company is concerned with the needs of the local people and with the heritage of the area as a first principle behind the redevelopment plans. The area is named as special on social and historical grounds, implying that the SDG themselves care about the impact of the redevelopment on the area, and have incorporated that concern into the plans from their earliest inception. Having set the area up as problematic, a second idea emerges. The statement in the quotation raises the question of finding a balance between the needs of capital and the needs of local people, which will then be resolved through careful design of the scheme and conscientious management of the redevelopment. This redevelopment, then, is offered as a solution to both the problems of Spitalfields as well as to the problems of the market. These ideas of a problematic area and a redevelopment solution were established in the initial SDG publicity as key concepts in the debate. As I show below, it was around these ideas that the Campaign's arguments were initially tied.

The first publicity material presented images of the proposed redevelopment using specific languages. For example, the various components of the redevelopment scheme were explained in literature which emphasised the range of purposes to be served and the needs to be met by the redeveloped site.

'The Spitalfields market site . . . will be developed to provide a lively new commercial and residential environment. There will be new streets, housing, open space, small business units, shopping and eating facilities; and in the middle of it all (helping to finance the entire scheme) three separate buildings which will provide international banking headquarters around a landscaped piazza with direct access from Bishopsgate' (SDG, 1986).

The redevelopment would serve commercial, retail, manufacturing, residential and leisure interests. It would be *lively*, one presumes bustling, international yet intimate. This international banking head-quarters would incorporate the requisite office blocks, yet the impact of these is played down. They are portrayed as impressive yet small scale:

'Building heights around the perimeter will be kept low (no more than four domestic storeys) to respect the scale of neighbouring architecture. . . . Only in the very centre of the scheme will heights rise to seven or eight storeys – a scale which will make the offices barely visible from thoroughfares that flank the site, but at the same time provide impressive views from streets approaching to the north and south' (SDG, 1986).

Redevelopment, these quotations imply, would constitute both a harmonious addition to the urban fabric of the area and an oppor-tunity to change that same fabric, modernising and renewing the area.

In addition, held within the portrayal of the redevelopment constructed by the SDG emerged the idea that this redevelopment scheme provided a blueprint for inner-city regeneration, having an impact beyond the confines of its location in East London:

'It has become clear that the future of Spitalfields is a matter of general as well as local concern. It is clear, too, that whatever is proposed here for the market site and for the neighbouring community could well become a blueprint for other inner city developments in Britain and perhaps abroad' (SDG, 1987a).

In their statements on the social and physical changes that would follow a redevelopment scheme, a constant feature was an emphasis on the existing social and environmental problems of Spitalfields and on the desirability of change through wholescale renewal. For example, in a preliminary statement on the redevelopment scheme, the area is described in these terms:

'Tower Hamlets, of which Spitalfields is an important constitutent element is part of a deprived Inner City area. It has a rich mixture of cultures whose conflicts are exacerbated by the intensity of Inner City living. The area is traditionally Working Class and has a high poverty level which stems from lack of investment and employment. There is an urgent need for a programme which will bring much needed cash . . . and social service provision to ease integration and bring better quality buildings, facilities and housing into the area' (SDG, 1987b).

The image we can 'read off' from this statement is of a classic 'inner city', a mythical place perhaps, but an idea with great currency naming the inner city as the seat of a range of social problems and a place requiring radical solutions. Their construction of Spitalfields as semi-derelict and without hope for the future was crucial to the SDG's manufacture of a case for the redevelopment of Spitalfields. The area was portrayed, time and time again, as derelict, run down, and problematic in its morphology, state of repair and social structure. Having constructed Spitalfields in this way, it was then relatively simple to produce a redevelopment scheme and in very straight-forward terms name it as the solution to these problems:

'The intention of the design is to mend a part of Spitalfields destroyed when the existing market buildings were erected, by extending the surrounding street pattern of previous centuries into the site, and so marrying the new fabric of Spitalifelds to the old' (SDG, 1986).

Spitalfields was constructed as an inner-city area, and using the connotations associated with that term the SDG was able to present its scheme as a solution. The issue of finding a balance between all interests is also raised within the scheme as inner-city renewal. Given the then government's insistence on the use of private sector finance in the regeneration of urban areas, it is not surprising to find an emphasis on the 'appropriate' method of financing for the scheme. Note the use of the word 'realistic' in the following quotation:

'[I]t is vital that whoever redevelops Spitalfields gets the proposal absolutely right. Right for local people, architecturally sympathetic to the Spitalfields environment and at the same time economically *realistic*. . . . The only practical solution is to find a middle ground between what the Corporation wants and what Tower Hamlets wants; and to ensure that this middle ground offers the people who live and work in Spitalfields the best deal possible in all the circumstances. (SDG, 1987a).

Note also the ways in which the question of differing interests in the redevelopment scheme is raised, and then resolved neatly using a vocabulary of consensus.

In summary, two images were constructed by the SDG, one of a rather derelict inner-city area with social and environmental problems, and one of the redevelopment scheme which emphasised its role as solution to the problems of the area. These two images were tackled head-on by the Campaign, reversed and reconstituted.

A first task for the Campaign was to construct a fundamentally negative view of redevelopment. This was, after all, a Campaign who during the first two years of its inception had only one goal, the prevention of the removal of the market and the halting of the redevelopment scheme. Whereas for the SDG redevelopment would constitute a harmonious addition to the fabric of Spitalfields, for the Campaign redevelopment entailed destruction of the entire area:

'If developers get their way, Spitalfields Market will disappear into a hole in the ground. . . . In its place will rise the monster office blocks of the international banking centre planned by Spitalfields Development Group' (CSSD, 1988b).

The SDG had constructed an image of the area as socially and environmentally problematic. The Campaign's representation still portrayed the Spitalfields that they knew as problematic, but constructed around this image ideas of community and activism:

'[D]eprivation certainly permeates the place but people have a fighting spirit and they form a vibrant community. The strengths of our community are too often overlooked by misguided, well-intentioned outsiders, as well as those who are ill-intentioned, and sensationalist journalism' (Maxwell, 1989).

The market, for the Campaign, rather than being a symbol of the dereliction of Spitalfields, represented a past and present that gave the place its life; redevelopment would obliterate the area.

'"You can't stop progress" . . . but what kind of progress? Spitalfields as an international banking centre, a microdot on the satellite screens linking Wall Street and Tokyo?' (CSSD, 1988b).

Redevelopment and its alternative (leaving the area as it was) were portrayed as stark opposites. The choice, the Campaign tried to emphasise, was clear:

'With 20% of the area up for redevelopment, and the hungry-eyed City of London looking for land for expansion, Spitalfields is threatened with a face lift. The new look for Spitalfields can be either the dry, drab, grey face of the City or the image of a vibrant community enhanced by the injection of much needed, community directed cash' (Maxwell, 1989).

Whilst the SDG's representation of the redevelopment scheme sidestepped the question of the need for office developments in that part of London, presenting them as some sort of natural requirement for the finance sector that would be inhabiting the area in future years, the Campaign's representation of redevelopment tackled head-on the need for office blocks in that part of London at all. The very presence of offices in Spitalfields, it was asserted, would destroy the area:

'We must *unite* to *protect* ourselves from more and more offices, trendy little shops and wine bars. *If we don't*, we will find ourselves surrounded by a *Berlin Wall* of OFFICES. This will *force us* OUT of OUR HOMES and WORK – *we won't be able to afford to live in Spitalfields*' (CSSD, 1989).

From this quotation we can see that a whole image of redevelopment is constructed around the idea of incursion of office space into an area such as Spitalfields. With the encroachment of offices, the quotation implies, comes associated services such as wine bars and boutiques. The images that the Campaign uses here are drastic; office developments equated to the Berlin Wall, the presence of offices in Spitalfields entailing eviction of residents. Unsurprisingly, they are the total opposite of the ideas of the SDG. Office development is portrayed as automatically destructive of the area and its heritage, depersonalising the area and rendering it sterile:

'The office development in Artillery Passage and Widegate Street [streets in the area] has destroyed what ten years ago was a thriving manufacturing district. Paxton the scalemakers, an ancient family business which occupied the Georgian shop in Artillery Passage is now a faceless computer centre' (CSSD, 1988b).

The SDG had used the notions of consensus and balance of interests in their construction of the idea of a beneficial redevelopment scheme for the area. The Campaign used languages of eviction, displacement and invasion. For example, the increase in office space in the area as a result of redevelopment was portrayed as the incursion of the rich and powerful in an unequal battle with the relatively poor and helpless:

'An office area gobbles up shops – which become office fronts. In the competition for space, small shopkeepers don't have a chance against multimillionaire corporations' (CSSD, 1988b).

Furthermore, in this portrayal shopkeepers were not the only ones to be affected by this invasion. The need for housing on the market site in a chronically badly-housed area was used in arguments against the redevelopment:

'This Campaign has always been opposed to the City of London moving eastwards ... WE NEED HOMES NOT OFFICES. Our precious land is being taken for use for PROFIT – not for the benefit of PEOPLE!' (CSSD, 1990).

Important to note here is the language of this quotation, an example of a style of communication present throughout the Campaign's publicity material. The Campaign were careful to associate themselves with the people of the area, and took on the role of spokespeople for Spitalfields. In contrast to the SDG's statements declaring in detached tones what the company could do for the area, the Campaign presented their ideas and arguments in personal terms; the implication being that they, as residents of the area, were articulating the opposition that they reckoned to share with the rest of the population.

The Campaign used the issues of housing and employment in their arguments, attempting to undermine the SDG's claims with regard to both. The redevelopment scheme was to incorporate a number of housing units for sale, as well as 118 units for rent which would be built for local needs and managed by local housing associations. The SDG had been keen to emphasise this aspect of the scheme as indicative of their concern for the social problems of the area, particularly those arising from housing shortages. The Campaign were scathing in their comments, using this issue of 118 housing units to discuss the wider problems of housing in the area and the implications for this housing crisis of redevelopment schemes such as that planned for Spitalfields Market:

'*Only 118* residential units for fair rent by housing associations will be built on the site – this won't even make a dent in the long waiting list for housing locally, with large numbers of the families in bed and breakfast and poor housing conditions. As we have seen in the Docklands, the knock-on effects will be privatisation of whole council estates and *land and house prices rocketing.* The average local wage is £100 per week, or less, so whole families will be *driven out* of the area' (CSSD, 1987).

As this quotation shows, local knowledges were vital for the Campaign in stating their case. By incorporating into their arguments evidence and examples from local situations, the Campaign were constructing an image that they hoped would both appeal to local residents and appear to outsiders as the representative voice of the population of the area.

This tactic of the incorporation of local knowledges into arguments aimed at undermining the claims of the SDG was also used to effect with regard to the question of the impact of the redevelopment scheme on employment in the area. The SDG had used the idea of employment creation as a means of publicising and gaining support for the redevelopment scheme:

'The new development will generate 6,000 extra jobs, not only in white collar banking work but in clerical, maintenance and other service functions. Jobs available to local people of all cultures and backgrounds' (SDG, 1986).

Again, the Campaign's representation of the redevelopment plans undermined this idea by identifying in its own portrayal inaccuracies in the SDG's account of employment creation and of the possible impact of the redevelopment on the labour market in the area in more general terms:

'Local industries will be the first to suffer from the redevelopment. The rise in land values will *wipe out* the local garment and clothing industry – local employment for many Bengali men and women. There is already over 20% local unemployment. 63% of Tower Hamlets residents are manual or semi-skilled workers; the developers say there will be 6,000 new office jobs created, many of which will probably be traditional women's work – secretaries etc., but they also say that local people would have to be trained in new technology. BUT their miserable £50,000 p.a. over five years for training can only train a handful, so the workers will have to come from *elsewhere*' (CSSD, 1987).

The two portrayals of a redeveloped Spitalfields at times mirrored each other completely. The SDG's representation of redevelopment, as outlined above, emphasised that the benefits to the social life of the area brought by the presence of the fruit and vegetable market were to be replaced with a busy thriving office and retail centre. The Campaign in turn constructed an image of a redeveloped Spitalfields devoid of its present inhabitants, who in the Campaign's representation gave the area its very life and vitality:

'[The redevelopment] will change the face of Spitalfields, *driving out* local people whilst bringing in well-off City workers and the type' (CSSD, 1987).

For example, both representations raised questions concerning the mechanisms by which urban areas gained their vitality and streetlife. The SDG stated that:

'The intention is to create a *living* environment: something that doesn't just die at 5 o'clock when office workers leave. Because this will be so much more than an office development it will be used by the whole community in Spitalfields, and life will continue as it would in any other thriving mixed environment' (SDG, 1986).

The Campaign answered that:

'Unless we act now, . . . Spitalfields *will* develop into an 8.00am–4.00pm office worker's community; an area which will be sterile, desolate landscape outside of ordinary office hours and days' (CSSD, Open Letter, 14 January, 1988).

The SDG represented redevelopment ultimately as the opportunity to revitalise an area that they portrayed as derelict and unsightly:

'For the environment – a real improvement, free from the congestion and rubbish of the market at work and from the desolate wasteland that remains when the market is closed during the day' (SDG, 1986).

Whilst the SDG portrayed the existing Spitalfields in terms of desolation, the Campaign's construction was of a wasteland to come:

'Residents fear that the new business people will show little loyalty. They will leave at the end of the day, turning the area into a ghost town at night' (*City Recorder*, 14 January, 1988).

FIG. 1. An Image from the Campaign to Save Spitalfields from the
Developers.

The wasteland that the SDG were so keen to improve would, after redevelopment, be landscaped, offering open space and public gardens for relaxation and enjoyment in an area sorely lacking in such facilities:

'The square will have direct access out to Commercial Street and across the Christ Church where the Spitalfields Development Group proposes to improve and landscape the church gardens for public enjoyment' (SDG, 1986).

The Campaign was scornful of this, pointing in their own publicity material to a possible misuse of such space after redevelopment:

'"Public Gardens" . . . Who will use them? They will be thronged with thousands of office workers at lunchtime and deserted, or no-go areas, at night' (CSSD, 1988b).

'Office areas, witness the City of London, have no room for family life. Children wouldn't be welcome in a district crowded with international executives and no one is going to play football in the "atrium" of an office tower' (CSSD, 1988b).

Finally, the logos and graphics used respectively by each group contrast each other neatly. The Campaign used the image of a hand,

FIG. 2. An Image from the Spitalfields Development Group.

seemingly stopping the advance of a rather sinister-looking cityscape (see **Figure 1**). The message is clear enough. In constrast, the SDG produced a logo caricaturing the buildings in one section of the market, presenting a homely, cosy image and using to the full the one part of the old market with any architectural merit, the Horner Buildings constructed by Sherrin in the 1890s in the Arts and Crafts Style (see **Figure 2**).

Concluding Thoughts

Ultimately, the Campaign was unsuccessful in their efforts to prevent the relocation of the fruit and vegetable market from Spitalfields. In May 1991 it closed, although plans for the redevelopment scheme have still to be finalised at the time of writing. The two parties are still engaged in debate over the future of the area, still constructing different images of Spitalfields as it is, and Spitalfields as it could be. Unsurprisingly, from two different agendas for the area we have two opposed sets of images. The SDG, a development company, needed to portray Spitalfields as derelict and problematic in order to highlight the ways in which their scheme for the fruit and vegetable market could prove beneficial. The Campaign, with an interest in preventing redevelopment so as to preserve the community and character of the area, constructed representations of redevelopment which emphasised the way that such schemes would destroy the area and in no way serve the best interests of the local residents, however much the area required investment.

I feel it is important to conclude this chapter with some sort of evaluation of the power of these two representations of Spitalfields. Did either story, for example, have any use in persuading public opinion in Spitalfields concerning the desirability or otherwise of redevelopment? Such evaluations are difficult judgements to make, certainly for myself, given that my research into the redevelopment scheme, for various reasons, focused exclusively on the CSSD and SDG, to the exclusion of any attempts to investigate in similar detail 'other voices' and 'other stories' about Spitalfields and its future. I shall, however, draw the following conclusions.

The CSSD was ultimately much more concerned with eliciting public support for its arguments and its representations of Spitalfields than was the SDG. The latter had no real need for public support for its development proposals, given that the decisions on the possibility and the form of the redevelopment were taken outside of the public sphere and with, ultimately, little regard to the wishes of the residents of Spitalfields. The SDG constructed an image of Spitalfields for local consumption, as I have related here; its construction was a formality undertaken because it needed to show itself as concerned with public opinion. Whilst the SDG did distribute glossy leaflets and conduct so-called public consultation exercises, its survival as an institution did not depend on the logic of its arguments when put before local residents. In contrast, the CSSD had as its main purpose and function the task of persuading local residents of the dangers of the redevelopment scheme for Spitalfields. The Campaign knew that it could not exist legitimately as an organisation without the support of at least some sections of the local population. The high profile of the Campaign in Spitalfields was part of this attempt to gain support. The language of opposition, as detailed above, was also an important strategy. Although at times perhaps simplistic and almost crude, it is also in turns drastic, polemical and hard-hitting. The portrayal of Spitalfields as a vibrant and active community was crucial also. I cannot state with any certainty how the Campaign's representation in this respect was received within Spitalfields – I cannot evaluate its power as such – but I would nonetheless argue that the *idea* of opposition was a major contribution made by the Campaign. The notion of constructing a representation of the area in this manner hence fell on sympathetic ground, even if the Campaign's precise message did not. The Campaign's construction of Spitalfields was believed by some, but had an impact on many more in terms of the impetus it gave to other groups to start thinking seriously about the redevelopment of Spitalfields and its possible impacts. Although the stories told of Spitalfields by the Campaign may or may not have had currency amongst the local population, the fact of the Campaign's existence and its involvement in the redevelopment debate was important as an example of the ways in which local people might, potentially, take part in the development process. In due course there were consequences of this process entirely unforeseen for the Campaign, and ultimately unfortunate. But that is another story.

13

'Tackintosh': Glasgow's Supplementary Gloss[1]

ERIC LAURIER

Tacking

'For a long time I suppose that an average intellectual like myself could and should join the struggle (even if it was only with regard to himself) against the tidal wave of collective images, the manipulation of affects. This was called demystification. I still struggle, now and then, but deep down I really do not believe in it anymore. Now that power is everywhere (a great and sinister discovery – even if a naïve one – of people of my generation), in whose name do we demystify? Denouncing manipulation itself becomes part of a manipulation system: *recuperated*, such would be the definition of the contemporary *subject*. The only thing left to do would be to make heard a voice *to one side*, an *oblique* voice: a voice *unrelated*' (Barthes, 1985, p. 97).

How can I *write obliquely*, as Barthes suggests, but then I am doing so already by writing in *italics*. Yet, outside of this calligraphic pun, how do I begin to explain something in a 'voice unrelated'? In answering I should avoid one of the traps of demystification, which is to pretend by clearing away the fog which surrounds an image to have exposed what it is really all about. For instance, it would be to say that Glasgow's extravagant year of culture in 1990 was not about focusing on a rich and vibrant cultural milieu, but was about hiding a grim 'working class' history from tourists and captains of industry. In other words, I must not say 'it is not this! it is that!'

To write 'to one side' is not to write above nor to dig below to expose the foundations of reality; it is to attempt to be non-hierarchical. And this is a problem in the first instance, since to talk about 'tack' (about things considered 'tacky') is to look *down* one's nose. Oh dear: let me leave tack until later. Returning to the problems of presenting, what is to be avoided is the creation of a distance rather than a proximity between what one is writing about and the writing itself – as much for the form as for the content. One depends on the other, the two cannot be pulled apart. They exist only in their relation, not in their lonely and truthful independence; not as one empirical dirty town and next door the residential suburb of theory.

You may hence be unfortunate enough, if you decide to read on, to become involved in a commentary often going astray, authored by myself in argumentation with itself. To return to Roland Barthes:

'What is proposed, then, is a portrait – but not a psychological portrait; instead, a structural one which offers the reader a discursive site . . .' (Barthes, 1990, p. 3).

This site was my own site which is now the reader's: my own internal dialogues are here already and you can to a certain extent alight on them as you choose.

Ad Temptation

'I can resist everything but temptation' said Oscar.

In the past and even now, there *has* been and is advertising, and the idea of supplying the public with relevant information was part and parcel of an old (a tried and tested) economic idea of the world. In this advertised world individuals needed to be supplied with information about products to make rational decisions about 'value for money'. Whichever product most effectively fitted the objective characteristics of cost, utility, efficiency, availability and so on, would be purchased. And this form of advertising lingers on, perhaps most recognisably in washing-up liquid adverts – Fairy still washes *that* many more dishes for just a few more pence. Such a form of advertising refers to advertising's own history: it is stylistically and so perhaps also ideologically conservative. Fairy's has become just one strategy amongst many. Yet the message is founded not only on advertising's own conservatism: to a critical viewer (and we are legion, even if I cannot say anything else about *us*) there is an awareness that sometimes men *are* now seen with their hands in the sink. Concurrent with the appearance of these 'men' there seems to have been a disappearance of the old guard of white-coated experts who tested the purity of clothes (which only frames ago had suffered ground-in dirt, greasy stains and blood – sounds like the kind of thing a mind suffers in encountering the dirty world out there).

Early forms of advertising made very direct reference to the product, and so to sell a bar of soap a bill-board poster pictured the soap. In semiotic terms these would be considered denotative signs (iconic), in which the product for sale is signified by a direct reference to it. *Too easy.* A packet of cigarettes is advertised by a picture of a packet of cigarettes, not by a rip in purple silk, but the former fails to engage my attention (except as a novelty for its blatant demand): it is *easy to refuse*. Advertising's language, or more appropriately its image-repertoire, has hence grown in recent times to the extent where denotative language has become a sign of itself (Fairy Liquid's value for money) and where direct representation has become a strategy amongst many others (rather than the only way of proceeding). *Not so easy.* There is an abundance of other forms of signification which derive their meaning from *metonymy* and *connotation*. Metonymic advertising uses a part to represent a whole; for example, The Eiffel Tower for Paris. For the purposes of this essay I would like to use connotative advertising in the sense of a metaphor, of using one thing to represent another. Part of Glasgow's reformulation was to use Mr. Happy (from the children's *Mr. Men* books, 'and what about the gendering of abstract terms for first readers' asked Mr Right-On) underlined by the slogan 'Glasgow's Miles Better'. Mr. Happy was used connotatively to suggest that 'Glasgow is like *happy*'. This is a particularly useful development as far as I the consumer am concerned, because I can now purchase metaphoric and metonymic adverts of Glasgow while the city is for the time being out of my price range.

Would it be possible to advertise Glasgow denotatively? The whole of Glasgow is not easily pictured: it can be captured in its spatial totality by satellite imagery, but unfortunately for the advertiser such a city bears little relation to the city of my imagination. Being a city it will likely wish to 'advertise itself', and this wish will have to be catered for connotatively or metonymically, much as my own memory caters for me.

And now for a 'commercial break' briefly to consider the mixing of ad(vert)s and representation, and to ask some questions. Ads as a particular form of representation, obeying certain rules? Ads are uncritical, whereas art is critical? Ads may be more talked about than an art exhibition, but what does this mean? Publics? Art believes in its autonomy; it can fake its disinterest, whereas ads cannot and would not want to. What about the brevity of adverts? And at the same time, what about their constant reproduction and so the importance of their resilience – they must stand up to constant use on billboards and in television slots? What is the relation to 'tack'? Art, ads and capital? Art is a direct investment, a fixed asset, but ads are expendable in that an ad has value that quickly disappears (it is the product which must sustain the source of capital). Adverts for advertisers and adverts for art? Posters for an exhibition and the value of the exhibits? Still there is also 'tack' in ads; are they always so complex and do I as a consumer always make the effort to reconstruct that meaning? Or do I, as Baudrillard might claim, show resistance in ignoring, in silence, in mimicry?

Eric Laurier

Ad Analysis

'As a professional designer and creator of advertising I have a nagging respect for semiotics – not because of its dissection of "lies", but because of its emphasis on sign, symbol and myth, the hard stuff of mass communication – and indeed of high art. That's why I've always been entranced by Picasso's startling observation that "art is the lie that tells the truth" – especially relevant to advertising' (Lois, quoted in Blonsky, 1985, p. 312).

Advertisers are very interested in semiotics and semiotics has always been very interested in advertising. In Marshall Blonsky's excellent collection of essays on semiotics *On Signs* there are essays by semioticians on advertising and essays by advertisers on semiotics: all of which is slightly worrying, since it is increasing advertising's and semiotics' sense of their own self-importance. Nevertheless, semiotics is frequently used to 'decode' adverts, so it is particularly appropriate that if I am going to use some of the tools provided by semiotics it should be in relation to advertising and concurrently to the part that a particular form of *high* art plays in the charade. There is not the space here to provide a quick example of bill-board poster decoding at work. 'Decoding' is a phrase that I have a problem with since it signifies that same notion of demystification discussed by Barthes, and it is perhaps more useful to write *recoding* since what normally occurs is a shift from a combination of an image and a short phrase to an extended piece of writing which seeks to *re-write* the advert. The re-write's purpose is: firstly, to allow us to see semiotics in action (just what exactly is it doing?); secondly, to fragment the sign; and thirdly to point to significant absences.

An advert (or any sign) provides us with a seeming wholeness: it is there, the advert has presence. Let us briefly revisit my passing reference to men with their hands in the sink. In looking at this advert, I forget what it is not: it is not a Bengali male, it is not an old male, it is not female (yet I noticed that), he is not discontented, he is not poor, there are remarkably few dishes beside the sink. I do not notice what has been suppressed, what was rejected before this particular actor was selected. I do record that this advert is no longer shown. Part and parcel of discovering the absences in a sign is also to fragment the sign, to realise it is a position in a chain of signifiers rather than only an advert in and for itself.

Death of the Advertised

Advertising persuades us to like it, to like its product, to desire the advertised. These are its intentions, and to understand it I think we should try to discover these intentions in detail. Or do I? No, I disagree: as if I should find out why several major city centre buildings were commissioned by real estate agents and investors, or as if I should establish what the architects were trying to do in designing those buildings, whilst you just get a thrill out of the experience. Within a building there are a multiplicity of intentions: there, struggle resulted in the raising of the block of concrete, steel and glass. And, in the meantime, *what is there to say to myself about the reasons for my constant glancing up at this office when I pass down this street.*

Consider the notion of 'the death of the author', which is frequently misused or misinterpreted (appropriately enough) to convict Roland Barthes of 'critical fascism' – murdering The Author so that Barthes can do what he likes with her or his writings. Texts (whether they be novels, adverts, offices) can be played with/without permission of the author, or even without an acknowledgement of her or his purpose. This point is stretched further on the rack to say that 'from the author-killer's view there are only texts', and then after this brutal interrogation it too dies. The floor is gone from beneath our feet, and we are left held up only by a web of intertextuality: the relations between *Crime and Punishment,* a rip in purple silk and the Trump Tower (Morris, 1988a). But *I am still looking up at the building, any answers?*

The problem is not so much one of the author's intentions as of a critic's search for The Origin (most commonly artistic genius). To circumnavigate this second problem, Barthes showed the author to be a myth: a myth not to be ignored, but like all other myths one to be subjected to criticism not subjugated by it. Barthes continued to begin with an author and a text, the specifities of these things remained, and Barthes re-wrote them into another set of texts called theory. And so, *to examine the building I have to rebuild it.*

And to examine advertising of any sort is to look away from the origin of advertising and to look toward consumers, to myself and to ask how re-advertising occurs.

Ad Consumption

In view of past Marxist analyses of production processes, the common assumption of contemporary cultural studies is that consumption processes have been largely ignored and that all eyes should now turn to them, much like the turn of literary studies from examining the author's intentions to considering what the reader does with texts. Concentrating on consumerism does allow various *valorising* conceptions of the human subjects of mass culture to make an appearance in our analyses (Morris, 1988b). To work through the relevant points outrageously rapidly, this manoeuvre allows the receivers of products to be considered as active, critical users rather than as passive 'couch-potatoes' waiting to be manipulated by media.

As I outlined earlier, a consumer's activity is not readily defined by economic activity models: akin to any form of meaning construction, it is coded but multiple, fragmented, and frequently contradictory. And, related to these ideas, the everyday practice of consumption cannot be reflected as a mirror image of production, nor can it be said to be derived from production studies. The implication here for cultural studies can almost be summed up as follows: have fun with consumption; get to look at adverts; think about sex and notions of fragmented subjectivity; and you can forget about production because it is *passé* ('it's been done to death darling and we just don't need to hear anything more about it').

Yet to formulate consumption in such a way – as an opposite and independent position from production – is to be tripped up by one of the fundamental tenets of difference (Morris, 1988b). To build binary opposites is to make one dependent on the other, and so there cannot be consumption without production. Thinking through their difference rather than their argued independence, it is apparent that they merge in many places and that each process certainly does have effects on the other (in both senses of *other*), even if they are causal or may never ever be explicable. In the words of cultural analysis, production has been projected as The Other, suffering from an oppositional function which imagines it to be controlled by corporate power, factories, 'fat cats' and the determinator economics of capitalism. And such an account conveniently forgets the abundance of changes in production which have come about through capital's own new 'virtual reality', and through the economy's own reproducing of itself in the to-ing and fro-ing between consumption and production processes. Indeed, to think for a moment of corporate raiding as a form of consumption guided by a mysterious set of aesthetics seemingly derived from Viking mythology is to realise just how intertwined production and consumption actually are.

Now this is my parachute point (my access hatch to the street) to investigate 'tack', since 'tack' seems inexplicable without reference to production as well as to consumption. But I still want to defer my treatment of tack, because first I would like to consider Glasgow and my attempt to tell myself something about it.

Let Glasgow Flourish

'Let Glasgow flourish' is the phrase which lies at the bottom of Glasgow's coat of arms, and during the nineteenth century and early-twentieth century the city certainly did. Referred to as the 'second city of the empire', Glasgow held Great Exhibitions, built admired buildings and *fathered* great artists. Or such is the tale told by the

Glasgow District Council (GDC) which views itself as a new city *father* proud of its 'old' offspring. A city invented as such, dominated by the Left, by a traditional form of civic socialism, this city has a certain tale to tell. And at its simplest this is a narrative of the great days of the city, then the decline of heavy industry with the ensuing struggle but ultimate death of the working class, and then the rise of cultural industry in a post-industrial hyper-real world where greatness and grittiness float free on the Clyde, reflecting more cleanly than ever – another Utopia.

My own unease regarding this narrative arose when Glasgow was nominated Europe's 'City of Culture for 1990', on the basis of exchanges between political groups, the European Economic Community (EEC) and the Glasgow District Council (GDC). Culture was a big 'C' Culture, a *strategy* (in de Certeau's sense, from the top down: see de Certeau, 1984) of those political groups, a single authorial intent immediately splitting and forming into a chorus of strategies larger than their authorial intentions within them, strategies with cities as their texts – labyrinthine manifestoes. During 1990, while I worked in Glasgow, there was for me a sense of strain: not of the eyes, for there was plenty to see, but of the ears as I listened for other stories, for lost voices, resisting the larger narrative.

To articulate lost voices could be said to be, by definition, impossible: they are, after all, *lost*. Let me slip smartly by that hurdle by *being* one of those lost voices, since I was someone who was relatively silent throughout 1990. I was part of an audience and a populace seen but not heard. What should I say about Glasgow, its culture? I worked there, but lived just outside it in an 'upper class enclave on the Clyde' (which sounds like a general metaphor for cultural studies), so why should I not listen to myself? What, then, were my tactics to create a brief moment of expression against the engulfing spectacle of 1990 in Glasgow? How did I make do? How do I make do now?

You use anecdotes? My anecdotes are intended to be referential, to be models less abstract in their language than theories, so although they are not true (which is a claim that I evade at all points), they will be functional. More like journalism than science, they carry news of 'what's going on' and they are *debatable* in a positive sense.

Perfectly normal, I never owned a car-sticker saying 'Glasgow's Miles Better', not even in Italian, nor even in German or French. Then again, I was out of reach of this campaign through not owning a car either. But now to the 'tack'.

Tack and Kitsch

In the train of signifiers 'tack' and 'kitsch' share the same carriage: they travel third class and no one else travels with them. Milan

Kundera claimed not to be able to find an equivalent for kitsch in English, leading me to suspect that he meant English speakers would be unable to speak about such a thing. Either he did not know many forms of English or was being rather smug over something he felt English speakers were too kitschy to identify. And yet a television series like *Signs of the Times* demonstrates just how sensitive British home-owners are to questions of taste rather to ones of pure economics (Barker and Parr, 1992). This is how Kundera explains kitsch, which he feels is misunderstood and untranslatable:

'[K]itsch is something other than simply a work in poor taste. There is a kitsch attitude. Kitsch behaviour. The kitsch-man's (*kitschmensch*) need for kitsch: it is the need to gaze into the mirror of the identifying lie and to be moved to tears of gratification at one's own reflection' (Kundera, 1990, p. 135).

This definition is from the 'sixty-three words' that are Kundera's keywords; those he fears readers are most likely partially rather than fully to comprehend. In the *Jerusalem Address* Kundera returns to kitsch to redefine it:

'[T]he word "kitsch" describes the attitude of those who want to please the greatest number, at any cost. To please, one must confirm what everyone wants to hear, put oneself at the service of received ideas. Kitsch is the translation of the stupidity of received ideas into the language of beauty and feeling. It moves us to tears of compassion for ourselves, for the banality of what we think and feel. . . . Given the imperative necessity to please and thereby to gain attention of the greatest number, the aesthetic of the mass media is inevitably that of kitsch; and as the mass media come to embrace and to infiltrate more and more of our life, kitsch becomes our everyday aesthetic and moral code' (Kundera, 1990, p. 163).

Tack is less an emotional response than a quality of the subject, and it is a tricky word to use when all around are celebrating the consumer's ability creatively to reconstitute this very mass culture. I would have to be pretty dopey to buy something that is tacky, unless the tackiness becomes a joke. Except the joke is never on me. Kundera also has a sense of the process of tackiness: in the first moment of production all creations are free of poor taste, but as the distance grows the creation loses its sense until at its azimuth before oblivion (or reprieval) it becomes tacky nonsense. Tack is as fleeting as cultural analysis itself, and it is constantly on the move at the margins of mass media. Clothing of the sixties which has been 'out' has passed through the margins of tackiness where it was ascribed minimum value, and now it comes back into fashion, suddenly recaptured by the forces of production to be reconsumed as a historical reference of a time that has gone beyond embarrassing unfashionableness to become the *avant-garde* of heritage (there must be another way?). This is the society that Kundera forgot, since he considered tack as the decline of the sovereignty of intention. Perhaps he was unfamiliar with the work of Andy Warhol: I judge from Warhol's diaries that his intent was banal, yet his pop art has become infused with the gravity of an elite interpretation.

Money is Taste
Poverty is Tack

In suggesting an attitude of those who want to please the greatest number, kitsch could be over-production, which is not at any cost but rather a process of producing at the least possible cost economically (for Kundera this assumes the highest 'cost' artistically) the greatest number *and* a sense – kitsch – that still allows tears of compassion (fears of . . .). Fordism is kitsch: buy it now and have a laugh through the fashionable frames of cultural consumption. Does this all sound like advertising again? To perfect kitsch would be to use received ideas such as 'great art comes from great suffering' to sell something to the maximum number of people, in which case the history of art become kitsch and ends up part of the package (and part of the place).

In the decline of Glasgow's traditional sector of industrial production great suffering occurred, then, and what better raw material to turn the flywheels of cultural creativity? Or such are the lines from the advert sent out to convince people that Glasgow could well be a place outstandingly blessed with culture! But is this really the right sort of suffering raw material: is it not just too tacky, a form of working-class culture suitable for selling abandoned cranes, industrial museums, gritty pulp romances such as *No Mean City* and a diminutive mean detective called Taggart, but not one suitable for selling inner-city real estate? Is this not the point where tack and kitsch depart company, incorporate rather different forms of suffering, become rather different forms of advertisement, and operate differently in the coming together of culture and capital in the selling of the city?

Another part of the story which I would like to speak from 'the side' is that which comes before the fall: that part of the story which tells of a time when Glasgow was flourishing, when all was well in the city. This period led on to a second type of received idea at work here: the culture of success, the bourgeois trappings, the 'Great' period, the Florence of the Medicis – to be more specific, *the Glasgow of Charles Rennie Mackintosh*, a Glasgow of glossy places that a supplement reader might like to live in. And it is these glossy places as present (and maybe as *re*presented) in the current Glasgow cityscape that are the centrepiece of this essay, and around which I want to walk.

As an additional note here reference should be made to the problem (for the developers and city fathers) arising from the inertia of substance: the problem that the cranes, warehouses and tenements of old, suffering Glasgow have to be *empty* before they can be refilled with meaning. There is nothing more useless to a city-seller than a working-class city that is still working class. And all the massaging of images in the world will be to no avail if the gritty places cannot be cleared ready for being 'glossed up'.

Dog Stories

In the days when my sister and I regularly walked Jason the dog, our chosen path all too often ran along the street on to which Hill House faces in all its menacing rockiness. Hill House was designed and its construction supervised by Charles Rennie Mackintosh. His controversial work had come to the attention of the Blackie Family (owners of a publishing house), who commissioned the building of the house. Currently the owners of Hill House are the National Trust, and yesterday (Sunday 12 April, 1992) it was visited by 206 people. My sister works at the house, she is there as I write this, her job being to guide visitors around the house.

And when we walked Jason we would make our way past Hill House, then onward to pass Lovell Homes' latest product: a row of half-a-dozen designer houses. These were houses that my father had decided were very tacky, not to say shocking, but about which my sister and I were ambivalent (the dog, being non-visual, was disinterested in modernist references and expressed no opinion). Thinking through de Certeau's (1984) notion of walking rhetorics, ours were thoughtful pauses in our pedestrian phrases, pauses to reconsider – verbally as often as not – these houses. We were passing them by, an act which de Certeau considered normally to be omitted when pedestrian routes are mapped around the city: moments outside of science and inside our socio-poetic lives.

Sister, dog and I repeat our route over and over, the pauses we make are predictable, and this route could be thought of as one of our sayings or perhaps more romantically as a musical phrase. It is performative and rightly so: we dawdle outside the Lovell estate, and it becomes a situation where we can escape the Glaswegian Spectacle, fleetingly creating our culture and then it is gone.

A moment snatched from where?

Lovell Homes are sharply contrasted black and white houses. Each is a variation on the general theme (in this case a generative architectural grammar), a theme that my sister and I think of as 'Mackintosh'. Distinctive elements of this style entail a square divided into nine smaller squares, with a brightly coloured sliver of stained glass using an *art-nouveau* pattern in the centre of each black door.

What is recognisable is a system of meaning, a clear set of ingredients, and each house combines the elements differently. Yet Lovell Homes display an unwillingness to abandon a traditional overall structure to the house, two up-two down, and inside the houses are like any other suburban residence: this means that the Mackintosh is merely 'tacked on' the outside (a case of kitsch into tack?).

In our conversations passing along in front of the houses, we were pleased with their distinctive style and bemused by their obvious borrowings from Hill House just along the road. Our trip past them became a lesson and a laugh over the aspirations of the middle classes, and also over our own aspirations, since we would quite happily have bought one – except that Lovell Homes commanded such high prices. Even more problematically for our own taste, though, the estate later quadrupled from a dozen to fifty houses (an axis of tack in its own right), and our opinion changed of the houses: they had now become tacky. As Kundera would have said, they were trying to please the greatest number at any cost. Here was style over-production occurring, distinction dwindling, and my sister and I did not add the new houses to our route.

Within our walks there were other significant absences, paths not taken. We were not interested in the suburban estates just around the corner from us. Our walks 'aspired'. What, then, *of* the non-Mackintosh? Their randomness, their inarticulacy, their blandness and their banality: we in *our* way silenced them. Yet the same thing was occurring to the piece of pedestrian verse that we had created: it was growing apart from its origin in a symbolisation of the desire to capitalise, to produce and to consume. Our rerouting of the houses was to dismiss the houses as tacky, and tack became our tactic.

What of the woman I encountered as I paced along the same route, using it as a reminder of what my sister, dog and I were doing? What strange bricolage was she involved in, using her oriental dog which farted constantly (those were the sounds I heard), her expensive coat, her designer estate? Walking out of her Mackintosh house to Hill House and back, equipped with a walkman; to hide the noises of the dog or to create a slow and sensuous ballet with the dramatic backdrop of the Lovell estate, the harmony of the houses in front and the turning point of the composition as she encountered Hill House, then triumphantly returning to her sitting room and laying dog and walkman aside until the next performance? Then again, I could not tell if it was Wagner or Manilow that I overheard.

Hill House

As well as attributing a signified for the signifier of Lovell Homes (an authentic for the inauthentic by juxtaposition), a guarantee of the price (a pegged currency), Hill House serves a further function for me. I use it to advertise myself: not only as I am doing just now to display my cultural assets, but also as an explanatory trip for foreign friends who stayed weekends. For my visiting friends (although they of course did not know it) Sunday afternoon visits to Hill House were inevitable, and in such a fashion I ended up visiting Hill House almost a dozen times. All of this was over two years ago, when the house had not yet known Mackintosh fever because the 'City of Culture' extravaganza was yet to come.

It was perhaps a question posed to my guests of whether they regarded Hill House as important as I did, whether they enjoyed its *art nouveau* interiors, its play of light and shadow, its display of the best and worst of modernism. Those were my prepared versions of what Hill House meant, leaving me rooted passively in the house's rooms, a knowing (know-it-all) subject waiting patiently to point out certain things and forcing my friends to use me as the reference point for all that was beautiful, good and true about the house. A selfish need to play the part of a guide, of an expert and of a patron (father?) and bore.

Here was something to discuss, to get to know Glasgow through, to understand my culture. As upon a special stage, like an intellectual round table, privileged judgements were to be made here about the realm of *identities*. With an Australian friend who studied Fine Arts we spent a summer's afternoon smelling the roses in the garden instead of admiring the restoration of the original rose wallpaper. Tactics? My sister tells me that several elderly ladies have approached her to ask questions about the Glasgow Girls instead of about Charles Rennie Mackintosh, and does this imply that the discourse built into the house might now be changing? If I were to take you round the house today, I would say 'that is not what I meant at all'.

Four Photographs of a Journey

Glasgow School of Art. Hidden, it has to be discovered, perched on a steep slope behind Sauciehall Street (see **Figure 1**). After taking this photograph I became embarrassed about taking another since the building was being photographed by a second photographer who appeared to be altogether more serious. Actually the problem was not so much the other photographer as the building in the first place. My relationship with it is not one which makes the memorisation by photograph necessary. The building is part of my own origin, not somewhere I am going to or where I am just visiting. To photograph this populist architecture is, I realise, as tacky as tourism; it is a given, it has been advertised. This conclusion is inescapable, since I am

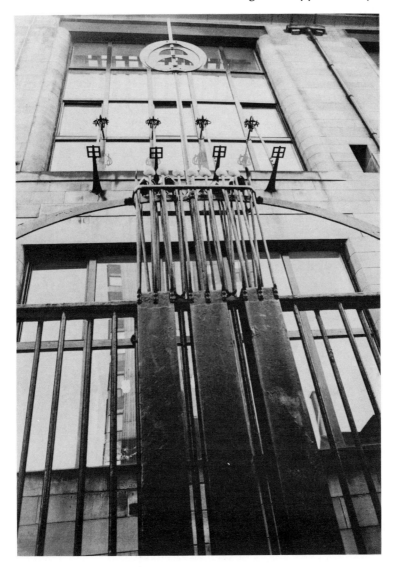

FIG. 1. Glasgow School of Art.

witnessing somebody else doing it and would be witnessed by that person in all of my own cumbersome technique. It is curious not to know what else to photo other than a building which has been aggrandised by the *authorities* rather than by me. I sneak away with one frame which I hope suggests its proximity, my own intimate encounter rather than a distant and total view of the building which is a tourist's vision.

FIG. 2. Catherine Shaw.

There are typical elements of Mackintosh in the photo, notably the use of wrought iron to add and to emphasise the geometric qualities of the building whilst also apparently stretching its verticals. My eye moved up, and behind the railings to the subdivided square of the window: Mackintosh would be looking for the light effect as that grid crossed the interior hour by hour. Next I regarded the organically shaped remainder, the abstracted peacock feather, a romance of iron work and a common feature of the 'Glasgow Style'. Finally, I noted the intrusion of the building opposite, its reflection in the window – what is all too often now called Modernism, yet I think of it as the triumph of tack manifested in Modernism's nemesis.

Catherine Shaw defiantly stamps her name on the door (see **Figure 2**). 'I am not Kitsch, I am quality'. One of the aspects of Mackintosh's style which I dislike is that it is total, and this total aesthetic is the reason that the shop *is* kitsch. Shop-window, door handles, the very displays are not in Mackintosh's style and so they are unacceptable, are tacky. Unrepentantly I stare through the window, saying 'I know better'. My own contradictions fail to worry me: disliking Mackintosh's style for its intolerance of other styles. (In Hill House the Blackie Family were allowed one room to cram all of their normal yet transgressive furniture into: the dining room, dark, conservative, sharing only its fireplace and its lightshades with the rest of the house.)

At the same time all of the simulcra Mackintosh, bar the buildings, are contained in this treasure trove. When I enter the shop, the objects all go quiet. Of course they are dumb, they know nothing, they have to be caught in the act, adorning someone's wrist, filled with wine, lighting a room or perched in the heart of a house on the mantelpiece.

Global Crossing. Just to the right of the frame Union Street becomes Jamaica Street and they are crossed by Argyle Street (see **Figure 3**). Local legends are here signified by street names: a memory that is a crack in the global economic consciousness which has claimed this crossing. Why global economic consciousness? On the other three corners squat Pizza Hut, McDonalds and Dunkin' Doughnuts. As part of a process sometimes referred to as 'global localisation', the Boots Store which was on this site was demolished and replaced by a 'building' indicative of an aspect of localisation, the facade. A building becoming a bill-board. Mackintosh's style on the exterior is part of an advertising strategy, *and it is blatant and I could not help laughing as I read the* Kentucky Fried Chicken *sign*. Then, whilst waiting for the lights to change, I overheard 'I think that new building across the road is horrible.'

Without some local knowledge, the dislike of this apparently aesthetically-pleasing building is hard to comprehend. Let me try to fill in the gaps by putting the Boots Store back on to the corner spot. In style, it was typical of the 1950s, architecturally dated, *passé*, out of fashion with a certain group of people. Fortunately it had escaped those architectural authors, and it had become a corner where *passing by* stopped: a site where friends and lovers waited in their free moments, their breaks, their lunch-hours and all other time outside of work (except leisure is work, as Guy Debord pointed out). Initially this meeting place may have grown up around the critical signifier

Fig. 3. Global Crossing.

for a meeting place – the giant clock on top of the building, the adjudicator for those who arrived too early and too late. It is a corner that then became infused with meaning because people paused there; it is haunted by memories, personal memories, perhaps expressible as 'under the Boots clock I waited; I was stood up; I was met and was happy; I was late and it was scarey; we were both late'. These are the moments of the crisis in our everyday lives, and also of the subject's surrender in its free time to a system.

Now the meeting place has been yanked out with all the dismay of a healthy tooth, the gauze of a dominant reproduction has been stuck over it. Boots was not advertising itself there, nor was the GDC advertising Glasgow: the economics of meeting friends, of waiting for lovers, are mysterious to capital. And if they were easily translatable, then 'my dear', we would be in jeopardy.

In Princes Square on the spiral stair, the bannisters are polished, tactile, the steps are acoustic wood (see **Figure 4**). I climbed them every morning on my way into work. In Princes Square the stairs are foregrounded, and this is unexpected since in contemporary shopping centres escalators form the dominant mode of passing from level to level: escalators which force a system of travel upon one, given that there is little choice in how to deal with an escalator because it even has a motorway-like lane system saying 'please stand on the right, please stand on the left'. No one dares sit down on an escalator, or jump back and forward pensively from step to step. Across the open interior of the square there are its escalators, and they are the closest to the main entrances. Whilst walking into the square, if a visitor wishes to be direct they can take the escalators, shun the open space of the main square. By riding the escalators they become spectators, watching the other people on the grand stage of the stairs or finding their own stares returned in the passengers in the lifts which rise up in the corners on either side of the grand central staircase.

Within the square there are three forms linking the floors, all of which open dramatically on to one another. They are centred around the stairs, the least and most denotative of the three symbols of elevation. They are the *most* direct sign of rising or falling between the floors because stairs were the first, the Classical, the basic; no multi-storey is without them. And yet at the same time they are the *least* direct sign, since they are the most ornate with their twisted black iron railings, their lamps and ultimately their constant reference to the 'Glasgow Style'. Upon them on every day of the week people sit or stand, witnessing other people's shopping (in the broadest sense): they are almost a public forum within the square. These stairs have become a sign of their function (as all functional things do), and once their direct relationship is given up we can use them as we like. They are a form and a function full of play.

What makes the stairs' symbolic value still more apparent is that concealed within the corners of the surrounding building of the square are two other sets of stairs, and these are back-stairs to the front-stairs. These back-stairs are rudely functional; plain-tiled,

Fig. 4. Princes Square.

windowless, straight railings. None of the shops are tied directly to them, and rather they are linked with the staff passages, the toilets, the back-entrance: all of the shitty aspects of the square's functioning.

As I walk though the Square I am offered these three potential motions in walking-gesture, in my narrative of *potential* consumption. Here is a medium which accepts a complex, contradictory consumer who might even want to use the stairs. It is an open text, and what do I mean by open text? Let me refer to the floors of Lewis', Fraser's and other department stores, where there are smooth line-paths to direct one around the store and carpeted sections schematicising where I should pause to purchase. Purchase here is in both senses, as I can find a place to rest my weight and I can also rest my self further on the thing I intend to buy. The designer has an intention which he or she forces upon me: *follow the path*, no matter how winding it gets. I am left crossing the shopfloor obliquely, hoping to find my way through the jungle. But Princes Square itself is open: it is all tiled, without floor codes to bind me to a particular path, and the tiles are evenly or abstractly patterned. The light used is not pinpointed by spotlights to alert my eye to certain goods, calligraphy and so on, and the light is daylight through the glass-roof. There is a myth at work nevertheless in its use of natural light, unobtrusive night-light, tiles and the like: the Square could be said to be presenting a myth of outdoor street life, albeit outdoor life enclosed.

There is a further catch: within this square there are Katherine Hamnett, Crabtree and Evelyn, Monsoon and so on, the shops of the affluent. To shop here one has to be rich. By a form of revisited industrial benevolence, the rest of the populace are allowed to come in and play in this place of freedom whilst the rich support it and enjoy some pearly grit on the way. Perhaps it is the museum of our *fin-de-siècle*, with living exhibits.

The Glasgow Style

Catherine Shaw, the Kentucky Fried Chicken Building and Princes Square are part of a rediscovery of the so-called 'Glasgow Style', the reproduction of a form of representation popular in Glasgow between 1895 and 1920 that is now recognisable as part of the *art nouveau* style and more generally of Modernism.

To consider tack for a moment, what appears to be the case is that straightforward reproduction as in the case of Catherine Shaw or diluted reproduction as in the case of Kentucky Fried Chicken incites the revulsion of those apparently being expressed by it (whilst *kitschmensch* heart swells with pride). Where re-interpretation has been carried out, as in Princes Square, the possibility is opened up of slipping out from underneath the swell of tack. But is this not just the old quote about lesser artists copying and great artists borrowing?

The appearance of the Glasgow Style from 1895 to 1920 was heralded by the introduction of an asymmetrical, rectilinear style which diverged from the dominant *art nouveau* motifs as exemplified in the whiplash style of Art Nouveau in France and Belgium (Burkhauser, 1990). Most similar to the Glasgow Style (also known as the 'Scotto-Continental') were the Viennese and German Secession movements. The allure of tracing out Art Movements draws me into its *high* circles, which is not necessarily a bad thing, but I think that the important point to be made here is Glasgow's European rather than British orientation during this period. Margaret and Frances MacDonald and Charles Rennie Mackintosh exhibited to greater acclaim in Vienna than they ever did in Great Britain (or even Glasgow), and Gustav Klimt was influenced by the work of the MacDonald Sisters.

The Glasgow Style came out of the Glasgow School of Art, and was almost an industrial art movement, since in common with the 1880s cult of Design strong links were emphasised between industrial production and artistic production. Objects were to advertise themselves (and note in this respect advertising being conceived as a field of aesthetics hived off from the product into a middle domain between product and consumer). Rather than being a painting-based phenomenon, this artistic current manifested itself in the creation of clothing, furniture, metalwork, interior design, the areas referred to at the turn of the century as 'Arts and Crafts'. This form of art, *not* painting and therefore gallery-based, allowed the movement of its motifs into everyday life instead of its procession from gallery to accumulation and exclusion in board rooms, mansions and the storage vaults of museums, and it was certainly a vital shift for the style to become identified – and for it to be able to become so associated – with a city (an everyday city in motion) rather than with a particular section of the art world.

'Nowhere else has the modern movement of art been entered upon more seriously than at Glasgow; the church, the school, the house, the restaurant, the shop, the poster, the book with its printing, illustrating and binding, have all come under the spell of the new influence' (*The Studio* (art periodical), 1906, quoted in Burkhauser, 1990, p. 37).

The signifiers of this movement were elongated, organic symbols: for instance, the Glasgow Rose, the Peacock feathers on the railings of the Art School. Female figures as used by the movement were elongated, with stretched arms and fingers, resulting in the artists involved being referred to as the 'Spook School' (Burkhauser, 1990). Again, crucial to the process whereby the style came to symbolise Glasgow was its lack of acceptance in the rest of Britain, thereby restricting its spread and echoing both the independence of the rest of the city's everyday culture (of which the style quickly became a part) and its exchange outside of Britain along axes undefined by London.

As well as the Glasgow Style's European links, its oriental influences are remarkable, and the open, light spaces of the style's interiors were intended to be developments of oriental interiors which used paper walls, sliding panels and uncluttered space. When I consider Glasgow's future, its dreamed of utopias, then the fact that these should contain elements of the Far East seems ironic; or perhaps it is suspicious that they should now be involved in the strategies of city selling?

One For All and All For One

Although the Glasgow Style was a movement – a group of artists, architects and designers all working together in developing a style – Charles Rennie Mackintosh has been singled out metonymically to signify the whole movement. Returning briefly to the idea of absences, the rest of the movement has in effect disappeared. The myth of The Author returns in another form: the Glasgow Style symbolically requires one author, a point of singular origin for it to be advertisable (to be brief, to be tacky, to be too distant from its origin for us to realise the *width* of that local point). Rather than mention Margaret MacDonald, Frances MacDonald, Jessie M. King, Annie French, the Walton Sisters, Herbert MacNair and so the list could go on, there is one *man*. Consequently he becomes a genius: to be made responsible for the success, the diversity of the whole of the Glasgow Style, *well he must have been a genius.*

This process whereby *one* stands for a whole movement – and in turn becomes part of a discourse of male authors, individual geniuses, urban culture, advertising – is not unopposed, and each of the signs within itself bears the marks of a struggle. My sister was asked by those elderly ladies visiting Hill House 'what particular Glasgow Girl helped here, and is there any of her artwork on display?' And I refuse to take Mackintosh as sole creator, which makes this essay part of the struggle against Mackintosh's imposition as the Glasgow Style, as Glasgow's art and culture.

Like Roland Barthes's methods, part of the altering of this myth will come from close attention to Charles Rennie Mackintosh's life – his social surroundings. What does a brief look reveal? His relationship with Margaret MacDonald, who became his wife? She was in fact co-author of much of what is now referred to as Mackintosh. There is a series of postcards and greetings cards which use botanical watercolours as their illustrations, and on the back of the cards they are labelled 'Charles Rennie Mackintosh'. On at least one of the illustrations, however, there are the initials 'C. R. M. M. M.' which stand for *both* partners in the marriage.[2]

The interiors of Hill House were designed by Margaret MacDonald, and she worked on the fireplaces, the wallpaper, tiles, in fact most of the detail. If 'Tackintosh' today is considered, I think it is notable that concern over facades and ignorance of the interiors is a defining feature of Lovell Homes, the Kentucky Fried Chicken Building, all of which I declaim as tack for ignoring the detail of the first productions – a detail which was Margaret MacDonald's as well as Charles Rennie Mackintosh's. To use the latter is to gloss over the other. In putting the Mackintosh advertising strategy into practice, what has become noticeable is how tactics have begun to unravel the tack (to sort out tack from kitsch from something else again?).

The Glasgow Girls

In 1988 there was an exhibition held at the Glasgow School of Art on the so-called 'Glasgow Girls', and in 1990 a major exhibition was held at the Kelvinhall Gallery. Produced at the same time as the 1990 exhibition, the book *The Glasgow Girls: Women In Art and Design 1880–1920* has been a runaway success in Glasgow. I gleaned most of my knowledge of the Glasgow Girls from it. The history of this group of female artists and designers has been suppressed since 1920 to the extent that the book and the exhibition were hailed as rediscoveries. One further but small personal rediscovery was my sister's delving into our great-aunt's photograph album; our great-aunt, it turned out, was one of the minor figures in the Glasgow Girls and the album contains class photos which had never been stored (oddly enough) in the Art School's archives. Jude Burkhauser notes, in her introduction to her *Glasgow Girls* book (1990), that 'there was almost no critical work published on Scottish Women in Art and Design until 1980', although women had played an important part in the field for the last two centuries. Art history's construction of its object, art and artists, has instituted a mythology wherein artists are male, lone geniuses.

It is not only the operations of art history which have lost the voices of women; there are also the current operations of institutions. In these walls there are the *major* arts such as painting, sculpture and architecture and the *minor* arts (interior design, the decorative arts), and in this binary division the majors arts tend toward the masculine and the minor arts toward the feminine. In the school itself at the turn of the century this meant that, although women were encouraged to attend art school, they were allotted to the decorative arts classes. Furthermore, although the decorative arts in themselves are no less important than the other arts, part of their diminution has come from the connotations of the private and the domestic as the site of the feminine.

Glasgow's urban advertising strategy has muscled into a masculine discourse of publicity, of individual genius, and of the major arts.

Reconsidering Glasgow's narrative, it has been conspicuously masculine, moving from the (production) hey-days of the last century to the industrial decline (with its dramatic effects on the male work force, as linked to the gritty aggression of male working-class strife) and now onward: it has almost reinvented itself out of this story as a new man's city. It becomes a city selling itself through its manly narrative and through the images of this manliness. And yet the figure of Charles Rennie Mackintosh in the advertising photos looks like a woman's man with his soft bow tie and his stance to one side: after the rough production of Glasgow's *culture*, its consumption has turned out to be much more gentle-manly.

The Glasgow Rose

Although normally attributed to Charles Rennie Mackintosh, the distinctive insignia of the Glasgow Style was formed under Jessie Newbery's hand, its innovative style arising out of her knowledge of gardening combined with her artistic medium, *appliqué*. And, as Jude Burkhauser points out, the Glasgow Rose's various significations cannot be ignored. Past religious connotations of the flower have laid with Christian evocations of the Virgin, and also with pre-Christian symbolisms as part of Aphrodite's sexual mysteries. In the works of Frances and Margaret MacDonald the rose became vaginal in illustrations, centred on women's bodies. In the painting *The Heart of the Rose* it contains a child. This thorny symbol has its place in the struggle over Glaswegian identity: a place full of birth, sexuality, femininity as well as containing Great Men and Great Buildings and Great Advertising. Margaret MacDonald used an imprint of the Glasgow Rose to produce the wallpaper for the interior of Hill House, and it is now printed on postcards, painted on the walls of flats, printed on mugs, worked in silver on ear-rings and woven into jumpers and appears on many more products. It is not *just wallpaper*.

Meanwhile

My imagined readers may hold notions of Glasgow as a city undergoing a culturally-driven revival, yet still shackled to its working-class heavy industrial past. The working-class past is notably absent from this essay, and I am illustrating a middle-class obsession with architecture and art rather than with other definitions of culture which would be more enabling. Then again, it is also the advertising you have been watching in your colour supplements. As Jessie Newbery put it:

'I believe that the design and decoration of a pepper pot as important in its degree as the conception of a cathedral' (Newbery, quoted in Burkhauser, 1990).

Eric Laurier

Notes

1 I would like to acknowledge the help of several semi-fictitious people – Chris Positive, Phil Nice, Mark Nebulous, the excellent Issy Possibility, Mary Contrary, the atypical Maeve Cha', Jason Woof, Dad Trad, Eric Pretend and Lara Syncratic – without who I would never have finished my homework.
2 The card in question is entitled *Larkspur, Walberwick 1914*, from 'The Mackintosh Collection', *Camsail Fine Art*, Dumbartonshire, Scotland.

Consolidated Bibliography

'350 Millions pour Cory.' *Libération* 13 July (1989).

ABERCROMBIE, N., HILL, S. and TURNER, B. S. *Sovereign Individuals of Capitalism*, London: Allen and Unwin, 1986.

'Action Directe: le juge fait un pas.' *Le Figaro* 20 July (1989).

ADSETTS, N. 'The Sheffield Partnership.' Paper presented at the Annual Conference of the Institute of British Geographers, University of Sheffield, 1991.

AGUDO, P. 'C'est possible.' *L'Humanité* 19 July (1989).

AGULHON, M. (ed.) *Histoire de la France urbaine. IV. La ville de l'âge industriel. Le cycle Haussmannien*, Paris: Seuil, 1983.

AGULHON, M. 'Quelle stratégie pour le bicentenaire?' *Revue Politique et Parlementaire* July–August (1987) pp. 14–16.

AGULHON, M. 'Pourquoi célébrer 1789?' *L'Histoire* 113 (1988) p. 5.

ALLAIN-DUPRE, E. 'Pyramid du Louvre: les structures de l'invisible.' *L'Architecture d'Aujourd'hui* 253 (1987) pp. 62–69.

AMALVI, C. 'Le 14 juillet: du Dies Irae à Jour de fête.' In P. Nora (ed.), *Les lieux de mémorie. I. La République*, Paris: Gallimard, 1984.

AMBROSE, P. *Whatever Happened to Planning?*, London: Methuen, 1986.

ANDERSON, J. 'The "New Right", Enterprise Zones and Urban Development Corporations.' *International Journal of Urban and Regional Research* 14 (1990) pp. 468–489.

Architectures Capitales, Paris: Electa Moniteur, 1987.

ASHWORTH, G. J. and VOOGD, H. *Selling the City: Marketing Approaches in Public Sector Urban Planning*, London: Belhaven, 1990.

BACQUE, R. and MURAT, L. 'La bataille de Paris: la préparation des fêtes du bicentenaire.' *Profession-Politique* 5 June (1989) pp. 6–7.

'Badoit fait pétiller la révolution.' *Le Figaro*, 7 October (1987).

BAILEY, J. T. *Marketing Cities in the 1980s and Beyond*, Cleveland: American Economic Development Council, 1989.

BAKER, R. and THOMAS, J. (eds.) *Writing the Past in the Present*, St. David's University College, Lampeter, Wales: Department of Archaeology, 1990.

BALFOUR, J. *Rockefeller Center: Architecture as Theatre*, New York: McGraw Hill Book Company, 1978.

BARKER, M. and PARR, M. *Signs of the Times: A Portrait of the Nation's Tastes*, Manchester: Cornerhouse Publications, 1992.

BARRE, ZEVI and NOUVEL. 'Commentary and conclusions.' *Architectural Design Profile. Les Halles* (1980) pp. 75–76.

BARTHES, R. *A Lover's Discourse*, London: Penguin, 1990.

BARTHES, R. 'Day by day with Roland Barthes.' In M. Blonsky (ed.), *On Signs*, Oxford: Basil Blackwell, 1985.

BASSETT, K. 'Economic Restructuring, Spatial Coalitions and Local Economic Development Strategies: A Case Study of Bristol.' *Political Geography Quarterly* 5 (1986) pp. S163–S178.

BASSETT, K. *et al.* 'Living in the Fast Lane: Economic and Social Change in Swindon.' In P. Cooke (ed.), *Localities: The Changing Face of Urban Britain*, London: Unwin Hyman, 1989, pp. 45–86.

Consolidated Bibliography

BATTERY PARK CITY AUTHORITY. *Battery Park City Draft Summary Report and Master Plan*, BPCA, One World Financial Center, New York, NY 10281–10971, 1979.

BATTERY PARK CITY AUTHORITY. *Battery Place Residential Area Design Guidelines*, BPCA, One World Financial Center, New York, NY 10281–10971, 1985.

BATTERY PARK CITY AUTHORITY. *Analysis of Tenant Survey, 6 April* (provided courtesy of John McMillan), BPCA, One World Financial Center, New York, NY 10281–10971, 1988.

BAUDRILLARD, J. *America*, London: Verso, 1988a.

BAUDRILLARD, J. 'Consumer Culture.' In M. Poster (ed.), *Jean Baudrillard: Selected Writings*, Cambridge: Polity Press, 1988b.

BEGUE, B. 'Vos guelles les pauvres.' *Politis. Le Citoyen* 69, 7–12 July (1989) pp. 10–13.

BEIGEL, F. 'A Prototype for a Public Hall: Isle of Dogs Neighbourhood Centre.' *A3 TIMES* 5:11 (1989) pp. 13–16.

BENJAMIN, D. 'Job Well Done: Brothers set up Work for 2000.' *Dudley Evening Mail* 17 March (1989a).

BENJAMIN, D. 'Merry Hill 5000 Jobs Boost Tarmac's 40m Contract.' *Dudley Evening Mail* 4 July (1989b).

BENNETS, L. '16 Tenements to become Artists Units in City Plan.' *New York Times* 4 May (1982).

BENOIST, J.-M. 'Bicentenaire: non la Révolution n'est pas un bloc.' *Le Figaro* 13 July (1989).

BERET, C. 'Une révolution formaliste.' *Art Press* spécial (1988) pp. 100–115.

BERGER, F. 'Le protocole dans tous ses états.' *Libération* 8 July (1989).

BERTIER, E. 'A son panthéon personnel: Carnot et Danton.' *Libération* 15 July (1989).

BEYNON, H. 'Regulating Research: Politics and Decision-Making in Industrial Organisations.' In A. Bryman (ed.), *Doing Research in Organisations*, London: Routledge, 1988

BIANCHINI, F. 'Urban Renaissance? The Arts and the Urban Regeneration Process in 1980s Britain.' Working Paper No. 7, Centre of Urban Studies, University of Liverpool, 1989.

'Bicentenaire 89: Finlay annulé.' *Libération* 30 March (1988).

'Bicentenaire: Jarre première victime.' *Le Quotidien de Paris* 26 September (1988).

Bienvenue à la manifestation et au concert du 8 Juillet, Leaflet distributed by the organisers of the 8 July demonstration, Paris (1989).

BIETRY-RIVIERE, E., NATAF, I. and SARRE, V. 'TV étrangères: l'oeil des stars.' *Le Figaro. Fig-Eco* 15 July (1989).

BLONSKY, M. 'Endword.' In M. Blonsky (ed.), *On Signs*, Oxford: Basil Blackwell, 1985.

BLUMIN, S. M. *The Emergence of the Middle Classes: Social Experience in the American City, 1760–1900*, Cambridge: Cambridge University Press, 1989.

BLUNKETT, D. *Local Enterprise: How it Can Help the Alternative Economic Strategy*, Nottingham: Institute for Worker Control Pamphlet No. 79.

BONLAURI, B. 'Ce qu'Edgar Faure veut faire.' *Le Figaro* 27 May (1986).

BONLAURI, B. 'Vive Robespierre.' *Le Figaro* 22 September (1987).

'Bonne fête?' *Le Figaro* 13 July (1989).

BONNET, F. 'D'est en ouest de Paris, l'avenues des chantiers-Élysées.' *Libération* 12 July (1989a).

BONNET, F. 'Un trou dans les comptes du Cube.' *Libération* 12 July (1989b).

BORSAY, P. 'The English Urban Renaissance: The Development of Provincial Culture.' *Social History* 5 (1977) pp. 581–603.

BORSAY, P. *The English Urban Renaissance: Culture and Society in the Provincial Town, 1660–1770*, Oxford: Clarendon Press, 1989.

BORSAY, P. 'The Emergence of a Leisure Town or an Urban Renaissance? Comment.' *Past and Present* 126 (1990) pp. 189–196.

BOUFFIN, Y. 'Les dix de Billancourt ramenés à la poste.' *Libération* 10 July (1989).

BOULDING, P., HUDSON, R. and SADLER, D. 'Consett and Corby: What Kind of New Era?' *Public Administration Quarterly* 12 (1988) pp. 235–255.

BOULY, M.-L. 'Jean-Michel Jarre.' *Le Figaro. Madame Figaro* 15 July (1989).

BOURDIEU, P. *Distinction: A Social Critique of the Judgement of Taste*, London: Routledge and Kegan Paul, 1984.

BOYER, M. C. 'Pretty as a Picture.' Unpublished manuscript, available from author.

BOYER, M. C. 'The Return of Aesthetics to City Planning.' *Society* 25:4 (1988) pp. 49–56.

BOYLE, M. and HUGHES, G. 'The Politics of the Representation of "the Real": Discourses from the Left on Glasgow's Role as European City of Culture, 1990.' *Area* 23 (1991) pp. 217–228.

BRADLEY, D. 'Reviving the Lower Don Valley.' *The Planner* 76 (1990a) p. 23.

BRADLEY, D. 'Regeneration through Sport and Leisure.' *The Planner* 76 (1990b) p. 23.

BRADLEY, D., MERCER, D. and STURCH, L. 'Regeneration through Shopping.' *The Planner* 76 (1990) p. 23.

BRAITBERG, J.-M. 'Paris-barrières.' *Le Quotidien de Paris* 14 July (1989).

BREEN, T. H. 'An Empire of Goods: The Anglicisation of Colonial America, 1690–1776.' *Journal of British Studies* 25 (1986) pp. 467–499.

BREEN, T. H. '"Baubles of Britain": The American and Consumer Revolutions of the Eighteenth Century.' *Past and Present* 119 (1988) pp. 73–104.

BRINDLEY, T., RYDIN, Y. and STOKER, G. *Remaking Planning: The Politics of Urban Change in the Thatcher Years*, Hemel Hempstead: Allen and Unwin, 1989.

BROWNILL, S. *Developing London's Docklands: Another Great Planning Disaster?*, London: Paul Chapman, 1990.

BUCK-MORSS, S. *The Dialectics of Seeing: Walter Benjamin and the Arcades Project*, Cambridge, Mass.: MIT Press, 1989.

BURGESS, J. 'Selling Places: Environmental Images for the Executive.' *Regional Studies* 16 (1982) pp. 1–17.

BURGESS, J. 'News From Nowhere: The Press, the Riots and the Myth of the Inner City.' In J. Burgess and J. R. Gold (eds.), *Geography, the Media and Popular Culture*, London: Croom Helm, 1985.

BURGESS, J. and WOOD, P. 'Decoding Docklands: Place Advertising and Decision-Making Strategies of the Small Firm.' In J. Eyles and D. M. Smith (eds.), *Qualitative Methods in Human Geography*, Oxford: Polity Press, 1988.

BURKHAUSER, J. 'Introduction.' In J. Burkhauser (ed.), *Glasgow Girls: Women in Art and Design, 1880–1920*, Edinburgh: Canongate, 1990a.

BURKHAUSER, J. 'The Glasgow Style.' in J. Burkhauser (ed.), *Glasgow Girls: Women in Art and Design, 1880–1920*, Edinburgh: Canongate, 1990b.

C. T. 'Les États-Unis rejettent l'idée d'organiser une conférence Nord-Sud.' *Le Monde* 16 July (1989).

CAME, F. 'Jacques Attali: "Comment faire du Nord-Sud sans le Nord?".' *Libération* 10 July (1989a).

CAME, F. 'Comment les sherpas avaient ficelé le sommet.' *Libération* 17 July (1989b).

CAMPBELL, B. 'New Games People Play.' *The Guardian* 23 August (1989).

CAMPBELL-BRADLEY, I. *Enlightened Entrepreneurs*, London: Weidenfeld and Nicolson, 1987.

CANARY WHARF DEVELOPMENT CONSORTIUM. *Canary Wharf: The Global Market Place* (1986 publication: no longer available, but copy obtained from Savills Estate Agents, Wapping Wall, London E3).

CANN, B. 'The Park of La Villette: Urban Park as Building.' *Places* 4:3 (1987) pp. 52–55.

'Cap maintenu.' *L'Humanité* 15 July (1989).

CARR, C. 'Night Clubbing. Reports from Tompkins Square Park Police Riot.' *Village Voice* 16 August (1988).

CARR, L. G. and WALSH, L. S. 'The Standard of Living in the Colonial Chesapeake.' *William and Mary Quarterly* 45 (3rd series) (1988) pp. 135–159.

CARR, M. and WEIR, S. 'Sunrise City.' *New Socialist* September (1986) pp. 7–10.

CARROLL, M. 'A Housing Plan For Artists Loses in Board of Estimates.' *New York Times* 11 February (1983).

CARTER, H. 'Whose City? A View From the Periphery.' *Transactions of the Institute of British Geographers* 14(NS) (1989) pp. 4–23.

CHAMPENOIS, M. 'Planner Extraordinaire.' *The Guardian Weekly* 22 December (1991) p. 13.

Consolidated Bibliography

CHASLIN, F. *Le Paris de François Mitterand*, Paris: Gallimard, 1985.

CHASLIN, F. 'Progress on the "Grands Projets".' *Architectural Review* 1076 (1986) pp. 27–30.

CHASLIN, F. 'Paris, Capital of the Pharoah's Republic.' In J.-L. Cohen and B. Fortier (eds.), *Paris: La ville et ses projets. A city in the making*, Paris: Babylone, 1988.

CHAUNU, P. 'Vérités cachées.' *Le Figaro* 22 September (1986).

CHOAY, F. 'Pensées sur la ville, arts de la ville.' In M. Agulhon (ed.), *Histoire de la France urbaine. IV. La ville de l'âge industriel. Le cycle Haussmannien*, Paris: Seuil, 1983.

CHRISTOPHER, A. J. 'From Flint to Soweto: Reflections on the Colonial Origins of the Apartheid City.' *Area* 15 (1983) pp. 145–149.

CLARK, P. *The English Alehouse: A Social History, 1200–1830*, London: Longman, 1983.

CLARK, T. J. *The Painting of Modern Life: Paris in the Art of Manet and his Followers*, London: Thames and Hudson, 1984.

CLARKE, D. B. 'Towards a Geography of the Consumer Society.' In C. Philo (ed.), *New Words, New Worlds: Reconceptualising Social and Cultural Geography*, St. David's University College, Lampeter, Wales: The Social and Cultural Geography Study Group of the Institute of British Geographers, 1991.

CLARKE, D. B. and BRADFORD, M. G. 'The Uses of Space by Advertising Agencies within the United Kingdom.' *Geografiska Annaler* 17B (1989) pp. 139–151.

CLAY, C. G. A. *Economic Expansion and Social Change: England, 1500–1700* (2 volumes), Cambridge: Cambridge University Press, 1984.

CLOKE, P. (ed.) *Policy and Planning in Thatcher's Britain*, Oxford: Pergamon, 1992.

CLOKE, P., PHILO, C. and SADLER, D. *Approaching Human Geography: An Introduction to Contemporary Human Geography*, London: Paul Chapman, 1991.

COCHRANE, A. 'Restructuring Local Politics: Sheffield's Economic Policies in the 1980s.' Paper presented to the 8th Urban Change and Conflict Conference, University of Lancaster, September 1991.

COHEN, J.-L. and FORTIER, B. (eds.) *Paris: La ville et ses projets. A city in the making*, Paris: Babylone, 1988.

COLLS, R. 'The North Reborn: Why We're Proud.' Paper presented at the Annual Conference of the Institute of British Geographers, University of Glasgow, 1990.

COLQUHOUN, A. 'On Modern and Postmodern Space.' In J. Ockman (ed.), *Architecture, Criticism and Ideology*, Princeton: Princeton University Press, 1985.

COLSON, M.-L. 'Sept "nains" qui veulent bousculer les Grands.' *Libération* 15 July (1989).

Commémoration de bicentenaire de la Révolution française, Le dossier information de la culture, 15 June (1987).

CONFEDERATION NATIONALE DU TRAVAIL, ASSOCIATION INTERNATIONALE DES TRAVAILLEURS., *1789. La bourgeoisie prend le pouvoir. 1989. Elle l'a toujours*, Leaflet distributed at 8 July demonstration, Paris (1989).

CONIL, D. 'AD: le juge Bruguière mis à l'isolement.' *Libération* 19 July (1989a).

CONIL, D. 'Les quatre d'AD poursuivent leur grève de la faim.' *Libération* 21 July (1989b).

COOMBES, A. 'Annual Kevin Lynch Memorial Lecture,' reported in *Architects Journal* 4 July (1990) p. 15.

CORFIELD, P. 'Class by Name and Number in Eighteenth-Century Britain.' *History* 72 (1987a) pp. 38–61.

CORFIELD, P. 'Small Towns, Large Implications: Social and Cultural Roles of Small Towns in Eighteenth-Century England and Wales.' *British Journal for Eighteenth-Century Studies* 10 (1987b) pp. 125–138.

CORFIELD, P. 'Walking the City Streets: Social Role and Social Identification in the Towns of Eighteenth-Century England.' *Journal of Urban History* 16 (1990) pp. 132–174.

CORRIGAN, P. 'My Place or Yours? Particular Philosophies from Whose Stories (Vernacular Values Revisited)?' *Journal of Historical Geography* 17 (1991) pp. 313–318.

COSGROVE, D. 'Venice as Cultural Capital: Pre-Modern and Post-Modern Perspectives.' Paper presented at the Annual Conference of the Institute of British Geographers, University of Glasgow, 1990.

COSGROVE, D. and DANIELS, S. (eds.) *The Iconography of Landscape: Essays on the Symbolic Representation, Design and Use of Past Environments*, Cambridge: Cambridge University Press, 1988.

COUSINS, J. M., DAVIS, R. L., PADDON, M. J. and WATON, A. 'Aspects of Contradiction in Regional Policy: The Case of North East England.' *Regional Studies* 8 (1974) pp. 133–144.

COWEN, H. *et al.* 'Cheltenham: Affluence Amid Recession.' In P. Cooke (ed.), *Localities: The Changing Face of Urban Britain*, London: Unwin Hyman, 1989.

COX, K. R. and MAIR, A. 'Uneven Growth Machines and the Politics of Local Economic Development.' *International Journal of Urban and Regional Research* 13 (1989), pp. 137–146.

COX, K. R. 'The Politics of Turf and the Question of Class.' In J. Wolch and M. Dear (eds.), *The Power of Geography: How Territory Shapes Social Life*, London: Unwin Hyman, 1989.

CRILLEY, D. 'The Enchanting Mountain: Olympia and York and the Contemporary Megastructure.' In P. Knox (ed.), *The Restless Urban Landscape*, Englewood Cliffs, N.J.: Prentice Hall Publishers, 1992.

CSSD. 'What Will This Development Mean for the Women of Spitalfields?' Information leaflet (1987).

CSSD. 'It's Not That Easy!' Information leaflet (1988a).

CSSD. 'Threat to Our Future.' *Spitalfields News* (1988b) p. 8.

CSSD. 'The House of Lords Rejects the Spitalfields Community!' Information leaflet (1989).

CSSD. 'Spitalfields and Weavers – Up for Grabs?' Information leaflet (1990).

CULLER, J. 'Semiotics of Tourism.' *American Journal of Semiotics* 1 (1981) pp. 127–140.

D. L. 'République et perestroïka.' *Le Figaro* 12 July (1989).

D. R. 'Cérémonie pour les droits de l'homme.' *Le Monde* 14 July (1989).

DARKE, B. 'Gambling on Sport: Sheffield's Regeneration Strategy for the 90s.' Paper presented to the 8th Urban Change and Conflict Conference, University of Lancaster, September 1991.

DARMAILLACQ, S. 'On était resté pour le défile. . . .' *Libération* 17 July (1989).

DATEL, R. E. and DINGEMANS, D. J. 'Environmental Perception, Historic Preservation, and Sense of Place.' In T. F. Saarinen, D. Seamon and J. L. Sell (eds.), *Environmental Perception and Behaviour: An Inventory and Prospect*, University of Chicago: Department of Geography Research Paper No. 209, 1984.

DAVIDOFF, L. and HALL, C. 'The Architecture of Public and Private Life: English Middle-Class Society in a Provincial Town, 1780–1850.' In D. Fraser and A. Sutcliffe (eds.), *The Pursuit of Urban History*, London: Edward Arnold, 1983.

DAVIES, R. and HOWARD, E. *Retail Change on Tyneside: Third Consumer Survey at Metrocentre*, Oxford: Oxford Institute of Retail Management.

DAVIS, M. *City of Quartz: Excavating the Future in Los Angeles*, London: Verso, 1990.

DE CERTEAU, M. *The Practice of Everyday Life*, Los Angeles: University of California Press, 1984.

DE KERGORLAY, H. 'Une bonne affair pour Helmut Kohl.' *Le Figaro* 17 July (1989).

DE ROUX, E. 'Un ballet gracieux et gratuit.' *Le Monde* 16 July (1989).

DE TAUNEY, P. 'AD: toujours la grève de la faim.' *Le Figaro* 21 July (1989).

DE VRIES, J. *The Dutch Rural Economy in the Golden Age*, California: Berkeley University Press, 1974.

DE VRIES, J. 'Peasant Demand Patterns and Economic Development: Friesland, 1550–1750.' In W. Parker and E. Jones (eds.), *European Peasants and Their Markets*, New Jersey: Princetown University Press, 1975.

DE VRIES, J. *European Urbanisation, 1500–1800*, London: Methuen, 1984.

DELSOL, C. 'Les droits de l'homme n'ont fait pas recette.' *Le Figaro* 14 July (1989a).

DELSOL, C. 'Heureusement Jessye.' *Le Figaro* 17 July (1989b).

DENIS, S. 'Le prince.' *Le Quotidien de Paris* 15 July (1989).

DERBYSHIRE, I. *Politics in France from Giscard to Mitterand*, London: Chambers, 1988.

DEUTSCHE, R. 'Uneven Development: Public Art in New York City.' *October* 47 Winter (1988) pp. 3–53.

Consolidated Bibliography

DEWS, P. *Logics of Disintegration: Post-Structuralist Thought and the Claims of Critical Theory*, London: Verso, 1987.

DOCKLANDS BOROUGHS. *Local Democracy Works*, London: Docklands Joint Committee (1979).

DOCKLANDS CONSULTATIVE COMMITTEE. *Six Year Review of the LDDC*, LDCC, Unit 4, Stratford Office Village, Romford Road, London E15, 1988.

DOCKLANDS CONSULTATIVE COMMITTEE. *The Docklands Experiment: A Critical Review of Eight Years of the London Docklands Development Corporation*, LDCC, Unit 4, Stratford Office Village, 4 Romford Road, London E15, 1990.

DODGSHON, R. A. *The European Past: Social Evolution and Spatial Order*, London: Macmillan, 1987.

DONCELIN, J. 'L'Opéra-Bastille à l'épreuve.' *Le Figaro* 14 July (1989).

'Droits de l'homme: "un sujet de préoccupation internationale légitime".' *Le Monde* 18 July (1989).

DUNCAN, S. S. and GOODWIN, M. 'The Local State and Local Economic Policy: Why the Fuss?' *Policy and Politics* 13 (1985a) pp. 227–253.

DUNCAN, S. S. and GOODWIN, M. 'Local Economic Policies: Local Regeneration or Political Mobilisation.' *Local Government Studies* 11:6 (1985b) pp. 75–96.

DUNCAN, S. S. and GOODWIN, M. *The Local State and Uneven Development*, Cambridge: Polity Press, 1988.

DUNFORD, M. and PERRONS, D. *The Arena of Capital*, London: Macmillan, 1983.

DUPIN, E. 'A l'Élysée, le soulagement après le doute.' *Libération* 17 July (1989).

DUPONCHELLE, V. 'La saga de l'Opéra-Bastille.' *Le Figaro* 13 July (1989a).

DUPONCHELLE, V. 'La Grande Arche de A à Z. . . .' *Le Figaro* 18 July (1989b).

EARLE, P. *The Making of the English Middle Class: Business, Society and Family Life in London 1660–1730*, London: Methuen, 1989.

ECO, U. *Travels in Hyperreality*, London: Picador, 1986.

ECO, U. *Foucault's Pendulum*, London: Weidenfeld and Nicolson, 1989.

European Marketing Data and Statistics 1992 (27th ed.), London: Euromonitor, 1992.

Evening will Come, They will Sew the Blue Sail, Edinburgh: Graeme Murray, 1991.

EXPRESS AND STAR. Merry Hill Supplement, *Express and Star* November (1988).

FABRA, P. 'La France favorable à un sommet entre pays riches et pauvres.' *Le Monde* 15 July (1989).

FAUJUS, A. 'Un entretien avec M. Olivier Stirn.' *Le Monde* 16 July (1989).

FAURE, E. 'Prendre la Bastille.' *Projet* 213 (1989) pp. 5–9.

FAWCETT, T. 'Eighteenth-Century Shops and the Luxury Trade.' *Bath History* 3 (1990) pp. 49–75.

FEATHERSTONE, M. *Consumer Culture and Postmodernism*, London: Sage Publications, 1991.

FEDERATION DES ASSOCIATIONS DE SOLIDARITE AVEC LES TRAVAILLEURS IMMIGRES. Leaflet circulated at 8 July demonstration, Paris (1989).

FERGUSON, S. 'The Boombox Wars.' *Village Voice* 16 August (1988).

FINE, B. and LEOPOLD, E. 'Consumerism and the Industrial Revolution.' *Social History* 15 (1990) pp. 151–179.

FISHER, T. 'Building the New City: Battery Park City.' *Progressive Architecture* March (1988) pp. 86–93.

FOLEY, P. 'The Impact of the World Student Games on Sheffield.' Paper presented at the Annual Conference of the Institute of British Geographers, University of Sheffield, 1991.

FONTAINE, A. 'Après la fête.' *Le Monde* 19 July (1989).

FOSTER, H. '(Post) Modern Polemics.' *Perspecta* 21 (1984) pp. 144–153.

FURET, F. 'French Intellectuals from Marxism to Structuralism.' In F. Furet (ed.), *In the Workshop of History*, Chicago: University of Chicago Press, 1984 (essay originally published in French in 1967).

FURET, F. 'La Révolution française est terminée.' In F. Furet (ed.), *Penser la Révolution française*, Paris: Gallimard, 1978.

G. B. 'Renaud et ses potes désespèrent l'Élysée.' *Libération* 7 July (1989).

G. Lq. 'La Régie prête à négocier un reclassement des Dix hors de Renault.' *Libération* 20 July (1989).

GAILLARD, J. *Paris. La ville. 1852–1870*, Lille: Réproduction des Thèses, 1976.

GALLO, M. 'L'Histoire ne date pas.' *Le Matin* 25 February (1986) p. 20.

GARCIA, P. 'La révolution momifiée.' *Espaces Temps* 38/39 (1988) pp. 4–12.

GARCIAS, J.-C. and MEADE, M. 'Politics of Paris.' *Architectural Review* 1078 (1986) pp. 23–26.

GARDNER, C. and SHEPPERD, J. *Consuming Passion: The Rise of Retail Culture*, London: Unwin Hyman, 1989.

GAUTHIER, N. and VALLADAO, A. 'La brochette des 33 passée sur le grill des droits de l'homme.' *Libération* 14 July (1989).

GENE, J. P. 'Le croisade néo-chouanne de Philippe de Villiers.' *Libération* 14 July (1989).

GENEVEE, F. 'Le Bicentenaire-départ.' *L'Humanité* 19 July (1989).

GERMAINE-ROBIN, F. 'Sommet au pied de la crise.' *L'Humanité* 15 July (1989).

GERRITZ, L. 'Slam Dancer at NYPD.' *Village Voice* 6 September (1988).

GHIRADO, D. 'A Taste of Money: Architecture and Criticism in Houston.' *Harvard Architectural Review* 6 (1987) pp. 87–97.

GHIRADO, D. 'The Deceit of Postmodern Architecture.' In G. Shapiro (ed.), *After the Future: Postmodern Times and Places*, Albany: State University of New York Press, 1990.

GIDDENS, A., *A Contemporary Critique of Historical Materialism, Vol. 1: Power, Property and the State*, London: Macmillan, 1981.

GIDDENS, A. 'Structuralism, Post-Structuralism and the Production of Culture.' In A. Giddens and J. Turner (eds.), *Social Theory Today*, Cambridge: Polity Press, 1987.

GIDDENS, A. *Consequences of Modernity*, Cambridge: Polity Press, 1990.

GIDDENS, A. *Modernity and Self-Identity*, Cambridge: Polity Press, 1991.

GIESBERT, F.-O. 'Colossal.' *Le Figaro* 15 July (1989a).

GIESBERT, F.-O. 'On a célébré une idée révolutionnaire en crise!' *Le Figaro* 17 July (1989b).

GILL, B. 'The Skyline: The Malady of Gigantism.' *The New Yorker* 9 January (1989) pp. 73–74.

GILL, B. 'Battery Park City.' *The New Yorker* 20 August (1990) pp. 69–78.

'Gilles Perrault.' *Politis. Le Citoyen* 69, 7–12 July (1989) p. 14.

GITLIN, T. 'Postmodernism: Roots and Politics.' *Dissent* 36:1 (1989) pp. 100–108.

Glasgow's Great British Art Exhibition, Glasgow: Glasgow Museums and Art Galleries, 1990.

GLENNIE, P. D. 'Industry and Towns, 1500–1730.' In R. Butlin and R. Dodgshon (eds.), *An Historical Geography of England and Wales* (2nd ed.), London: Academic Press, 1990.

GLENNIE, P. D. 'Early Mass Consumption in Europe and North America.' Paper delivered at 8th International Conference of Historical Geographers, Vancouver, August 1992.

GLENNIE, P. D. and THRIFT, N. 'Modernity, Urbanism and Modern Consumption.' *Environment & Planning D: Society & Space* 10 (1992) pp. 423–443.

GOLDBERGER, P. 'Battery Park City is a Triumph of Urban Design.' *New York Times* 31 August (1986).

GOLDBERGER, P. 'Winter Garden at Battery Park City.' *New York Times* 12 October (1988a).

GOLDBERGER, P. 'Public Space gets a New Cachet in New York.' *New York Times* 22 May (1988b).

GOLDBERGER, P. 'Times Square: Lurching towards a Terrible Mistake?' *New York Times* 9 September (1989).

GOODWIN, M. 'Locality and Local State: Sheffield's Economic Policy.' University of Sussex, Working Paper in Urban and Regional Studies No. 52, 1986a.

GOODWIN, M. 'Locality and Local State: Economic Policy for London's Docklands.' University of Sussex, Working Paper in Urban and Regional Studies No. 53, 1986b.

GOODWIN, M. 'The Politics of Locality.' In A. Cochrane and J. Anderson J (eds.), *Politics in Transition*, London: Open University/Sage, 1989.

Consolidated Bibliography

GOODWIN, M. 'Replacing a Surplus Population: The Policies of the London Docklands Development Corporation.' In J. Allen and C. Hamnett (eds.), *Housing and Labour Markets: Building the Connections*, London: Unwin Hyman, 1991.

GOTTDEINER, M. 'Recapturing the Center: A Semiotic Analysis of Shopping Malls.' In M. Gottdeiner and A. Lagopoulos (eds.), *The City and the Sign: An Introduction to Urban Semiotics*, New York: Columbia University Press, 1986.

GRANT, E. 'The Sphinx in the North: Egyptian Influences on Landscape, Architecture and Interior Design in Eighteenth- and Nineteenth-century Scotland.' In D. Cosgrove and S. Daniels (eds.), *The Iconography of Landscape: Essays on the Symbolic Representation, Design and Use of Past Environments*, Cambridge: Cambridge University Press, 1988.

GREATER LONDON COUNCIL. *East London File*, London: County Hall, 1982.

GREER, W. R. 'The Fortunes of the Lower East Side are Rising.' *New York Times* 4 August (1985).

GRIFFITHS, R. A. (ed.) *Boroughs of Medieval Wales*, Cardiff: University of Wales Press, 1978.

GROUD, G. 'Paris et les grands hommes.' In *Quand Paris dansait avec Marianne, 1879–1889*, Paris: Paris-Musées, 1989.

GRUNDMANN, P. 'Le 'Goudovision' pour sept cents millions de téléspectateurs.' *Libération* 13 July (1989).

GUILBERT, P. 'La contre-fête.' *Le Figaro* 10 July (1989a).

GUILBERT, P. 'Le théâtre de la solidarité.' *Le Figaro* 12 July (1989b).

GUILBERT, P 'Un sommet désorienté.' *Le Figaro* 17 July (1989c).

GUTMAN, R. *Architectural Practice: A Critical Review*, Princeton: Princeton Architectural Press, 1988.

HALFORD, S. 'Spatial Divisions and Women's Initiatives in British Local Government.' *Geoforum* 20 (1989) pp. 161–174.

HALL, S. 'Brave New World.' *Marxism Today* October (1988) pp. 24–29.

HALSALL, M. 'City of Steel Cuts a Fresh New Dash.' *The Guardian* 9 June (1989).

HARDY, D. 'Historical Geography and Heritage Studies.' *Area* 20 (1988) pp. 333–338.

HARRIES, P. *et al.* 'The Marketing of Meaning: Aesthetics Incorporated.' *Environment and Planning B: Planning and Design* 9 (1982) pp. 457–466.

HARRIS, D. 'World of Shopping in the Bag.' *Express and Star* Black Country Supplement, November (1989).

HARRISON, M. *Crowds and History: Mass Phenomena in English Towns, 1790–1835*, Cambridge: Cambridge University Press, 1988.

HARVEY, D. *Social Justice and the City*, London: Edward Arnold, 1973.

HARVEY, D. 'Monument and Myth: The Building of the Basilica of the Sacred Heart.' *Annals of the Association of American Geographers* 69 (1979) pp. 362–381 (reprinted in HARVEY, D. *Consciousness and the Urban Experience: Studies in the History and Theory of Capitalist Urbanisation, Vol. I*, Oxford: Basil Blackwell, 1985).

HARVEY, D. *The Limits to Capital*, Oxford: Basil Blackwell, 1982.

HARVEY, D. *Consciousness and the Urban Experience: Studies in the History and Theory of Capitalist Urbanisation, Vol. I*, Oxford: Basil Blackwell, 1985.

HARVEY, D. 'Flexible Accumulation through Urbanisation: Reflections on the Transition to Postmodernism in the American City.' Paper presented at Symposium on Developing the American City: Society and Architecture in the Regional City, Yale School of Architecture, 1987.

HARVEY, D. 'From Managerialism to Entrepreneurialism: The Transformation in Urban Governance in Late Capitalism.' *Geografiska Annaler* 71B (1989a) pp. 3–17.

HARVEY, D. *The Condition of Postmodernity: An Enquiry into the Origins of Cultural Change*, Oxford: Basil Blackwell, 1989b.

HARVEY, D. 'Downtowns.' *Marxism Today* January (1989c) p. 21.

HASKI, P. 'Les critiques sur les "Riches" sont amères au Président.' *Libération* 13 July (1989).

HATTERSLEY, R. 'A Canny Lad.' *The Listener* 19 January (1989).

HAUSSMANN, G. *Mémoires de Baron Haussmann. II. Préfecture de la Seine*, Paris: Victor-Havard, 1989.

HAUTER, F. 'Accord pour aider l'Europe de l'Est.' *Le Figaro* 17 July (1989).

298

HEBDIDGE, D. *Subculture: The Meaning of Style*, London: Methuen, 1979.

HEBDIDGE, D. *Hiding in the Light*, London: Comedia, 1988.

HECK, S. 'Pei's Paris Pyramid.' *Royal Institute of British Architects Journal*, 92:8 (1985) pp. 42–49.

HENRY, M. 'Solidarité en Vendée sur le pointe des pieds.' *Libération* 10 July (1989).

HERLICH, G. 'Le crépuscule de Billancourt.' *Le Monde* 20 July (1989).

HEWISON, R. *The Heritage Industry: Britain in a Climate of Decline*, London: Methuen, 1987.

HMSO. *Employment for the 1990s*, London: Cmnd. 540.

HOBSBAWM, E. J. *Echoes of the Marseillaise: Two Centuries Look Back on the French Revolution*, London: Verso, 1990.

HOLCOMB, B. 'Purveying Places: Past and Present.' New Brunswick, N.J., Centre for Urban Policy Research Working Paper No. 17, 1990.

HOLSTON, J. *The Modernist City*, Chicago: University of Chicago Press, 1989.

HOWARD, D. 'France-USA: l'art des bicentenaires.' *Libération*, 8 June (1988).

HRH THE PRINCE OF WALES. 'A Vision of Britain.' BBC TV transmission, 28 October (1988).

HUDSON, R. 'Uneven Development in Capitalist Societies: Changing Spatial Divisions of Labour, Forms of Spatial Organisation of Production and Service Provision, and their Impacts upon the Localities.' *Transactions of the Institute of British Geographers* 13(NS) (1988) pp. 484–496.

HUDSON, R. and SADLER, D. 'Contesting Works Closures in Western Europe's Old Industrial Regions: Defending Place or Betraying Class?' In A. J. Scott and M. Storper (eds.), *Production, Work, Territory: The Geographical Anatomy of Industrial Capitalism*, London: Allen and Unwin, 1986.

HUDSON, R. and SADLER, D. 'Myths, Mirages, Miracles and the Cold Light of Day: Manufacturing Employment Change in Derwentside District in the 1980s.' University of Durham, Department of Geography Working Paper No. 1, 1990.

HUDSON, R. and SADLER, D. *The International Steel Industry: Restructuring, State Policies and Localities*, London: Routledge, 1992.

HUXTABLE, A. L. 'Times Square Renewal (Act II), A Farce.' *New York Times* 14 October (1989).

HUYSSEN, A. 'Mapping The Postmodern.' *New German Critique* 33 (1984) pp. 5–52.

IMBERT, D. 'Paris et la République: les enjeux d'une statue.' In *Quand Paris dansait avec Marianne, 1879–1889*, Paris: Paris-Musées, 1989.

Institute for Worker Control, Pamphlet No. 79 (Nottingham).

International Financial Statistics, Yearbook 1990. Washington: International Monetary Fund, 1990.

IRIS, Transnational Network for Social Liberation, *Quart Etat*, Leaflet distributed at 8 July demonstration, Paris (1989).

J.-F. R. 'Le dollar revient au sommet.' *Le Quotidien de Paris* 18 July (1989).

J. H. 'Le feu d'artifice de Chevènement.' *Libération* 17 July (1989).

JACK, I. 'The Repackaging of Glasgow.' *London Sunday Times Magazine* 2 December (1984).

JACKSON, P. *Maps of Meaning: An Introduction to Cultural Geography*, London: Unwin Hyman, 1989.

JACKSON, P. 'The Crisis of Representation and the Politics of Position.' *Environment and Planning D: Society and Space* 9 (1991) pp. 131–134.

JACOBS, J. *The Death and Life of Great American Cities: The Failure of Town Planning*, Harmondsworth: Penguin Books, 1985 (1961 orig.).

JACQUEMART, C. and MONET, J.-P. 'Dans l'opposition, chacun son Goude. . . .' *Le Figaro* 17 July (1989).

JACQUES DERRIDA in discussion with CHRISTOPHER NORRIS. 'Discussion and Comments.' In A. Papadakis, C. Cooke and A. Benjamin (eds.), *Deconstruction: Omnibus Volume*, London, Academy, n.d.

JAMESON, F. 'Postmodernism or the Cultural Logic of Late Capitalism.' *New Left Review* 146 (1984) pp. 53–92.

JARMAN, D. *Modern Nature: The Journals of Derek Jarman*, London: Century, 1991.

Consolidated Bibliography

'Jean Kaspar.' *Libération* 12 July (1989).

JENCKS, C. *What is Post-Modernism?*, London: Academy, 1986.

JENCKS, C. *The Language of Post-Modern Architecture*, London: Academy Editions, 1987 (5th. ed.).

JENCKS, C. 'Néo-classicisme et post-modernisme.' *Art Press* spécial (1988) pp. 106–111.

JOHN, D. 'Disney takes the Mickey . . . to Paris.' *The Guardian* 30 May (1989).

JOHNSON, K. 'Poetry and Public Service.' *Art in America* March (1990) pp. 161–163, 219.

JOHNSON, P. 'Conspicuous Consumption Amongst Working-Class Consumers in Victorian England.' *Transactions of the Royal Historical Society* 38 (5th series) (1988) pp. 27–42.

JONES, A. 'Japanese Investment – Welcome to Wales.' *Welsh Economic Review* 1(2) (1988) pp. 60–64.

JONQUET, F. 'Christian Dupavillon: monsieur cent idées.' *Le Quotidien de Paris*, 14 July (1989).

JOSSIN, J. 'Les Français jugent leur histoire.' *Le Figaro* 29 August (1983)

JULY, S. 'Le miroir français.' *Libération* 12 July (1989a).

JULY, S. 'Le Paris de bi-sommet.' *Libération* 15 July (1989b).

JUMET, D. 'Quelle histoire?' *Le Quotidien de Paris* 17 September (1986).

KAJMAN, M. 'Sous le signe de la mondialisation médiatique.' *Le Monde* 25 November (1987).

KALFLECHE, J.-M. 'Le grand dépit de l'Afrique.' *Le Quotidien de Paris* 18 July (1989).

KEATING, D., KRUMHOLZ, N. and METZGER, J. 'Cleveland: Post-Populist Public Private Partnerships.' In G. Squires (ed.), *Unequal Partnerships: The Political Economy of Urban Redevelopment in Postwar America*, New Brunswick, N.J.: Rutgers University Press, 1989.

KEITH, M. 'Racial Conflict and the "No-Go" Areas of London.' In J. Eyles and D. M. Smith (eds.), *Qualitative Methods in Human Geography*, London: Macmillan, 1988.

KIERAN, S. 'The Architecture of Plenty: Theory and Design in the Marketing Age.' *Harvard Architecture Review* 6 (1987) pp. 102–113.

KING, A. D. *Colonial Urban Development: Culture, Social Power and the Environment*, London: Routledge and Kegan Paul, 1976.

KLAUSNER, D. 'Behind the Dock Walls: Popular Planning in Newham.' London: London School of Economics, Geography Discussion Paper No. 17(NS), 1985.

KLAUSNER, D. 'Infrastructure, Investment and Political Ends: The Case of London's Docklands.' *Local Economy* 1:4 (1987) pp. 47–59.

KLOTZ, H. *The History of Postmodern Architecture*, Cambridge, Mass.: MIT Press, 1987.

KOPTIUCH, K. 'Third Worlding at Home: Transforming New Frontiers in the Urban U.S.' *Cultural Anthropology* (forthcoming).

KRIEDTE, P., MEDICK, H. and SCHLUMBOHM, J. *Industrialisation Before Industrialisation: Industry in the Genesis of Capitalism*, Cambridge: Cambridge University Press, 1981.

KRISTEVA, J. 'L'homme, le citoyen, l'étranger.' *Art Press* spécial (1988) pp. 21–27.

KUNDERA, M. *The Art of the Novel*, London: Faber, 1990.

KUPFERMAN, F. 'Bastille, tout le monde descend.' *L'Express* 17 July (1986) pp. 66–67.

'L'opposition boycotte le show des Champs-Élysées.' *Le Figaro* 12 July (1989).

'La communion du Puy-de-Fou.' *Le Figaro* 12 July (1989).

'La foule Goude.' *Le Quotidien de Paris* 15 July (1989).

La Lettre d'Arguments des Communistes Démocratiques 34, July (1989).

'La ruche élyséene.' *Le Monde* 14 July (1989).

LABERITT, G., NAJMAN, M. and SILBERSTEIN, P. 'Pour un nouvel internationalisme.' *2A/Rouge et Vert*, July 8 (1989).

LABOUR PARTY. 'Municipal Enterprise, Jobs and Training.' Paper for 26th Local Government Conference of the Labour Party, Sheffield, February 1982.

LACAN, J.-F. 'Débâcle télévisée.' *Le Monde* 16 July (1989).

LAKE, R. W. *Real Estate Tax Delinquency: Private Disinvestment and Public Response*, Piscataway N.J.: Centre for Urban Policy Research, Rutgers University, 1979.

LANGTON, J. and HOPPE, G. *Town and Country in the Development of Early Modern Western Europe*, Norwich: Geo Books, 1983.

LASH, S. and URRY, J. *Economics of Signs and Spaces: After Organised Capitalism*, Cambridge: Polity Press, 1992.

LASWELL, H. *The Signature of Power*, New York: Transaction Books, 1979.

LAURENT, P. 'Cela n'a pas duré plus de dix minutes. . . .' *L'Humanité* 18 July (1989).

LAVAU, G. 'Choisir les événements fondateurs.' *Projet* 213 (1989) pp. 150–159.

LAWLESS, P. 'Regeneration in Sheffield: From Radical Intervention to Partnership.' In D. Judd and M. Parkinson (eds.), *Leadership and Urban Regeneration, Cities in North America and Europe, Urban Affairs Annual Reviews Vol. 37*, London: Sage, 1990.

LAWLESS, P. 'Public–Private Sector Partnership in a Major Industrial Conurbation: The Case Study of Sheffield.' *East Midland Geographer* 14 (1991) pp. 4–13.

LAWLESS, P. and RAMSDEN, T. 'Land Use Planning and the Inner Cities: The Case of the Lower Don Valley, Sheffield.' *Local Government Studies* January/February (1990) pp. 33–47.

'Le GATT ouvre le dossier tourisme.' *Libération* 18 July (1989).

'Le mur de çon.' *Le Canard Enchaîné* 19 July (1989) p. 1.

LEISS, W., KLINE, P. and JHALLY, S. *Social Communication Through Advertising*, London: Methuen, 1987.

LEITNER, H. 'Cities in Pursuit of Economic Growth: The Local State as Entrepreneur.' *Political Geography Quarterly* 9 (1990) pp. 146–170.

LEMON, J. T. *'The Best Poor Man's Country in the World': A Geographical Study of Early Southern Pennsylvania*, New York: W. W. Norton & Co., 1976.

'Les "contre-revolutionnaires" veulent manifester.' *Le Monde* 19 October (1987).

'Les quatre chefs d'Action Directe et l'isolement.' *L'Humanité* 20 July (1989).

'Les quatre dirigeants d'Action Directe cessent leur grève de la faim.' *Le Monde* 22 July (1989).

'Les quatre dirigeants d'Action Directe sont "en train de mourir" selon leurs avocats.' *Le Monde* 15 July (1989).

'Les sept exaspèrent Pékin.' *Le Quotidien de Paris* 18 July (1989).

'Les travaux des sept pays les plus riches du monde.' *Le Figaro* 17 July (1989).

LEWIS, A. 'Urban Image and Revitalization: The Cleveland Strategy.' Unpublished Senior Thesis, New Brunswick, NJ: Rutgers University, 1988.

LEY, D. and OLDS, K. 'Landscape as Spectacle: World's Fairs and the Culture of Heroic Consumption.' *Environment and Planning D: Society and Space* 6 (1988) pp. 191–212.

LEY, D. 'Fragmentation, Coherence and the Limits to Theory in Human Geography.' In A. Kobayashi and S. Mackenzie (eds.), *Remaking Human Geography*, London: Unwin Hyman, 1989.

Liberty, Terror and Virtue. Southampton: Southampton Art Gallery, 1984.

LIFFRAN, H. 'Qui prendra la Bastille en 1989?' *Obs. de Paris* 8 July (1988).

LIGON, R. *A True and Exact History of the Island of Barbadoes*, London, 1673.

LIMERICK, P. N. *The Legacy of Conquest: The Unbroken Past of the American West*, New York: W. W. Norton & Co., 1987.

LLOYD, M. G. and NEWLANDS, D. A. 'The "Growth Coalition" and Urban Economic Development.' *Local Economy* 3 (1988) pp. 31–39.

LOCAL AUTHORITY ASSOCIATIONS. *Crisis and Opportunity: Local Authorities' Response to Decline of the Major Traditional Industries*, London, 1986.

LOGAN, J. R. 'Growth, Politics and the Stratification of Places.' *American Journal of Sociology* 84 (1978) pp. 404–415.

LOGAN, J. R. and MOLOTCH, H. L. *Urban Fortunes: The Political Economy of Place*, Los Angeles: University of California Press, 1987.

LOGEART, A. 'La chancellerie cherche un moyen de mettre un term à la grève de la faim.' *Libération* 21 July (1989).

LOJKINE, J. *La politique urbaine dans la région parisienne*, Paris: Mouton, 1972.

LOMBARD, M.-A. 'Une interview de Christian Dupavillon.' *Le Figaro* 12 July (1989).

LONCHAMPT, J. 'Soirée heureuse, acoustique radieuse.' *Le Monde* 15 July (1989).

LONDON DOCKLANDS DEVELOPMENT CORPORATION. *Corporate Plan*, LDDC, Great Eastern Enterprise, Millharbour, London E14, 1982.

Consolidated Bibliography

LONDON DOCKLANDS DEVELOPMENT CORPORATION. *Protecting the Heritage*, LDDC, Great Eastern Enterprise, Millharbour, London E14, 1987.

LONDON DOCKLANDS DEVELOPMENT CORPORATION. *Annual Report and Accounts, 1988–1989*, LDDC, Great Eastern Enterprise, Millharbour, London E14, 1989.

LONDON DOCKLANDS DEVELOPMENT CORPORATION. *Creating a Real City: An Arts Action Programme for the London Docklands*, LDDC, Great Eastern Enterprise, Millharbour, London E14, 1990a.

LONDON DOCKLANDS DEVELOPMENT CORPORATION. *Architectural Review*, LDDC, Great Eastern Enterprise, Millharbour, London E14, 1990b.

LOONEY, J. J. 'A Cultural Life in the Provinces: Leeds and York, 1720–1820.' In A. Beier, D. Cannadine and J. Rodenheim (eds.), *The First Modern Society: Essays in English History in Honour of Lawrence Stone*, Cambridge: Cambridge University Press, 1989.

LOOSELY, D. 'Jack Lang and the Politics of Festival.' *French Cultural Studies* 1 (1991) pp. 5–19.

LOPGATE, P. 'Whose Times Square?' *7 Days Magazine* 7 March (1990) pp. 43–47.

LORTIE, A. 'Le renouveau des "promenades de Paris".' In *Parcs et promenades de Paris*, Paris: Babylone, 1989.

LOWE, M. 'Trading Places: Retailing and Local Economic Development at Merry Hill, West Midlands.' *The East Midlands Geographer* 14 (1991) pp. 31–48.

'M. Jean-Noël Jeanneney, président de la mission du bicentenaire.' *Le Monde*, 26 May (1988).

'M. Juppé (RPR) réplique au critiques.' *Le Monde*, 21 July (1989).

MAIGNE, J., BERGER, F., REYNAERT, F. and VINCENDON, S. 'Bicentenaire: le jour du melting-pot est arrivé.' *Libération* 14 July (1989).

MAIR, A. and COX, K. 'Locality and Community in the Politics of Local Economic Development.' *Annals of the Association of American Geographers* 78 (1988) pp. 307–325.

MALAURIE, G. 'Bicentenaire: quelle fête commence?' *L'Express* 24 June (1988).

MALCHOW, H. L. 'Public Gardens and Social Action in Late-Victorian London.' *Victorian Studies* 29 (1985) pp. 97–124.

MANN, M. *The Sources of Social Power, Vol. I: A History of Power from the Beginning to A.D. 1760*, Cambridge: Cambridge University Press, 1986.

MARCHETTI, X. 'Apparences.' *Le Figaro* 17 July (1989).

MARCOVICI, P. 'Chine: les balles du 14-juillet.' *Le Quotidien de Paris* 14 July (1989).

MARCUSE, P. 'Gentrification, Residential Displacement and Abandonment in New York City.' Report to the Community Services Society, 1984.

MARCUSE, P. 'The Grid as City Plan: New York and *Laissez-Faire* Planning in the Nineteenth Century.' *Planning Perspectives* 2 (1987) pp. 287–310.

MARREY, B. *Le fer à Paris-Architecture*, Paris: Picard, 1989.

MARTIN, R. 'Thatcherism and Britain's Industrial Landscape.' In R. Martin and R. Rowthorn (eds.), *The Geography of De-industrialisation*, London: Macmillan, 1986.

MARTIN, R. 'The Political Economy of Britain's North–South Divide.' *Transactions of the Institute of British Geographers* 13 (1988) pp. 389–418.

MARX, K. and ENGELS, F. *The German Ideology*, New York: International Publishers Edition, 1970.

MASSEY, D. 'Flexible Sexism.' *Environment and Planning D: Society and Space* 9 (1991a) pp. 31–57.

MASSEY, D. 'A Global Sense of Place.' *Marxism Today* June (1991b) pp. 24–9.

MAXWELL, P. *Save Spitalfields: A Photographic Exhibition by Phil Maxwell*, Exhibition catalogue, p. 1.

MCCRACKEN, G. *Culture and Consumption: New Approaches to the Symbolic Character of Consumer Goods and Activities*, Bloomington: Indiana University Press, 1987.

MCDOWELL, L. 'Towards an Understanding of the Gender Division of Urban Space.' *Environment and Planning D: Society and Space* 1 (1983) pp. 59–72.

MCGARR, P. 'The Great French Revolution.' *International Socialism* 43 (1989) pp. 15–110.

MCINNES, A. 'The Emergence of the Leisure Town: Shrewsbury, 1660–1760.' *Past and Present* 120 (1988) pp. 53–87.

MCINNES, A. 'The Emergence of a Leisure Town or an Urban Renaissance? Reply.' *Past and Present* 126 (1990) pp. 196–202.

MCKENDRICK, N., BREWER, J. and PLUMB, J. H. *The Birth of a Consumer Society: The Commercialisation of Eighteenth-Century England*, London: Hutchinson, 1982.

MCLEOD, M. 'Architecture and Politics in the Reagan Era: From Post-modernism to Deconstructivism.' *Assemblage* 8 (1988) pp. 23–59.

MENNELL, S. *All Manners of Food: Eating and Taste in England and France from the Middle Ages to the Present*, Oxford: Basil Blackwell, 1985.

MERRINGTON, J. 'Town and Country in the Transition to Capitalism.' In R. H. Hilton (ed.), *The Transition from Feudalism to Capitalism*, London: New Left Books, 1976.

MILIESI, G. 'Les Américains se paient notre dette.' *Le Quotidien de Paris* 15 July (1989a).

MILIESI, G. 'Réaliste, Lui?' *Le Quotidien de Paris* 18 July (1989b).

MILIESI, G. 'Cette dette qui laisse les banquiers de marbre.' *Le Quotidien de Paris* 18 July (1989c).

MITCHELL, T. 'The World as Exhibition.' *Comparative Studies in Society and History* 31 (1989) pp. 217–236.

MOLOTCH, H. 'The City as a Growth Machine.' *American Journal of Sociology* 82 (1976) pp. 309–332.

MOORE, R. 'Where Serious Design plays Second Fiddle to the Music.' *The Independent* 5 June (1991).

MORRIS, J. 'The Japanese are Here: For Better or Worse?' *Welsh Economic Review* 1(1) (1988) pp. 45–47.

MORRIS, M. 'Apologia: Beyond Deconstruction, Beyond What?' In M. Morris (ed.), *The Pirate's Fiancée*, London: Verso, 1988a.

MORRIS, M. 'Banality in Cultural Studies.' *Block* 14 (1988b) pp. 15–26.

MORRIS, M. 'Things to do with Shopping Centres.' In S. Sheridan (ed.), *Crafts: Feminist Cultural Criticism*, London: Verso, 1988c.

MORUZZI, J.-F. 'L'escalade d'Action Directe.' *Le Quotidien de Paris* 21 July (1989).

MOUVEMENT RÉVOLUTIONNAIRE INTERNATIONALISTE. *Sommet des Riches. Bicentenaire des Riches*, Leaflet distributed at 8 July demonstration, Paris (1989).

MUKERJI, C. *From Graven Images: Patterns of Modern Materialism*, New York: Columbia University Press, 1983.

MURET, J.-P. 'Architecture and Politics: The Real Problem of Les Halles.' *Architectural Design Profile: Les Halles* (1980) pp. 77–80.

NATAF, I. 'Jean-Paul Goude: "Je ne crains que la pluie et les stars".' *Le Figaro* 10 July (1989).

NOIRIEL, G. 'Un modèle pour le mouvement ouvrier?' *Projet* 213 (1989) pp. 73–84.

NORA, P. (ed.) *Les lieux de mémoire. I. La République*, Paris: Gallimard, 1984b.

NORA, P. 'Entre mémoire et histoire.' In P. Nora (ed.), *Les lieux de mémoire. I. La République*, Paris: Gallimard, 1984b.

NORTHCUTT, W. 'François Mitterand and the Political Use of Symbols: the Construction of a Centrist Republic.' *French Historical Studies* 17 (1991) pp. 141–158.

OLYMPIA AND YORK. *Canary Wharf: the Untold Story*, Unpaginated brochure distributed by O&Y, 10 Great George Street, London SW1P 3AE, 1990a.

OLYMPIA AND YORK. *Ogilvy & Mather, No. 3 Cabot Square, Canary Wharf, London E 14*, 10 Great George Street, London SW1P 3AE, 1990b.

OLYMPIA AND YORK. *Canary Wharf: Vision of a New City District*, 10 Great George Street, London, SW1P 3AE, 1990c.

ORTOLI-LANOE, E. and LECOURT, E. '"Dix" de Renault: la marche calme.' *Le Figaro. Fig-Eco* 13 July (1989).

ORY, P. *Les expositions universelles de Paris*, Paris: Ramsay, 1982.

ORY, P. 'Le centenaire de la Révolution française.' In P. Nora (ed.), *Les lieux de mémoire. I. La République*, Paris: Gallimard, 1984.

ORY, P. 'La beauté du mort.' *Espaces Temps* 38/39 (1988) pp. 21–24.

OZOUF, M. 'Le Panthéon: L'École normale des morts.' In P. Nora (ed.), *Les lieux de mémoire. I. La République*, Paris: Gallimard, 1984.

OZOUF, M. 'Robespierre, mort ou vif.' *Le Nouvel Observateur* 23 October (1986).

OZOUF, M. 'Les fêtes révolutionnaires.' *Art Press* spécial (1988) pp. 8–14.

Consolidated Bibliography

PATOZ, J. 'Gauche: Mitterand veut reprendre l'initiative.' *Le Quotidien de Paris* 15 July (1989).

PAWLEY, M. 'First Simulate, then Stimulate.' *The Guardian* 4 June 4 (1990).

Pékins de tous les pays . . . unissons-nous, Leaflet distributed at the 8 July demonstration, Paris (1989).

PERRY, M. 'The Internationalisation of Advertising.' *Geoforum* 21 (1990) pp. 35–50.

PETITFRERE, C. 'Les rebelles de l'Ouest (1789–1989).' *L'Histoire* 113 (1988) pp. 79–85.

PEUCHAMIEL, B. 'Silence aux pauvres.' *L'Humanité* 14 July (1989).

PITTSBURGH. *Pocketful of Pittsburgh*, Pittsburgh: Neighborhoods for Living Center, 1989.

PITTSBURGH. *Pittsburgh Facts: 1989–1990 Statistical Guide to the Pittsburgh Metropolitan Region*, Pittsburgh: Greater Pittsburgh Office of Promotion, 1990.

PLANNING. 'Dudley Powerless to Resist Enterprise Zone Skyscraper.' *Planning* 17 November (1989).

PLANNING. 'Steelworks Project Secures First Grant.' *Planning* 23 October (1987).

PLANNING. 'Heseltine Reasserts Control on Tower.' *Planning* 10 May (1991).

POISSON, G. *Guides des statues de Paris: monuments, decors, fontaines*, Paris: Hazan, 1990.

POLITIS, N. 'The New Wave: From the East of Paris to the Suburbs, a New Relationship Between the Centre and the Outskirts.' In J.-L. Cohen and B. Fortier (eds.), *Paris: La ville et ses projets. A city in the making*, Paris: Babylone, 1988.

PORTELLI, A. *The Death of Luigi Trastulli and Other Stories: Form and Meaning in Oral History*, Albany: State University of New York Press, 1991.

PRED, A. 'Spectacular Articulations of Modernity: The Stockholm Exhibition of 1897.' *Geografiska Annaler* 73B (1991) pp. 45–84.

PRICE, P. 'The Sheffield Partnership.' Paper presented at the Annual Conference of the Institute of British Geographers, University of Sheffield, 1991.

Quand Paris dansait avec Marianne, 1879–1889, Paris: Paris-Musées, 1989.

RABAN, J. *Soft City*, London: Hamish Hamilton, 1974.

RAMOS, L.-M. 'Vive la Vendée, vive la Pologne.' *Le Figaro. Fig-Mag* 15 July (1989).

RANCE, C. '14 juillet: Bastille-Défense, en avant, Arche!' *Le Figaro. Fig-Mag* 15 July (1989).

RAUSE, V. 'Pittsburgh Cleans Up Its Act.' *New York Times Magazine* 26 November (1989).

RELPH, E. *Place and Placelessness*, London: Pion, 1976.

RELPH, E. *The Modern Urban Landscape*, London: Croom Helm, 1987.

REMOND, R. 'La tradition contre-révolutionnaire.' *Projet* 213 (1989) pp. 160–167.

'Renaud.' *Le Politis. Citoyen* 69, 7–12 July (1989) pp. 15–16.

RENFREW, C. *The Emergence of Civilization: The Cyclades and the Aegean in the Third Millennium B.C.*, London: Methuen, 1972.

REYNAERT, F. 'Le jour d'y croire est arrivée.' *Libération* 25 November (1987).

REYNAERT, F. 'Le bicentenaire de la Révolution est en plans.' *Libération* 5 April (1988).

REYNAERT, F. 'Les historiens de la Révolution font melting-pot en Sorbonne.' *Libération* 7 July (1989).

RHODES, P. 'Biggest and Best in Happy Fantasy built by Don and Roy.' *Express and Star* Merry Hill Supplement, November (1989a).

RHODES, P. 'When all the Map was Black.' *Express and Star* Black Country Supplement, August (1989b).

RIBAUD, A. 'Tonton a trouvé Maggie tout sucre, tout fiel – Non, Sire, c'est un apothéose.' *Le Canard Enchaîné* 19 July (1989).

RICHARDS, T. *The Commodity Culture of Victorian England: Advertising and Spectacle, 1851–1914*, California: Stanford University Press, 1990.

RISTAT, J. '"La Marseillaise" baillonnée dans son berceau.' *L'Humanité* 15 July (1989).

ROBINSON, F. '"It's Not Really Like That": Living With Unemployment in the North East.' Newcastle-on-Tyne. BBC(NE), 1987.

ROBINSON, F. and SADLER, D. 'Consett After the Closure.' University of Durham, Department of Geography Occasional Paper No. 19, 1984.

ROBINSON, F. and SADLER, D. 'Routine Action, Reproduction of Social Relations and the Place Market: Consett After the Closure.' *Environment and Planning D: Society and Space* 3 (1985) pp. 109–120.

ROBINSON, J. '"Progressive Port Elizabeth": Liberal Politics, Local Economic Development and the Territorial Basis of Racial Domination.' *Geoforum* 22 (1990) pp. 293–303.

ROLLAT, A. 'Le droit du citoyen devant la loi.' *Le Monde* 16 July (1989a).

ROLLAT, A. 'Lendemain des fastes.' *Le Monde* 18 July (1989b).

ROMAN, J. 'Représenter le peuple souverain.' *Projet* 213 (1989) pp. 29–38.

RONCAYOLO, M. 'La production de la ville.' In M. Agulhon (ed.), *Histoire de la France urbaine. IV. La ville de l'âge industriel. Le cycle Haussmannien*, Paris: Seuil, 1983.

ROSE, G. 'Imagining the East End of London in the 1980s.' Paper presented at the Annual Conference of the Institute of British Geographers, University of Glasgow, 1990.

ROSE, G. 'Review of Crosby's *The Ends of History* and Young's *White Mythologies*.' *Journal of Historical Geography* 17 (1991) pp. 344–346.

ROSE, J. and TEXIER, C. (eds.) *Between C&D. New Writings from the Lower East Side Fiction Magazine*, New York: Penguin, n.d.

ROSNER, A. 'Projects for Les Halles, 1950–1979.' *Architectural Design Profile: Les Halles* (1980) pp. 11–15.

ROSS, K. *The Emergence of Social Space: Rimbaud and the Paris Commune*, London: Macmillan, 1988.

SACK, R. 'The Consumer's World: Place as Context.' *Annals of the Association of American Geographers* 78 (1988) pp. 642–664.

SADLER, D. 'The Social Foundations of Planning and the Power of Capital.' *Environment and Planning D: Society and Space* 8 (1990) pp. 323–338.

SAID, E. W. *Orientalism*, New York: Pantheon Books, 1978.

SALINS, P. 'The Creeping Tide of Disinvestment.' *New York Affairs* 6:4 (1981) pp. 5–19.

SAMSON, M. 'Perrault: un résultat proche du néant sur la dette.' *Libération* 18 July (1989).

SAMUEL, R. 'Workshop of the World: Steam Power and Hand Technology in Mid-Victorian Britain.' *History Workshop Journal* 3 (1977) pp. 6–72.

SAMUEL, R. 'Introduction: Exciting to be English.' In R. Samuel (ed.), *Patriotism: The Making and Unmaking of British National Identity, Vol. 1: History and Politics*, London: Routledge, 1989.

SAVITCH, H. V. *Post-Industrial Cities: Politics and Planning in New York, Paris and London*, Princeton, NJ: Princeton University Press, 1988.

SBRAGIA, A. 'The Pittsburgh Model of Economic Development: Partnership, Responsiveness, and Indifference.' In G. Squires (ed.), *Unequal Partnerships: The Political Economy of Urban Redevelopment in Postwar America*, New Brunswick, N.J., Rutgers University Press, 1986.

SCHUDSON, M. *Advertising: The Uneasy Persuasion*. New York: Basic Books, 1984.

SCHUURMAN, A. 'Probate Inventories and Material Culture: Research Issues, Problems and Results.' *A.A.G. Bijdragen* 23 (1980) pp. 19–32.

SCHWARTZENBERG, E. '14 juillet: les coulisses de la grand parade.' *Le Figaro* 11 July (1989).

SCOTT BROWN, D. 'Architectural Taste in a Pluralist Society.' *Harvard Architectural Review* Spring (1980) pp. 41–51.

SCOTT, A. J. *The Urban Land Nexus and the State*, London: Pion, 1980.

SEED, J. and WOLFF, J. 'Class and Culture in Nineteenth-Century Manchester.' *Theory, Culture and Society* 2:2 (1984) pp. 38–53.

SENNETT, R. *The Conscience of the Eye: The Design and Social Life of Cities*, London: Faber and Faber, 1990.

SEYD, P. 'Radical Sheffield: From Socialism to Entrepreneurialism.' *Political Studies*, 38 (1990) pp. 355–44.

SHAMMAS, C. 'Consumer Behaviour in Colonial America.' *Social Science History* 6 (1982) pp. 67–86.

SHAMMAS, C. *The Pre-Industrial Consumer in England and America*, Oxford: Clarendon Press, 1990.

SHARPE, J. A. *Early Modern England: A Social History, 1550–1760*, London: Edward Arnold, 1987.

SHAW, K. 'The Politics of Public–Private Partnerships in Tyne and Wear.' *Northern Economic Review* 19 (1990) pp. 2–16.

SHEFFIELD CITY COUNCIL. *Alternative Economic Policies: A Local Government Response*, Sheffield: Corporate Management Unit, City Council, Town Hall (1981).

SHEFFIELD CITY COUNCIL. *Response to an Invitation by Central Government to Apply for an Enterprise Zone, and Recommendations for an Alternative Strategy*, Sheffield: Employment Department, City Council, Town Hall (1982).

SHOTTER, J. 'Social Accountability and the Social Constitution of You.' In J. Shotter and T. Gurgan (eds.), *Texts of Identity*, London: Sage, 1989.

SIBLEY, D. 'Purification of Space.' *Environment and Planning D: Society and Space* 6 (1988) pp. 409–421.

SIMOLO, N. 'Jean-François Vilar tient le Hébert fan-club.' *Politis. Le Citoyen* 69, 7–12 July (1989) p. 60.

SMITH, N. 'Toward a Theory of Gentrification: A Back to the City Movement by Capital not People.' *Journal of the American Planning Association* 45 (1979) pp. 538–548.

SMITH, N. *Uneven Development: Nature, Capital and the Production of Space*, Oxford: Basil Blackwell, 1984.

SMITH, N. 'Gentrification, the Frontier, and the Restructuring of Urban Space.' In N. Smith and P. Williams (eds.), *Gentrification of the City*, Boston: Allen and Unwin, 1986.

SMITH, N. 'Lower East Side as Wild West: New City as New Frontier.' Unpublished manuscript, New Brunswick, N.J., Rutgers University, Department of Geography, 1988.

SMITH, N., DUNCAN, B. and REID, L. 'From Disinvestment to Reinvestment: Tax Arrears and Turning Points in the East Village.' *Housing Studies* 4 (1989) pp. 238–252.

SMITH, S. J. 'Political Interpretations of "Racial Segregation" in Britain.' *Environment and Planning D: Society and Space* 6 (1988) pp. 423–444.

SMITH, S. J. *The Politics of 'Race' and Residence: Citizenship, Segregation and White Supremacy in Britain*, Cambridge: Polity Press, 1989.

SOJA, E. W. *Postmodern Geographies: The Reassertion of Space in Critical Social Theory*, London: Verso, 1989.

SORKIN, M. (ed.) *Variations on a Theme Park: The New American City and the End of Public Space*, New York: The Noonday Press, 1992.

SOULIER, P.-B. 'Bravo l'Opéra.' *Politis. Le Citoyen* 69, 7–12 July (1989) pp. 50–54.

SPEER, A. 'Foreword.' In L. Krier (ed.), *Albert Speer: Architecture 1932–1942*, Brussels: Archives d'Architecture Moderne, 1985.

SPIRE, A. 'Les braises de la Révolution française.' *Le Matin* 25 February (1986).

SPITALFIELDS DEVELOPMENT GROUP. 'A Solution for Spitalfields and What it Means to You.' Information leaflet (1986).

SPITALFIELDS DEVELOPMENT GROUP. 'Update.' Information leaflet (1987a).

SPITALFIELDS DEVELOPMENT GROUP. 'A Proposal for the Redevelopment of Spitalfields Market.' Information leaflet (1987b).

SPITERI, G. 'Allons enfants de la cohabitation.' *Le Quotidien de Paris* 17 September (1986).

SPITERI, G. '1989: aux larmes citoyens!' *Le Quotidien de Paris* 24 May (1988).

SPITERI, G. 'Comment Mitterand a vécu son sommet.' *Le Quotidien de Paris* 18 July (1989).

STALLYBRASS, P. and WHITE, A. *The Politics and Poetics of Transgression*, London: Methuen, 1986.

STAROBINSKI, J. 'Lecture-interprétation des images de 89.' *Art Press* spécial (1988) pp. 97–99.

STEIN, S. R. 'French Ferociously Debate the Pei Pyramid at the Louvre.' *Architecture: The American Institute of Architects Journal* May (1985) pp. 25–46.

STEINBACH, S. 'Une mise en scène à la mode.' *L'Humanité* 13 July (1989).

STENLI, J.-B. 'Bicentenaire: ça ira pas.' *L'Express* 3 June (1989).

STERNLEIB, G. and LAKE, R. W. 'The Dynamics of Real Estate Tax Delinquency.' *National Tax Journal* 29 (1976) pp. 261–271.

STRONG, R. 'Rus in Urbe (the Kevin Lynch Memorial Lecture, transcribed by P. Luck).' *Urban Design Quarterly* June (1989) pp. 45–48.

STYLES, J. 'Design for Large-Scale Production in Eighteenth-Century Britain.' *Oxford Art Journal* 11 (1988) pp. 10–16.

STYLES, J. 'Manufacture, Consumption and Design in Eighteenth-Century England.' In J. Brewer and R. Porter (eds.), *Consumption and Society in the Seventeenth and Eighteenth Centuries*, London: Routledge, 1992.

SUTCLIFFE, A. *The Autumn of Central Paris: The Defeat of Town Planning, 1850–1970*, London: Edward Arnold, 1970.

SWYNGEDOUW, E. 'The Heart of the Place: The Resurrection of Locality in an Age of Hyperspace.' *Geografiska Annaler* 71B (1989) pp. 31–42.

TAFURI, M. 'The Disenchanted Mountain.' In G. Ciucci *et al.* (eds.), *The American City From the Civil War to the New Deal*, Cambridge, Mass.: MIT Press, 1980.

TAUPIN, B. 'Un monsieur Bons-Offices pour les "dix" de Renault.' *Le Figaro. Fig-Eco* 11 July (1989a).

TAUPIN, B. 'Dix de Renault: première rencontre Régie-CGT.' *Le Figaro. Fig-Eco* 21 July (1989b).

THOMPSON, J. B. *Ideology and Modern Culture*, Cambridge: Polity Press, 1991.

THORAVEL, A. 'Dix noms pour un symbole.' *Libération* 11 July (1989).

THORNTON, P. J. *Seventeenth-Century Interior Decoration in England, France and Holland*, London: Yale University Press, 1978.

THRIFT, N. 'Images of Social Change.' In C. Hamnett, L. McDowell and P. Sarre (eds.), *Restructuring Britain: The Changing Social Structure*, London: Sage, 1989.

TIBBALDS, F. 'An Overview.' *Urban Design Quarterly* June (1989) pp. 40–44.

TIBBALDS, F. *et al. City of Birmingham City Centre Design Strategy*, Birmingham: Birmingham City Council, 1990.

TONKA, H. 'La révélation française.' *Art Press* spécial (1988) pp. 162–164.

TOURAINE, A. 'La révolution n'est plus ce qu'elle était.' *Projet* 213 (1989) pp. 20–27.

Tourism Policy and International Tourism in OECD Countries, Paris: OECD, 1986.

TOWN AND COUNTRY PLANNING ASSOCIATION. *Bridging the North–South Divide*, London, 1989.

TRICOT, H. 'Marathon diplomatique entre l'Élysée et l'Hôtel Marigny.' *Le Quotidien de Paris* 14 July (1989).

TSCHUMI, B. *Cinégramme folie: le Parc de la Villette*, London: Butterworth, 1987a.

TSCHUMI, B. 'Disjunctions.' *Perspecta* 23 (1987b) pp. 108–119.

'Un grand bazar.' *Libération* 9 June (1988).

'Un retraite de Russie.' *Libération* 12 July (1989).

'Une commission contestée.' *Le Monde* 23 March (1988).

URRY, J. *The Tourist Gaze: Leisure and Travel in Contemporary Societies*, London: Sage, 1990.

VALENTINE, G. 'The Geography of Women's Fear.' *Area* 21 (1989) pp. 385–390.

VARENNE, F. 'Bicentenaire: les Parisiens punis.' *Le Figaro* 25 September (1988).

VAREY, S. *Space and the Eighteenth-Century English Novel*, Cambridge: Cambridge University Press, 1990.

VENTURI, R. *Complexity and Contradiction in Architecture*, New York: Museum of Modern Art, 1966.

VENTURI, R. *et al. Learning From Las Vegas*, Cambridge, Mass.: MIT Press, 1972.

VERGO, P. (ed.) *The New Museology*, London: Reaktion Books, 1989.

VERNET, H. 'La naufrage touristique.' *Le Quotidien de Paris* 24 May (1988).

VERNET, H. 'Trocadéro: faites venir les pauvres.' *Le Quotidien de Paris* 14 July (1989).

VINCENDON, S. 'Goude, le saga d'un sauteur d'obstacles.' *Libération* 14 July (1989).

VOVELLE, M. '"La Marseillaise": la guerre ou la paix.' In P. Nora (ed.), *Les lieux de mémoire. I. La République*, Paris: Gallimard, 1984.

VOVELLE, M. 'La Révolution "une et indivisible".' *Revue Politique et Parlementaire* July–August (1987) p. 19.

VOVELLE, M. 'L'âge de la prise de conscience.' *Espaces Temps* 38/39 (1988) pp. 36–40.

Consolidated Bibliography

WARD, D. *Poverty, Ethnicity and the American City, 1840–1925*, Cambridge: Cambridge University Press, 1989.

WARD, S. V. 'Local Industrial Promotion and Development Policies, 1899–1940.' *Local Economy* 5(2) (1990) pp. 100–118.

WATSON, S. 'Gilding the Smokestacks: The Symbolic Representation of Deindustrialized Regions.' Paper presented at Urban Change and Conflict Conference, University of Bristol, 17–20 September 1989a.

WATSON, S. 'Gilding the Smokestacks: The New Symbolic Representations of Deindustrialised Regions.' *Environment and Planning D: Society and Space* 9 (1989b) pp. 59–71.

WEATHERILL, L. 'A Possession of One's Own: Women and Consumer Behaviour in England, 1660–1760.' *Journal of British Studies* 25 (1986) pp. 131–156.

WEATHERILL, L. *Consumer Behaviour and Material Culture in Britain, 1660–1760*, London: Routledge, 1988.

WEBER, M. *The City*, New York: Free Press, 1958 (orig. in 1921).

WERNICK, A. 'Advertising and Ideology: An Interpretive Framework.' *Theory, Culture and Society* 2 (1983) pp. 16–32.

WHEATLEY, P. *The Pivot of the Four Quarters: A Preliminary Inquiry into the Origins and Character of the Ancient Chinese City*, Edinburgh: Edinburgh University Press, 1971.

WHITE, J. 'Beyond Autobiography.' In R. Samuel (ed.), *People's History and Socialist Theory*, London: Routledge and Kegan Paul, 1981.

WHITT, J. A. 'Mozart and the Metropolis: The Arts Coalition and the Urban Growth Machine.' *Urban Affairs Quarterly* 23 (1987) pp. 15–36.

WILLIAMS, P. 'Constituting Class and Gender: A Social History of the Home, 1700–1900.' In N. Thrift and P. Williams (eds.), *Class and Space: The Making of Urban Society*, London: Routledge and Kegan Paul, 1987.

WILLIAMS, W. 'Rise in Values Spurs Rescue of Buildings.' *New York Times* 4 April (1987).

WILLIAMSON, J. *Decoding Advertising*, London: Marion Boyars, 1978.

WILLIS, P. *Common Culture: Symbolic Work at Play in the Everyday Cultures of the Young*, Milton Keynes: Open University Press, 1990.

WILLIS, P. *Profane Culture*, London: Routledge and Kegan Paul, 1978.

WISEMAN, C. 'A Vision with a Message, Battery Park City, New York.' *Architectural Record* March (1987).

WOMEN AND GEOGRAPHY STUDY GROUP OF THE IBG. *Geography and Gender: An Introduction to Feminist Geography*, London: Hutchinson, 1984.

WOODWARD, R. *Saving Spitalfields: The Politics of Opposition to Redevelopment in East London*, Unpublished PhD Thesis, University of London, 1991.

WOOLF, J. 'The Invisible *Flâneuse*: Women and the Literature of Modernity.' *Theory, Culture and Society* 2:3 (1985) pp. 37–46.

WOOLF, P. 'Symbol of the Second Empire: Cultural Politics and the Paris Opera House.' In D. Cosgrove and S. Daniels (eds.), *The Iconography of Landscape: Essays on the Symbolic Representation, Design and Use of Past Environments*, Cambridge: Cambridge University Press, 1988.

WRIGHT, P. *On Living in an Old Country: The National Past in Contemporary Britain*, London: Verso, 1985.

WRIGHTSON, K. *English Society, 1580–1660*, London: Hutchinson, 1982.

WRIGLEY, E. A. 'Urban Growth and Agricultural Change: England and the Continent in the Early Modern Period.' *Journal of Interdisciplinary History* 15 (1984) pp. 683–728.

YOUNG, R. *White Mythologies: Writing History and the West*, London: Routledge, 1990.

Index

Some terms are so common throughout the volume – such as 'city', 'culture', 'history', 'selling' and 'place marketing' – that they are not indexed. Figures in plain type (e.g. 23) refer to terms in the main text, whereas figures in italics (e.g. *24*) refer to terms in endnotes to chapters.